# MAUI

## A Paradise Guide

By Greg and Christie
Stilson

Cover design by Susie O'Day
Illustrations by Janora Bayot

Published by
Paradise Publications
Portland, Oregon

In affiliation with
Prima Publishing and Communications

Distributed by
St. Martin's Press

# *"Maui No Ka Oi"*
# *(Maui is the Best)*

Dedicated to Maren and Jeffrey, two terrific travelers.

Paradise Publications
8110 SW Wareham Circle
Portland, OR   97223   U.S.A.
(503) 246-1555

**MAUI, A Paradise Guide**
Copyright © 1990 Paradise Publications, Portland, Oregon

|  |  |  |  |
|---|---|---|---|
| First | Edition: | Dec | 1984 |
| Second | Edition: | May | 1986 |
| Third | Edition: | May | 1988 |
| Fourth | Edition: | May | 1990 |

ISSN # 0895-9609
Printed in U.S.A.

SPECIAL SALES
Paradise Publications is pleased to make available any of the Paradise Series Guides at special discounts for bulk purchases intended as sales premiums or promotions. Also available are special editions which may include personalized covers or business imprints for a variety of special needs. Excerpts of the existing Paradise Guides can also be arranged by contacting the publisher. Paradise Publications, 8110 S.W. Wareham Circle, Portland, OR   97223.

COVER DESIGN:  Susie O'Day.
PEN & INK SKETCHES AND MAPS:  Janora Bayot

Aerial photograph of the Makena area and the Maui Prince Resort is courtesy of the Maui Prince Hotel.

# TABLE OF CONTENTS

# V. BEACHES AND BEACH ACTIVITIES

# VI. RECREATION AND TOURS

## OCEAN ACTIVITIES

## LAND ACTIVITIES

## AIR TOURS

# VII. RECOMMENDED READING

# VIII. INDEX

# IX. READER RESPONSE - ORDERING INFORMATION

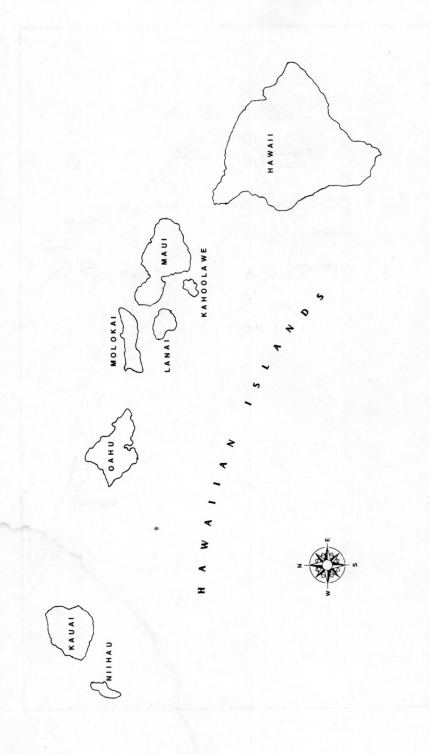

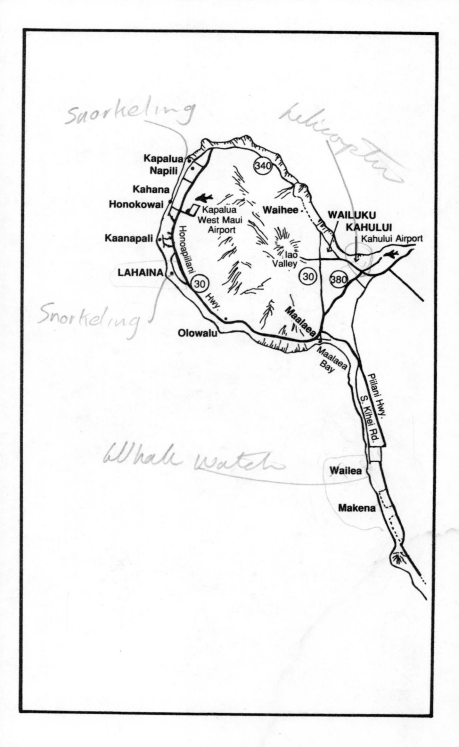

# MAUI

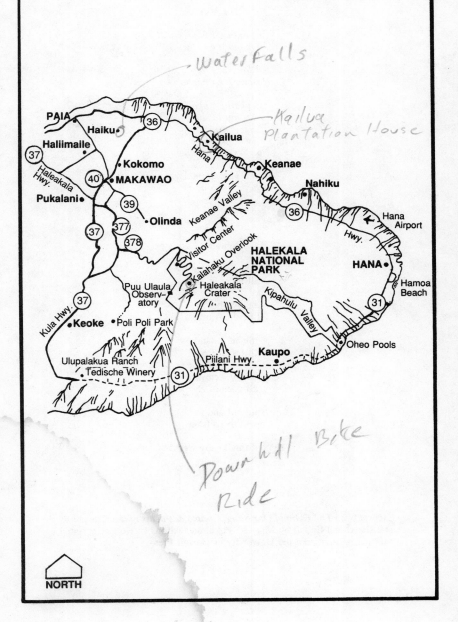

Waterfalls

Kailua Plantation House

Downhill Bike Ride

NORTH

### E'Ike Mai

I luna la, i luna
Na manu o ka lewa

I lalo la, i lalo
Na pua o ka honua

I uka la, i uka
Na ulu la 'au

I kai la, i kai
Na i'a o ka moana

Ha'ina mai ka puana
A he nani ke ao ne


Behold

Above, above
all birds in air

below, below
all earth's flowers

inland, inland
all forest trees

seaward, seaward
all ocean fish

sing out and say
again the refrain

Behold this lovely world


Excerpt from *The Echo of Our Song, Chants & Poems of the Hawaiians*
Translated and Edited by Mary K. Pukui and Alfons L. Korn. Reprinted
with permission from the University of Hawaii Press.

# Introduction

Congratulations on choosing Maui as the site of your vacation. You will soon see why it has the deserved slogan, Maui No Ka Oi (Maui is the Best). The sun and lush tropicalness, and some of the finest accommodations, blend sublimely together to create a perfect holiday paradise - a place both magical and beautiful.

Our personal perspective is that of the visitor. We continue to travel frequently to the island to update our information, discover new things, rediscover old things, make wonderful new friends, and to thoroughly enjoy the tropical energy and seductive charm of Maui. While first-time visitors will delight in the diversity of activities that Maui has to offer, those making a return visit can enjoy discovering new sights and adventures on this magnificent island.

The chapters on accommodations and sights, restaurants, and beaches are conveniently divided into areas with similar characteristics and indexes are provided for each chapter. This allows a better feel for, and access to, the information on the area in which you are staying, and greater confidence in exploring other areas. Remember that except for Hana, most of the areas are only a short drive and worth a day of sightseeing, beach exploring, or a meal at a fine restaurant.

Maui can be relatively inexpensive, or extravagantly expensive, depending on your preference in lodgings, activities, and eating arrangements. Therefore, we have endeavored to give complete and detailed information covering the full range of budgets. The opinions expressed are based on our personal experiences, and while the positive is emphasized, it is your right to know, in certain cases, our bad experiences. To aid in your selections, a **BEST BET** summary is included at the beginning of the General Information Chapter. Also refer to individual chapters for additional best bets and check for ★'s which identify one of our special recommendations.

Our guide is as accurate as possible at the time of publication, however, changes seem extremely rapid for an island operating on "Maui Time." Ownerships, managements, names, and menus do change frequently as do prices. For the latest information on the island, Paradise Publications has available *THE MAUI UPDATE*, a quarterly newsletter. We invite you to receive a complimentary issue or a yearly subscription see ORDERING INFORMATION at the back of this guide.

If your itinerary includes visits to the other islands, the Paradise Guide Series includes *Kaua'i, A Paradise Guide* by Don and Bea Donohugh, *O'ahu, A Paradise Guide* by Ken Bierly and *Hawai'i: The Big Island, A Paradise Guide* by John Penisten. These essential travel accessories not only contain information on all the condos, hotels, restaurants, along with specific recreational activity information, but also inside information that most visitors never receive, shared by an author who knows and loves the islands. Each of these titles also feature update newsletters. See ORDERING INFORMATION.

As this guide features pen and ink sketches, a couple of recommendations for picture books might be helpful. An inexpensive, 48-page, full-color photographic book *Maui The Romantic Island*, highlights the most memorable sights and gives you a good feel for the island. It is available through Paradise Publications; see ORDERING INFORMATION. A magnificent coffee table size publication, *Maui, On My Mind*, by Rita Ariyoshi contains outstanding photographs that depict the island at its best. The book is available at local island gift and book stores and at mainland travel bookstores.

We are confident that as you explore these islands, you too will be charmed by their magic. Keep in mind the expressive words used by Mark Twain nearly 100 years ago when he visited and fell in love with Hawaii.

> *"No alien land in all the world has any deep strong charm for me but that one, no other land could so longingly and so beseechingly haunt me, sleeping and waking, through half a lifetime, as that one has done. Other things leave me, but it abides; other things change, but it remains the same. For me its balmy airs are always blowing, its summer seas flashing in the sun; the pulsing of its surfbeat is in my ear; I can see its garlanded crags, its leaping cascades, its plumy palms drowsing by the shore, its remote summits floating like islands above the cloud wrack; I can feel the spirit of its woodland solitudes, I can hear the splash of its brooks; in my nostrils still lives the breath of flowers..."*

Although the islands have changed greatly during the century since his visits, there remains much to fall in love with. The physical beauty and seductiveness of the land remains despite what may seem rampant commercialism, and the true aloha spirit does survive.

Aloha,

Greg and Christie Stilson

# General Information

## *OUR PERSONAL BESTS*

BEST FOOD SPLURGE: Champagne Sunday brunch at the Prince Court (Maui Prince Resort in Makena), champagne Sunday brunch at Raffles' (Stouffer Wailea Beach Resort), or champagne Sunday brunch at the Sound of the Falls (Westin Maui at Kaanapali).

BEST SUNSET, AND COCKTAILS: West Maui - Kapalua Bay Lounge; South Maui - Maui Prince Resort, Molokini Lounge.

BEST DAILY BREAKFAST BUFFET: Swan Court at the Hyatt Regency

BEST SALAD BAR: Royal Ocean Terrace

BEST DINNER VALUES: Early bird specials are offered at a number of island restaurants. There are generally more specials offered during the summer months and the price may also reflect the time of year. Some of the better early bird offerings are at the Marriott's Moana Terrace at Kaanapali, Chuck's or Island Fish House in Kihei.

BEST RESTAURANT ATMOSPHERE: Swan Court at the Hyatt Regency Hotel or The Sounds of the Falls at the Westin.

MOST OUTRAGEOUS DESSERT: The Lahaina Provision Company's Chocoholic Bar at the Hyatt Regency Hotel.

BEST SHOPPING: Affordable and fun - Kahului Swap Meet each Saturday, and one in Kihei too!; Touristy - Lahaina waterfront; Practical - Kaahumanu Shopping Center in Kahului; Extravagant - Hyatt Regency and Westin Maui; Odds 'n Ends - Long's Drug Store, Kahului or Lahaina and opening soon in Kihei.

BEST HOTEL VALUE: The Stouffer Wailea Beach Resort with a 50% off coupon from the Entertainment Book.

MOST SPECTACULAR RESORT GROUNDS: Hyatt Regency and the Westin Maui at Kaanapali

BEST MAUI GET AWAY FROM IT ALL RESORT: Hotel Hana Ranch

ALOHA WEAR: The traditional tourist garb is available in greatest supply at the 17,000 square foot Hilo Hattie's factory in the new Lahaina Center at the Kaanapali end of Lahaina. The muumuu factory at the Maui Mall in Kahului has a good selection as do the stores in the Kaahumanu Shopping Center, also in Kahului. Gentlemen might be especially interested in Kula Bay. On Lahainaluna just off Front St. in Lahaina, this shop takes classic aloha shirt styles of the 1930's, 40's and 50's, updates the colors, and uses an all cotton fabric for easy care and comfort.

BEST EXCURSIONS: Most spectacular - a helicopter tour.
Most unusual - a bike trip down Haleakala.
Best adventure on foot - a personalized hike with guide Ken Schmitt.
Best sailing - a day-long snorkel and picnic to Lanai with the congenial crew of the Trilogy

BEST BEACHES: Beautiful and safe - Kapalua Bay and Ulua Beach. Unspoiled - Oneloa (Makena) and Mokuleia (Slaughterhouse) Beaches.

BEST BODY SURFING: Slaughterhouse in winter (only for experienced and strong swimmers)

BEST SURFING: Honolua Bay in winter (for experienced surfers)

BEST WINDSURFING: Hookipa Beach Park (for experienced windsurfers)

BEST SNORKELING: North end - Honolua Bay in summer. Kapalua area - Kapalua Bay and Namalu Bay. Kaanapali - Black Rock at the Sheraton. Olowalu at Mile Marker 14. Wailea - Ulua Beach. Makena - Ahihi Kinau Natural Reserve. Island of Lanai - Hulopoe Beach Park. And Molokini Crater

BEST POSTCARD HOME: Patty Halbrook, Sports Photo, will take a personalized photo of you and/or your family at any location and turn it into postcards in 24 hours.

UNUSUAL GIFT IDEAS: For the green thumb, be sure to try Dan's Green House, on Prison Street in Lahaina for a Fuku-Bonsai planted on a lava rock. They are specially sprayed and sealed for either shipping or carrying home. Perfumes made from all-island products are a "scentsational" gift from the original Waikiki Fragrance Factory at the Cannery in Lahaina. Chocolate Chips of Maui, "the ultimate munchie," is a combination of rich dark chocolate or milk chocolate and the original Maui potato chip. Try them frozen or use them as a scoop for ice cream.

BEST T-SHIRTS: A huge selection is available at the T-Shirt Factory near the Kahului Airport. Sizes range from infants to XL adults and, with plenty to choose from, it is easy to mix and match styles and sizes. Crazy Shirts are more expensive, but of excellent quality and beautiful designs. The Kihei Swap Meet and the Kahului Swap Meet have an assortment of inexpensive shirts as well.

**BEST FREE STUFF:** Free Friday night walk through Lahaina Galleries "art night." Free snorkeling, scuba and windsurfing instruction clinic at Ocean Activities Center at Stouffers (879-9969). Free introductory scuba instruction offered poolside at several major resorts including Kapalua. A self-guided tour of the Hyatt Regency Maui or Westin Maui Resorts. The Maui Zoo in Wailuku. All the public beaches with their free parking. Canoe races held at Honokaoo Park. Halloween Parade in Lahaina. Free bus tour of Lahaina sponsored by the area merchants that can be picked it up behind the Wharf Shopping Center. Classic movies "under the stars" at the Wailea Village Shopping Center each Wednesday. Hula show at the Kapalua Shops (currently Thursday mornings), Aloha night each Friday at the Wharf Shopping Center. The Makawao Parade held Fourth of July weekend. Fourth of July festivities at Kaanapali Beach. Sand Castle Contest in Kihei (November).

**BEST FLORIST:** A Special Touch in Lahaina has a magical way with floral arrangements. For best flower values visit the Kahului Swap Meet (for making your own arrangements). Leis sometimes available here also.

**BEST GIFT FOR FRIENDS TRAVELING TO MAUI:**
A copy of ***MAUI, A PARADISE GUIDE*** and
a subscription to the quarterly ***MAUI UPDATE*** newsletter!

**BEST NIGHT SPOTS:** Lively evenings are available at Spatts at the Hyatt. For Hawaiian melodies at their best, visit the Molokini Lounge evenings at the Maui Prince Resort for the mellow strains of George Paoa. The new Stardancer floating on the waters off Lahaina is a wonderful new addition to evening time entertainment. More nightlife information is listed at the conclusion of the restaurant chapter.

**BEST TAKE-HOME FOOD PRODUCTS:** Paradise Fruit in Kihei

**UNUSUAL VISUAL ADVENTURE:** The Omni-Experience Theater in Lahaina.

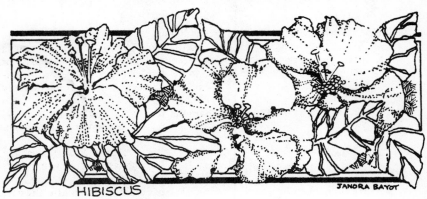

HIBISCUS                                          JANORA BAYOT

King Kamehameha I

# HISTORY

Far beneath the warm waters of the Pacific Ocean is the Pacific Plate, which moves constantly in a northwest direction. Each Hawaiian island was formed as it passed over a hot vent in this plate. Kaua'i, the oldest of the major islands in the Hawaii chain was formed first and has since moved away from the plume, the source of the lava, and is no longer growing. Some of the older islands even farther to the northwest have been gradually reduced to sandbars and atolls. The Big Island is the youngest in the chain and is continuing to grow. A new island called Lo'ihi (which means prolonged in time), southeast of the Big Island is growing and expected to emerge from the oceanic depths in about a million years.

It was explosions of hot lava from two volcanos that created the island of Maui. Mauna Kahalawai (Ma-ow-na Ka-HA-la-why) is the oldest, creating the westerly section with the highest point (elevation 5,788 ft.) known as Pu'u Kukui (Poo'oo koo-KOO-ee). The great Haleakala (HAH-leh-AH-kuh-LAH), now the world's largest dormant volcano, created the southeastern portion of the island. (The last eruption on Maui took place about 1789 and flowed over to the Makena area.) A valley connects these two volcanic peaks, hence the source of Maui's nick-name, "The Valley Isle."

The first Hawaiians came from the Marquesa and Society Islands in the central Pacific. (Findings suggest that their ancestors came from the western Pacific, perhaps as far as Madagascar). The Polynesians left the Marquesas about the 8th century and were followed by natives from the Society Islands sometime between the 11th and 14th centuries.

The Hawaiian population may well have been as high as 300,000 by the 1700's, spread throughout the chain of islands. Fish and poi were diet basics, supplemented by various fruits and occasionally meat from chickens, pigs and even dogs.

Four principal gods formed the basis of their religion until the missionaries arrived. The stone foundations of Heiaus, the ancient religous temples, can still be visited on Maui.

The islands were left undisturbed by western influence until the 1778 arrival of James Cook. He first spotted and visited Kaua'i and O'ahu and is believed to have arrived at Maui on November 25 or 26, 1778. He was later killed in a brawl on the Big Island of Hawai'i.

The major islands had a history of independent rule, with open warfare at times. On Maui, Kahului and Hana were both sites of combat between the Maui islanders and the warriors from neighboring islands.

Kamehameha the First was born on the Big Island of Hawaii about 1758. He was the nephew of Kalaiopi, who ruled the Big Island. Following the King's death, Kalaiopi's son came to power, only to be subsequently defeated by Kamehameha in 1794. The great chieftain Kahekili was Kamehameha's greatest rival. He ruled

Early Whaling Days

not only Maui, but Lana'i and Moloka'i, and also had kinship with the governing royalty of O'ahu and Kaua'i. King Kahekili died in 1794 leaving control of the island to his son. It was a bloody battle (more like a massacre since Kamehameha used western technology, strategy, and two English advisors) in the Iao Valley which resulted in the defeat of Kahekili's son, Kalanikupule, in 1795.

Kamehameha united all the islands and made Lahaina the capital of Hawaii in 1802. It remained the capital until the 1840's when Honolulu became the center for government affairs. Lahaina was a popular resort for Hawaiian royalty who favored the beaches in the area. Kaahumanu, the favorite wife of Kamehameha was born in Hana, Maui, and spent much of her time there. (Quiet Hana was another popular spot for vacationing royalty.)

Liholiho, the heir of Kamehameha the Great, ruled as Kamehameha II from 1819 to 1824. Liholiho was not a strong ruler so Kaahumanu proclaimed herself prime minister during his reign. She ended many of the kapus of the old religion thus creating a fortuitous vacuum which the soon to arrive missionaries would fill. These New England missionaries and their families arrived in Lahaina in the spring of 1823 at the invitation of Queen Keopuolani. They brought drastic changes to the island with the education of the natives both spiritually and scholastically. The first high school and printing press west of the Rockies was established at Lahainaluna. Built just outside of Lahaina, it now houses a museum, and is open to the public.

Liholiho and his wife were the first Hawaiian royalty to visit the United States. When their travels continued to Europe, they caught and succumbed to the measles while in London. Liholiho was succeeded by Kauikeaouli (the youngest son of Kamehameha the Great), who ruled under the title of Kamehameha the III from 1824 to 1854.

Beginning in 1819 and continuing for nearly 40 years, whaling ships became a frequent sight, anchored in the waters off Lahaina. The whalers hunted their prey north and south of the islands, off the Japanese coast and in the Arctic. Fifty ships were sometimes anchored off Lahaina, and during the peak year of whaling, over 400 ships visited Lahaina with an additional 167 in Honolulu's harbor. Allowing 25 to 30 seamen per ship you can quickly see the enormous number of sailors who flooded the area. While missionaries brought their Christian beliefs, the whaling men lived under their own belief that there was "No God West of the Horn." This presented a tremendous conflict between the sailors and missionaries, with the islanders caught right in the middle.

After months at sea, sailors arrived in Lahaina anxious for the grog shops and native women. It was the missionaries who put the island girls in muu-muus and set up guidelines that forbid them to visit the ships in the harbor. In 1832, a coral fort was erected near the Lahaina harbor following an incident with the unhappy crew of one vessel. The story goes that a captain, disgruntled when he was detained in Lahaina for enticing "base women," ordered his crew to fire shots at the homes of some Lahaina area missionaries. Although the fort was demolished in 1854, remnants of the coral were re-excavated and a corner of the old fort reconstructed. It is located harborside by the Banyan Tree.

An interesting fact is reported in the 1846 Lahaina census. The count included 3,445 Hawaiians, 112 foreigners, 600 seamen, 155 adobe houses, 822 grass houses, 59 stone and wooden houses, as well as 528 dogs!

A combination of things brought the downfall of the whaling industry. The onset of the Civil War depleted men and ships, (one Confederate warship reportedly set 24 whaling vessels ablaze) and the growth of the petroleum industry lessened the need for whale oil. Lastly, the Arctic freezes of 1871 and 1876 resulted in many ships being crushed by the ice. Lahaina, however, continues to maintain the charm and history of those bygone whaling days.

The whaling era strengthened Hawaii's ties with the United States economically and the presence of the missionaries further strengthened this bond. The last monarch was Liliuokalani, who ruled from 1891 to 1893. Hawaii became a territory of the United States in 1900 and achieved statehood in 1959.

Sugar cane brought by the first Hawaiians was developed into a major industry on Maui. Two sons of missionaries, Henry P. Baldwin and Samuel T. Alexander played notable roles, and their construction of a water pipeline to irrigate the arid central isthmus of Maui thus securing the future of the sugar industry and other agricultural development on the island.

Pineapple, another major agricultural industry, has played an important role in the history of Maui. Historians believe that pineapple may have originated in Brazil and was introduced to the modern world by Christopher Columbus on return from his second visit to the Americas. When it arrived in the islands is uncertain, but Don Francisco de Paula y Marin writes, in his diary on January 21, 1813 that "This day I planted pineapples and an orange tree."

The first successful report of pineapple agriculture in Hawaii is attributed to Captain James Kidwell, an English horticulturist. He brought the smooth cayenne variety of pineapple from Jamaica and began successfully raising and harvesting the fruit on O'ahu in 1886. Since the fresh fruits perished too quickly to reach the mainland, Captain Kidwell also began the first cannery, called Hawaiian Fruit and Packing Company, which operated until 1892 when it was sold to Pearl City Fruit Company. James Dole, a young Harvard graduate arrived on O'ahu from Boston in 1899, and by 1901 had established what has today become known as the Dole Pineapple Company.

Grove Ranch and Haleakala Ranch Company both began pineapple cultivation on Maui in 1906. Baldwin Packers began as Honolua Ranch and was owned by Henry Baldwin started in 1912.

The Grove Ranch hired David T. Fleming as company manager and began with several acres in Haiku which soon increased to 450 acres. W. A. Clark succeeded Fleming as Grove Ranch manager and while the acreage increased, for some unknown reason the pineapples failed. For ten years the fields were leased to Japanese growers who were successful. During these early years Haleakala Ranch Company continued to expand their acreage and to successfully produce pineapples. J. Walter Cameron arrived from Honolulu to become manager of Haleakala

Ranch Company in about 1925. In 1929 the ranch division was divided from the pineapple division and the company became Haleakala Pineapple Company. In 1932 the Haleakala Pineapple Company and Grove Ranch merged forming Maui Pineapple Company Limited and thirty years later in 1962, Baldwin Packers merged with Maui Pineapple Company to form what we know today as Maui Land and Pineapple.

Maui Land and Pineapple continues to raise pineapples as well as develop land into the fine resort area known as Kapalua.

The company owns 29,800 acres of land and uses 7,300 acres for company operations while employing approximately 1,800 people on a year-round or seasonal basis. While competition from abroad, particularly Thailand, has been fierce, Maui Land and Pineapple has chosen to maintain their market by supplying a quality product. Maui Land and Pineapple Company is the only 100% Hawaiian producer of canned pineapple in the world.

It was about 100 years ago that the first Macadamia Nut trees arrived from Australia. They were intended to be an ornamental tree, since they had nuts that were extremely difficult to crack. It was not until the 1950's that the development of the trees began to take a commercial course. Today, some sugarcane fields are being converted to Macadamia. It is a slow process, taking seven years for the grafted root (they do not grow from seed) to become a producing tree. However, beware of their hazards, 1/2 ounce of nuts contains 100 calories!

The Kula area of Maui has become the center for many delicious fruits and vegetables as well as the unusual Protea flower, a native of South Africa. Wineries are also making a comeback with the opening of the Tedeschi Winery a few years ago. They first began producing an unusual pineapple wine. In 1984, they introduced a champagne, and in 1985, a red table wine. Be sure to also sample the very sweet Kula onions (these are not the same as "Maui onions" that can be grown anywhere in Maui County) raised in this area and are available for shipping home. Coffee is currently being tested for commercial feasibility in Upcountry.

PINEAPPLE

JANORA BAYOT

# MAUI'S NAMES AND PLACES

Haiku  (HAH-ee-KOO)  abrupt break
Haleakala  (HAH-leh-AH-kuh-LAH)  house of the sun

Hali'imaile  (HAH-LEE-'ee-MAH-ee-leh)  maile vines spread
Hana  (HAH-nuh)  rainy land

Honoapiilani  (HOH-noh-AH-PEE-'ee-LAH-nee)  bays of Pi'ilani
Honolua  (HOH-noh-LOO-uh)  double bay

Iao  (EE-AH-oh)  cloud supreme
Kaanapali  (KAH-AH-nuh-PAH-lee)  land divided by cliffs

Kahana  (Kuh-HAH-nuh)  meaning unknown, of Tahitian origin
Kaho'olawe  (kuh-Ho-'oh-LAH-veah) taking away by currents

Kahului  (Kah-hoo-LOO-ee)  winning
Kapalua  (KAH-puh-LOO-uh)  two borders

Kaupo  (KAH-oo-POH)  night landing
Ke'anae  (keh-'uh-NAH-eh)  the mullet

Keawakapu  (Keh-AH-vuh-KAH-poo)  sacred harbor
Kihei  (KEE-HEH-ee)  shoulder cape

Lahaina  (LAH-HAH-ee-NAH)  unmerciful sun
Lana'i  (LAH-NAH-ee)  meaning lost

Maalaea  (MAH-'uh-LAH-eh-uh)  area of red dirt
Makawao  (mah-kah-wah-oh)  forest beginning

Makena  (Mah-KEH-nuh)  abundance
Napili  (NAH-PEE-lee)  pili grass

Olowalu  (oh-loh-wah-loo)  many hills
Paia  (PAH-EE-uh)  noisy

Pukalani  (poo-kah-lah-nee)  sky opening
Ulupalakua  (OO-loo-PAH-luh-KOO-uh)  ripe breadfruit

Waianapanapa  (WAH-ee-AH-NAH-puh-NAH-puh)  glistening water
Wailea  (WAH-ee-LEH-uh)  water Lea (Lea was the canoe maker's goddess)

Wailua  (WAH-ee-LOO-uh)  two waters
Wailuku  (WAH-ee-LOO-KOO)  water of slaughter

# HAWAIIAN WORDS/MEANINGS

alii  (ah-lee-ee)  chief
aloha  (ah-loh-hah)  greetings

hale  (Hah-lay)  house
hana  (HAHA-nah)  work

Heiau  (heh-ee-ah-oo)  temple
kai  (kye)  ocean

kahuna  (kah-HOO-nah)  teacher, priest
Kamaaina  (Kah-mah-ai-nuh)  native born

kane  (kah-nay)  man
kapu  (kah-poo)  keep out

keiki  (kayee-kee)  child
lanai  (lah-nah-ee)  porch or patio

lomi lomi  (loh-mee-LOH-mee)  to rub or massage
luau  (loo-ah-oo)  feast

makai  (mah-kah-ee)  toward the ocean
mauka  (mah-oo-kah)  toward the mountain

mauna  (MAU-nah)  mountain
mele  (MAY-leh)  Hawaiian song or chant

menehune  (may-nay-hoo-nee)  Hawaiian dwarf or elf
moana  (moh-ah-nah)  ocean

nani  (NAH-nee)  beautiful
ono  (oh-no)  delicious

pali  (PAH-lee)  cliff, precipice
paniolo  (pah-nee-ou-loh)  Hawaiian cowboy

pau  (pow)  finished
pua  (POO-ah)  flower

puka  (POO-ka)  a hole
pupus  (poo-poos)  appetizers

wahine  (wah-hee-nay)  woman
wiki wiki  (wee-kee wee-kee)  hurry

# WHAT TO PACK

When traveling to paradise, you won't need too much. Comfortable shoes are important for all the sightseeing. Sandals are the norm for footwear. Dress is casual for dining. Many restaurants require men to wear sport shirts with collars, but only one or two require a tie. Clothes should be lightweight and easy care. Cotton and cotton blends are more comfortable for the tropical climate than polyesters. Shorts and bathing suits are the dress code here! A lightweight jacket with a hood or sweater is advisable for evenings and the occasional rain showers. The only need for warmer clothes is if your plans should include hiking or camping in Haleakala Crater. While it may start out warm and sunny, the weather can change very quickly in Upcountry. Even during the daytime, a sweater or light jacket is a good idea when touring upcountry. (The cooler weather here is evidenced on the roofs of the homes where chimney stacks can be spotted.) Tennis shoes or hiking shoes are a good idea for the rougher volcanic terrain of Haleakala or hiking elsewhere as well. Sunscreens are a must. A camera, of course, needs to be tucked in. Many visitors are taking their memories home on video tape, and VHS rental units are available around the island. Binoculars are an option and may be well used if you are traveling between December and April when the whales arrive for their winter vacation. Special needs for traveling with children are discussed in the next section. Anything that you need can probably be purchased once you arrive. Don't forget to leave some extra space in those suitcases for goodies that you will want to take back home!

# TRAVELING WITH CHILDREN

Traveling with children can be an exhausting experience for parents and children alike, especially when the trip is as long as the one to Maui. There are an increasing number of direct flights to Maui out of Seattle, San Francisco, Los Angeles, Chicago and Dallas, which saves stopping over in Honolulu. These flights are very popular and fill up well in advance. Packing a child's goody bag for the long flight is a must. A few new activity books or toys that can be pulled out enroute can be sanity saving. Snacks (boxes of juice are a favorite with our children) can tide over the little ones at the airport or on the plane while awaiting your food/drink service. A thermos with a drinking spout works well and is handy for use during vacations. A change of clothes and a swim suit for the little ones can be tucked in your carry-on bag. (Suitcases have been known to be lost or delayed.) Another handy addition is a small nightlight as unfamiliar accommodations can be somewhat confusing for little ones during the bedtime hours.

Young children may have difficulty clearing their ears when landing. Many don't realize that cabins are pressurized to approximately the 6,000 foot level during flight. To help relieve the pressure of descent, have infants nurse or drink from a bottle, and older children may benefit from chewing gum. If this is a concern of yours, consult with your pediatrician about the use of a decongestant prior to descent.

**CAR SEATS:** By law, children under 4 must travel in child safety seats. While most rental agencies do have car seats for rent, you need to request them well in advance as they have a limited number. The one, and only, car seat we have rented had seen better days, and its design was only marginal for child safety. Prices run about $20- $25 per week or $4-$5 per day. You may wish to bring your own with you. Several styles are permitted by the airlines for use in flight, or it may be checked as a piece of baggage.

**BABYSITTING:** Most hotels have some form of babysitting service which runs about $7 an hour. *Babysitting Services of Maui* is an independent company that will send the sitter to your hotel or condominium. Their rates are $7 per hour with a three hour minimum. Fifty cents each additional child in the family and $2.50 an hour for each additional child from another family. (i.e. $9.50 an hour for two families with two children, $10.50 an hour for 2 families with two children each.) Contact Tony at 661-0558. We have used several of their people and found them all to be efficient and caring. Another agency is *Keiki Kids* which charges $7 for two children with a three hour minimum in the Kihei/Wailea/Makena areas. They can arrange sitters in West Maui for an additional charge. Phone Carla Gandy or Kerri DeBeer at 879-2522. With any of these agencies, or through your hotel, at least a 24 hour notice is requested. We suggest phoning them *as soon* as you have set up your plans. At certain times of the year, with the limited number of sitters available, it can be nearly impossible to get one. If you can plan out your entire vacation babysitting needs, it might also be possible to schedule the same sitter for each occasion.

Check with your condo office as they sometimes have numbers of local sitters. As you can easily figure from the above rates, spending much time away from your children can be costly. Consider the feasiblity of bringing your own sitter, it may actually be less expensive, and certainly much more convenient. This has worked well for us on numerous occasions. We have also tried contacting the Maui high schools, and even advertised in the local papers, with no success.

**CRIBS:** Most condos and hotels offer cribs for a rental fee that may vary from $2 to $10 per night. Companies such as Maui Rent (877-5827) charge $6 a day, $25 a week, and $35 for two weeks. For an extended stay you might consider purchasing one of the wonderful folding cribs that pack up conveniently. We tried the Fisher-Price version that weighs 20 pounds and fits into a small duffle bag. It checked easily as luggage and is very portable for car travel as well. We have heard recommendations about other types including the Snugli's carrier bed and diaper bag combo that zips easily to form a compact shoulder carryall, and the Houdini full-size playpen by Kantwent which weighs only 16 pounds and folds into a 8x8x42 inch space. Each of the units may be purchased for less than the cost of a ten-day rental fee. If you don't want to pack along a stroller, Fun Rentals (661-3053) in Lahaina rents them for $7 a day, $32 a week.

**EMERGENCIES:** There are several clinics around the island which take emergencies or walk-in patients. Your condominium or hotel desk can provide you with suggestions, or check the phone book. Kaiser Permanente Medical Care Facilities are located in Wailuku (243-6000) and in Lahaina (661-0081). See the section on Helpful Information for additional numbers.

**BEACHES/POOLS:** Among the best beaches for fairly young children are the Lahaina and Puunoa beaches in Lahaina, where the water is shallow and calm. Kapalua Bay is also well protected and has fairly gentle wave action. Remember to have children well supervised and wearing floatation devices, for even the calmest beaches can have a surprise wave. Several of the island's beaches offer lifeguards, among these are the Kamaole I, II, and III beaches in Kihei. Kamaole III Park also has large open areas and playground equipment. In the Maalaea area, follow the road down past the condominiums to the public access for the beach area. A short walk down the kiawe lined beach, to the small rock jetty with the large pipe, you'll discover a seawater pool on either side that is well protected and ideal for the younger child. Another precaution on the beach that is easily neglected is the application of a good sunscreen. (Remember to reapply after swimming.) A number of complexes have small shallow pools designed with the young ones in mind. These include the Marriott, Kaanapali Alii, Sands of Kahana and the Kahana Sunset. We recommend taking a life jacket or water wings (floaties) with you. Packing a small inflatable pool for use on your lanai or courtyard may provide a cool and safe retreat for your little one. Typically Maui resorts and hotels DO NOT offer lifeguard services.

**ENTERTAINMENT:** Currently there two theaters in Kahului. The Wharf Shopping Center in Lahaina has a tri-cinema with first run movies and a $1.99 early admission! There are a number of video stores which rent movies and equipment. Free "classic" movies are offered Wednesday nights at the Wailea Shopping Village under the stars on the Village Green. Bring a chair, mat or beach towel to sit on the ground. Call 879-4465 to verify the time and movie.

In the Lahaina/Kaanapali area, the colorful *Sugar Cane Train* runs a course several times a day along Honoapiilani Highway from Kaanapali to Lahaina. Transportation can be purchased alone or in combination with a trip on the Lin Wa, a glass bottom boat. Disembarking from the train at the Lahaina terminal, you will board a red, double decker bus for the short drive to the Lahaina Harbor. The Lin Wa trip lasts over an hour and includes fish viewing and the option of a brief swim before returning to port. (See Land Tours for additional details). There is time for a stroll or a visit to the missionary homes before returning to the train for the trip home or combine your train excursion with a trip to the Omni-Theater.

SUGAR TRAIN

Flashlights can turn the balmy Hawaiian evenings into adventures! One of the most friendly island residents is the Bufo (Boof-oh). In 1932 this frog was brought from Puerto Rico to assist with insect control in the cane fields. Today this large toad still emerges at night to feed or mate and seems to be easier to spot during the winter months, especially after rain showers. While they can be found around most condominiums, Kawiliki Park (the area behind the Luana Kai, Laule'a and several other condominium complexes with access from Waipulani Road off South Kihei Road) seems to be an especially popular gathering spot. We also enjoy searching for beach crabs and the African snails which have shells that may grow to a hefty five inches.

The other Hawaiian creature that cannot go without mention is the gekko. They are finding their way into the suitcases of many an island visitor, in the form of tee-shirts, sunvisors and jewelry. This small lizard is a relative of the chameleon and grows to a length of three or four inches. They dine on roaches, termites, mosquitos, ants, moths and other pesky insects. While there are nearly 800 species of geckos found in warm climates around the world, there are only about five varieties found in Hawaii. The house gecko is the most commonly found, with tiny rows of spines that circle its tail, while the mourning gecko has a smooth, satiny skin and along the middle of its back it sports pale stripes and pairs of dark spots. The mourning gecko species is parthenogenic. That means that there are only females which produces fertile eggs, no need for a mate! The stump-toed variety is distinquished by its thick flattened tail. The tree gecko enjoys the solitude of the forests, and the fox gecko, with a long snout and spines along its tail, prefers to hid around rocks or tree trunks. The first geckos may have reach Hawaii with early voyagers from Polynesia, but the house gecko may have arrived as recently as the 1940's, along with military shipments to Hawaii. Geckos are most easily spotted at night when they seem to enjoy the warm lights outside your door. We have heard they each establish little territories where they live and breed so you will no doubt see them around the same area each night. They are very shy and will scurry off quickly. Sometimes you may find one living in your hotel or condo. They're friendly and beneficial animals and are said to bring good luck, so make them welcome.

If you headquarter your stay near the Papakea Resort in Honokowai, you might take an adventurous nighttime reef walk. If an evening low tide does not conflict with your children's bedtime, put on some old tennis shoes and grab a flashlight. (Flashlights that are waterproof or at least water resistant are recommended.) The reef comes right into shore at the southern end of Papakea where you can walk out onto it like a broad living sidewalk. Try and pick a night when the low tide is from 9 - 11 pm (tide information is available in the Maui News or call the recorded weather report) and when the sea is calm. Searching the shallow water will reveal sea wonders such as fish and eels that are out feeding. Some people looked at us strangely as we pursued this new recreation, but our little ones thought it an outstanding activity. Shoes are a must as the coral is very sharp. Afterwards, be sure to thoroughly clean your shoes promptly with fresh water or they will become horribly musty smelling.

***The Maui Youth Theater***, having recently lost its performance hall in Puuene, will be on the move for the next couple of years until it relocates to the Kahului

Performing Arts Center, currently under construction. To find out where and when performances are scheduled, call 871-7484.

Also located in Kahului is the **Maui Zoo**. Admission is free and, while only a limited number of animals make their home here, it's a great stop off. Bring along a picnic lunch! For more information see WHERE TO STAY - WHAT TO SEE, Wailuku & Kahului.

Many restaurants offer a Keiki (children's) menu. There are also an assortment of Burger King's and McDonald's on the island. **Paradise Fruits** in Kihei has terrific yogurt shakes and healthy snacks or sandwiches along with a good selection of fresh fruits and vegetables. Inexpensive hamburgers and sandwiches are also available at the Azeka's Market take-out counter.

The annual Keiki Fishing Tournament is held sometime during July each year in Kaanapali. The large pond in the golf course is stocked with fish for the event.

The local bookstores offer a wealth of wonderful Hawaiian books for children. A young snorkeler might enjoy *Hawaiian Reef Fishes*, a coloring book by Lori Randall featuring forty reef fish with background information on each. The *Hawaiian Animal Life* coloring book by Sean McKeown has more than just pictures to color. The text presents interesting factual information designed to stimulate each child with a fundamental knowledge of Hawaii's birds, reptiles, amphibians and mammals. A number of colorful Hawaiian folk tales may be a perfect choice to take home for your children to enjoy, or as a gift. *My Travels in Hawaii* is a 56-page activity book that highlights scenic spots on all the islands.

Check with your resort concierge for additional youth activities. During the summer months, Christmas holidays and Easter, many of the resort hotels offer partial or full day activities for children. Rates range from free to $35 per day.

# TRAVEL TIPS FOR THE
# PHYSICALLY IMPAIRED

Make your travel plans well in advance and inform hotels and airlines when making your reservations that you are handicapped. Most facilities will be happy to accommodate. Bring along your medical records in the event of an emergency. It is recommended that you bring your own wheelchair and notify the airlines in advance that you will be transporting it. There are no battery rentals available on Maui. Other medical equipment rental information is listed below.

**ARRIVAL AND DEPARTURE:** On arrival at the Kahului airport terminal, you will find the building easily accessible for mobility impaired persons. Two parking areas are located in front of the main terminal for disabled persons. Restrooms with handicapped stalls (male and female) are also found in the main terminal. If you are unable to use the steps of the boarding ramps, you will need to be boarded with a special lift. Advance notification to the airlines is important.

***TRANSPORTATION:*** There is no public transportation on Maui and taxi service can be spendy. The only two car rental companies providing hand controls are Avis and Hertz. See the Rental Car listing for phone numbers. They need some advance notice to install the equipment. The Maui Economic Opportunity Center operates a van with an electric lift for local residents, however, visitors can make arrangements with them by calling (808) 877-7651.

***MEDICAL SERVICES AND EQUIPMENT:*** Maui Memorial Hospital is located in Wailuku (808-242-2036). There are also good clinics in all areas of the island. Check the local directory. Several agencies can assist in providing personal care attendants, companions, and nursing aides while on your visit. Maui Center for Independent Living (808-242-4966) provides personal care attendants, as does Aloha International Employment Service Health Care Registry (808-871-6373). Medical Personnel Pool (808-877-2676).

The following companies offer medical equipment rentals. Lahaina Pharmacy, Lahaina Shopping Center (808-661-3119) has wheelchairs, crutches, canes, and walkers with delivery by special arrangement. Maui Rent, 349 Hanakai, Kahului (808-877-5827) has walkers, wheelchairs, and shower chairs. It is again recommended that you contact them well in advance of your arrival.

***ACCOMMODATIONS:*** Each of the major island hotels offer one or more handicapped rooms including bathroom entries of at least 29" to allow for wheelchairs. Due to the limited number of rooms, reservations should be made well in advance. Information on condominium accessibility is available from the Maui Commission of the Handicapped Office.

***ACTIVITIES:*** None of the van tour companies currently offer wheelchair access. The Maui Easter Seal Society can provide information on recreational activities for the disabled traveler. Among the options are wheelchair tennis or basketball, bowling and swimming. Contact them in advance of your arrival at (808-877-4443). Wheelchair access to some of the tourist attractions may be limited. More information and phone numbers can be found in RECREATION AND TOURS.

Additional information can be obtained from the Commission of the Handicapped c/o State Department of Health, 54 High St., Wailuku, Maui 96793 (808-244-4441), or the Commission on the Handicapped, Old Federal Building, 335 Merchant St., #215, Honolulu, Hawaii 96813 (808-548-7606).

***Over the Rainbow, Inc.*** is the only travel, tour and activity agency on Maui that specializes in assisting the disabled traveler. "Imagination is your limit" they report when it comes to the activities they offer. They can assist in making reservations at a condominium, hotel or home to fit the needs of the traveler, make airport arrangements including van accessible wheelchair lifts and arrange for personal care such as attendants, pharmacists, or interpreters. As for recreation, how about snorkeling, bowling, golf, horseback riding, luaus, tennis (disabled opponent available), tours, jet skiing, or ocean kayaking! Wedding and honeymoon arrangements too. Write or call for their free brochure. 186 Mehani Circle, Kihei, Maui, HI 96753. (808) 879-5521.

# WEDDINGS - HONEYMOONS

If a Hawaiian wedding (or a renewal of vows) is in your dreams, Maui can make them all come true. While the requirements are simple, we have heard that some people have been given conflicting and confusing information. Here are a few tips, based on current requirements at time of publication, for making your wedding plans run more smoothly. The bride will need to have a rubella blood test and have the results certified by the lab and the physician. This can be done at home and brought with you. Both parties must be 18 years of age or have consent from both parents. Proof of age is required for anyone age 19 or under and evidence should be a certified copy of your birth certificate or baptismal record. A license must be purchased in person in the state of Hawaii. Call the Department of Health in Kahului (808-244-4313) for the name of a licensing agent in the area where you will be staying. The fee is currently $8. There is no waiting period once you have the license. Check with the Chamber of Commerce in Kahului for information regarding a pastor. Many island pastors are very flexible in meeting your needs, such as an outdoor location etc. The social director of the major resorts can often times assist you with your plans. For copies of current requirements and forms, write in advance to the State of Hawaii, Department of Health, Marriage License Section, 1250 Punchbowl St., Honolulu, Hi 96813 (808-548-5862).

Several independent companies are also available to handle all those important details. A basic package costs anywhere from $300 - $400. Although each company varies the package slightly, it will probably include assistance in choosing a location and getting your marriage license, a minister and a varying assortment of extras such as champagne, limited photography, cake, leis, and a bridal garter. Video taping, witnesses, or music are usually extra.

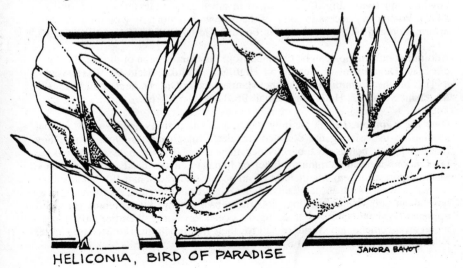

HELICONIA, BIRD OF PARADISE

JANORA BAYOT

*A Maui Wedding* (808-879-2355). 1993 S. Kihei Rd., Kihei, Maui, HI 96753. Contact Jan Lyle.

*A Wedding Made in Paradise*, PO Box 986, Kihei, Maui, HI 96753. (808-879-3444).

*Aloha Weddings* (808-244-1586) or 1-800-367-8047 ext. 540) P.O. Box 12091, Lahaina, Maui, HI 96761. Contact Cheryl Hutton.

*Arthur's Limousine Service* ★ 1-800-345-4667 or FAX 808-661-8673, P.O. Box 11865, Lahaina, Maui, HI 96761. They can provide the ultimate in wedding fantasies or simple ceremonies. Top this off with transportation in their 28-foot stretch limo to your honeymoon hotel.

*Beautiful Beginnings* (808-667-7555) 505 Front St., Suite 226, Lahaina, Maui, HI 96761. Contact Sandy Moore or Sandy Barker.

*Hawaiian Wedding Experience* (808-667-6689), P.O. Box 11093, Lahaina, Maui, HI 96761. Contact Patti O'Neill-Moore.

*Pierre of Lahaina Studios* (808-667-7988), 129 Lahainaluna Rd., Lahaina, Maui, HI 96761.

*Royal Hawaiian Weddings* 1-800-657-7857 U.S. (808-877-8811) P.O. Box 424, Puunene, HI 96784. Andrea Thomas and Janet Sosner have been putting together the ceremonies for the most special occasions since 1977. Choose from dazzling beachside sunsets, spectacular Iao Valley, sleek yachts or remote helicopter landings. Name your dream.

*Tropic Temptations* (808-667-7105) P.O. Box 1630, Lahaina, Maui, HI 96767-1630. Contact Teresa Kikkert or Chantel Berinot.

*Weddings the Maui Way* (808-877-7711) 353 Hanamau St., Suite 21, Kahului, Maui, HI 96732. Contact Libby Valley or Richard Dickinson.

Honeymoon packages are available from a number of Maui resorts. Prices vary depending on your length of stay and current air fares. Services might include your rental car, champagne, meals, sunset sails or sporting activities. Contact the complexes of your choosing for current honeymoon package rates.

## *HELPFUL INFORMATION*

**INFORMATION BOOTHS:** Booths located at the shopping areas can provide helpful information and lots of brochures! Brochure displays are everywhere.

**MAUI VISITORS BUREAU:** (808-871-8691), 380 Dairy Rd. Kahului, Maui, HI 96732

**TELEVISION:** KBPC Cable Television, Channel 7, is designed especially with tourists in mind. Information is provided on recreation, real estate, shopping, restaurants and other points of interest.

**RADIO:** KPOA 93.5 FM plays great old and new Hawaiian music, with a daily jazz program 8 pm to 1 am Tune in and catch the local disk jockeys "talking story"! Reception available only in West Maui.

**PERIODICALS:** Maui Star, Art to Onions, This Week Maui, Maui Gold, Rent A Car Drive Guide, Maui Island Guide, The Kaanapali Beach Guide, The Kihei/Wailea Beach Guide, The Makena Beach Guide, Lahaina Historical Guide, Real Estate Maui Style, The Maui Island Guide, Gold Coast News and A Taste of Maui are all free publications available almost everywhere. Most of these free publications offer lots of advertising, however, they do have coupons which will give you discounts on everything from meals to sporting activities to clothing. It may save you a bit to search through these before making your purchases.

*The Bulletin* - This is primarily a T.V. guide which is published newspaper style and available at no charge. It does have features on some local events and is very popular among the residents. You may have a little more trouble locating a copy of this one.

*Maui Beach Press* - This free newspaper format weekly has informative local stories, maps, entertainment, and restaurant information.

*Lahaina News* - A small free weekly newspaper. It contains a television guide and local news.

*Gold Coast News* - A free newspaper publication listing some interesting feature articles about Kihei's sunny south shores.

*Maui News* - This is the local Maui newspaper, published Monday thru Friday, and is available for 35 cents. A good source of local information.

**SUN SAFETY:** The sunshine is stronger in Hawaii than on the mainland, so a few basic guidelines will ensure that you return home with a tan, not a burn. Use a good lotion with a sunscreen, and reapply after swimming. Moisturize after a day in the sun. Wear a hat to protect your face. Exercise self-control and stay out a limited time the first few days, remembering that a gradual tan will last longer. It is best to avoid being out between the hours of noon and three when it is the hottest. Be cautious of overcast days when it is very easy to become burned unknowingly. Don't forget that the ocean acts as a reflector and time spent in it equals time spent on the beach.

**FOR YOUR PROTECTION:** Do not leave valuables in your car, even in your trunk. Many rental car companies urge you to not lock your car as vandals cause extensive and expensive damage breaking the locks. Many companies also warn not to drive on certain roads (Ulupalakua to Hana and the unpaved portion of Hwy. 34) unless you are willing to accept liability for all damages.

### *HELPFUL PHONE NUMBERS:*

EMERGENCY:  Police - Ambulance - Fire . . . . . . . . . . . . . 911
NON-EMERGENCY POLICE:
    Lahaina . . . . . . . . . . . . . . . . . . . . . . . . . . . . . . . 661-4441
    Hana . . . . . . . . . . . . . . . . . . . . . . . . . . . . . . . . . 248-8311
    Wailuku . . . . . . . . . . . . . . . . . . . . . . . . . . . . . . . 244-6400
Poison Control . . . . . . . . . . . . . . . . . . . . . . 1-800-362-3585
Helpline (suicide & crisis center) . . . . . . . . . . . . . . 244-7407
Red Cross . . . . . . . . . . . . . . . . . . . . . . . . . . . . . . 244-0051
Consumer Protection . . . . . . . . . . . . . . . . . . . . . . 244-4387
Directory Assistance:
    Local . . . . . . . . . . . . . . . . . . . . . . . . . . . . . . (1) 411
    Inter-island . . . . . . . . . . . . . . . . . . . . . . . 1-555-1212
    Mainland . . . . . . . . . . . . . . . . . . . . 1-(area code)-555-1212
Customs . . . . . . . . . . . . . . . . . . . . . . . . . . . . . . . 877-6013
Hospital (Maui Memorial):
    Information . . . . . . . . . . . . . . . . . . . . . . . . . . 242-2036
    Emergency . . . . . . . . . . . . . . . . . . . . . . . . . . . 242-2343
Camping Permits:
    State Parks . . . . . . . . . . . . . . . . . . . . . . . . . . 244-4354
    County Parks . . . . . . . . . . . . . . . . . . . . . . . . . 243-7230
Maui Visitors Bureau . . . . . . . . . . . . . . . . . . . . . . 871-8691
Time of Day . . . . . . . . . . . . . . . . . . . . . . . . . . . . 242-0212
Information - County of Maui . . . . . . . . . . . . . . . . 243-7866
Complaint Office - County of Maui . . . . . . . . . . . . . 243-7711
Haleakala National Park Information . . . . . . . . . . . . 572-7749
Haleakala Park Headquarters . . . . . . . . . . . . . . . . . 572-9306
Haleakala Weather . . . . . . . . . . . . . . . . . . . . . . . . 871-5054
Ohe'o Headquarters . . . . . . . . . . . . . . . . . . . . . . . 248-8251
Carthaginian Whale Watch Report . . . . . . . . . . . . . 661-8527
Weather:
    Maui . . . . . . . . . . . . . . . . . . . . . . . . . . . . . . . 877-5111
    Marine (also tides, sunrises, sunsets) . . . . . . . . . . 877-3477
    Recreational Area . . . . . . . . . . . . . . . . . . . . . . . 871-5054

ORCHIDS

**COSTS PER HOUR:** Did you ever wonder what something was costing in relation to the time spent. This is what we came up with based on approximate lengths of time with average prices.

| | |
|---|---|
| $190/hr | Parasail (based on $35 for a 10 - 12 min. ride) |
| $133 | Maui helicopter tour (1 1/2 hour trip) |
| $ 75 | Sailboat charter (usually 4 - 8 hrs.) |
| $ 60 | Rolls Royce limo service |
| $ 50 | Fishing boat charter (8 hrs.) |
| $ 47 | Round trip coach airfare LA to Maui (11 hrs.) |
| $ 43 | Dinner for two at a top restaurant (2 hrs.) |
| $ 20 | Horseback rides (up to $25 per hr.) |
| $ 17.50 | Introductory scuba dive (3 hrs.) |
| $ 17.50 | 18 holes of golf at a resort course (3 1/2 hrs.) |
| $ 11.85 | Deep sea fishing - Shared boat (8 hrs.) |
| $ 11.25 | Molokini snorkel trip (4 hrs.) |
| $ 11.25 | Lanai snorkel/sail/tour (8 hrs.) |
| $ 11.25 | Haleakala bike trip (8 hrs.) |
| $ 9.38 | First class hotel room ($225/day) |
| $ 6.25 | Diver certification course (36 hrs.) |
| $ 6.20 | Haleakala sunrise van tour (6 hrs.) |
| $ 4.80 | Hana van tour (10 hrs.) |
| $ 4.16 | Moderate condominium ($100/day) |
| $ 1.25 | Rental car ($30/day) |

# GETTING THERE

The best air prices can generally be arranged through a reputable travel agent who can often secure air or air with car packages at good prices by volume purchasing. Prices can vary considerably, so comparison shopping is a wise idea.

The major American carriers that fly from the mainland to The Honolulu International Airport on O'ahu, Hawaii are:

AIR AMERICA - 1-800-247-2475, in Honolulu 808-833-4433. This newest arrival on the scene has only one L-1011 for their Los Angeles - Honolulu route, so any mechanical difficulty will cause delays. One flight each Saturday except during Christmas season.

AMERICAN AIRLINES - 1-800-433-7300, Los Angeles 213-935-6045, Honolulu 808-523-9376. Direct flights to Maui from San Francisco and Los Angeles.

AMERICAN WEST AIRLINES - 1-800-247-5692; offers serve to Honolulu through its major mainland hubs of Las Vegas and Phoenix with connecting service to over 60 cities nationwide.

CANADIAN AIR LINES INTERNATIONAL - Eighteen weekly flights from Vancouver to and from Honolulu. Then connecting inter-island carriers to Maui.

CONTINENTAL - 1-800-525-0280; in Honolulu, 808-836-7730. Currently flights only to Honolulu with connect service to Maui, no direct Maui flights.

DELTA AIR LINES - 1-800-221-1212. They fly out of Atlanta, stopping in Los Angeles, then direct flights to Maui.

HAWAIIAN AIRLINES - 1-800-367-5320; in Honolulu, 808-537-5100. Based on acquaintances, friends, and our own experiences, Hawaiian Airs lower fares must be tempered by the increased risk of flight changes, delays, cancellations, and reroutings which occur all too frequently.

NORTHWEST AIRLINES - 1-800-225-2525; in Honolulu, 808-955-2255. No direct Maui flights.

PAN AMERICAN WORLD AIRWAYS - A very limited number of flights from New York, Los Angeles, and San Diego to and from Honolulu. Connecting service Miami, Philadelphia, N.Y., Baltimore, Boston, and Washington D.C.

TWA - 1-800-321-2000; in Honolulu, 808-241-6522.

UNITED AIRLINES - United has more flights to Hawaii from more U.S. cities than any other airline. They have no central 800 number, but do have one for each area in the United States. See your telephone directory. Their Honolulu number is 808-547-2211. They have a number of direct flights to Maui from Los Angeles, Denver, Chicago, Philadelphia and San Francisco. On Maui, phone 242-7911.

The direct flights available on United, Delta, and American Airlines save time and energy by avoiding the otherwise necessary stopover on Oʻahu. Travel agents schedule at least an hour and a half between arrival on Oʻahu and departure for Maui to account for any delays, baggage transfers, and the time required to reach the inter-island terminal. If you do arrive early, check with the inter-island carrier. Very often you can get an earlier flight which will arrive on Maui in time to get your car, and maybe some groceries, before returning to pick up your luggage when it arrives on your scheduled flight. Alert! We were foiled by this terrific plan when we hopped onto an earlier flight only to find that it was a prop-jet and the flight was enough longer that we arrived at the same time we would have on our scheduled jet! Oh well!

The inter-island carriers that operate between Honolulu and Maui are:

ALOHA AIRLINES - They fly only jets - mostly 737's. 1-800-367-5250 U.S., 1-800-663-9471 Canada, 1-800-663-9396 Alberta and B.C. Their Honolulu number is 808-836-1111, on Maui 808-244-9071. This airline tends to have more respect for its schedule than the others. They fly 1,200 flights weekly with their fleet of 15 Boeing 737's. Also weekly charter service to Christmas Island and long range charters upon request.

ALOHA ISLAND AIR - (formerly Princeville) Their fleet consists of 8 - 18 passenger twin engine deHavilland Dash 6 Twin Otters (turbo-prop) aircraft. They service the Kahului, Hana and Kapalua West Maui Airports on Maui as well as all other islands. From Hawaii the toll free number is 1-800-652-6541 or 1-800-323-3343 U.S., locally (808-877-5755). Charters available.

DISCOVERY AIRLINES - The new kid on the block began service in early 1990. Phone 808-946-1500 or 1-800-733-2525 in Hawaii. They operate more than 60 daily departures between O'ahu, Kahului and Lihue with a fleet of eight British Aerospace BAe-146's. They offer both club and first class inter-island service. Initially, flights have been scheduled between O'ahu, Kaua'i and Maui with service to be added later to The Big Island.

HAWAIIAN AIRLINES - The Honolulu number is 808-537-5100, in Maui 808-244-9111. Toll free 1-800-367-5320 U.S., 1-800-882-8811 Hawaii, 1-800-663-6296 Alberta, 1-800-663-2074 Canada B.C. They fly DC9's, and deHavilland Dash 7's. They service both the Kahului and Kapalua West Maui Airports. A recent press release notes that Hawaiian has added two new MD-80 jet aircraft with more to be added during the summer of 1990 to "assure Hawaiian's inter-island passengers of an on-time and reliable schedule."

UNITED - (See phone numbers above) Wide body DC-10's are now making 21 weekly flights between Lihue, Kona and Maui, to and from Honolulu.

Most visitors arrive at the Kahului Airport, via direct or inter-island flights. The The Kahului airport is currently undergoing some major and much needed expansion. Improvements in the parking area are also underway. A new rental car center has already opened.

From the airport it is only a 20-30 minute drive to the Kihei-Wailea-Makena areas, but a 45/60 minute drive to the Kaanapali/Kapalua areas. If your destination is West Maui from O'ahu, Kaua'i, or Hawai'i, it might be more convenient to fly into the new *Kapalua West Maui Airport*. However, the airport is serviced only by a few airlines. Restrictions allow only prop-jets to land here which means the flight is lower, slower, and louder, but more scenic.

One luxurious way to see the islands is aboard one of the American Hawaii Cruises ships, the *Independence* or *Constitution*. These comfortable 700-foot (800 passenger) ship's provide adequate accommodations and friendly service during the seven day sail around the islands. They come into port at each of the major islands for a day of touring. Also available are three and four day trips combined with land accommodations, which are especially convenient for honeymooners who often arrive Sundays after the ships Saturday departures. Another option is to extend your stay on land following a seven day cruise. About half the crew are from Hawaii. For additional information write American Hawaii Cruises and Land Vacations, 550 Kierny St., San Francisco, CA 94108. Phone 1-800-765-7000 U.S. (415-392-9400 in San Francisco) or from Canada phone collect (415) 392-9400.

# GETTING AROUND

***FROM THE AIRPORT:*** After arriving, there are several options. Taxi cabs, because of the distances between areas, can be very costly (i.e., $40 from Kahului to Kaanapali). There are several bus/limo services available also. The limited around-the-island public transportation that did exist has been terminated at this time with no plans to reinstate it. There are some local area shuttles. The best option may be a rental car unless your resort provides transportation.

The Grayline (877-5507) (1-800-367-2420) provides the Kahului Airport with service to Lahaina and Kaanapali. Baggage is charged per piece. The shuttle currently departs every hour from 7 am until 5 pm. Check in at the airport desk (call to verify current schedules and price). Service from Kahului to Kihei/Wailea is $9.50 per person and to Kapalua $17.50 per person.

Akina Bus Service Ltd. offers shuttle service from the Kahului Airport to the Kihei area every hour on the hour from 8 am until 4 pm $9.50. (879-2828)

Travel in style with one of the following limo services. Rates run $48 - $60 per hour with a minimum of 1 1/2 - 2 hours. ***Arthur's Limousine Service*** ★ 1-800-345-4667 or 808-661-5466. Arthur's offers several limosines including a super-stretch Lincoln limousine with VCR, two TV's, 3 bars, and 2 sunroofs. Also a stretch Cadillac, Lincoln Continentals, Lincoln Town cars and a stretch van. ***Silver Cloud Limousine Service*** (669-8580), ***Limousines by Roberts*** 808-871-6226, and ***Maui Limousine Service*** (669-7800) or (667-7800), offer a variety of limousines and Lincoln Town cars for transportation or island tours with service to the Kahului or Kapalua-West Maui airports. Also available are chauffers as drivers for your own car.

***LOCAL TRANSPORTATION:*** If you don't choose a rental car, you will find limited public transportation. The Kaanapali Trolley that runs through the Kaanapali Beach Resort area daily between 7 am and 11 pm is FREE! The trolley takes Kaanapali guests to Lahaina with pickups at Whalers Village and the resorts. Also free is the Lahaina Cannery Shuttle running from 9 am to 10 pm with stops at the Royal Lahaina, Sheraton Maui, Kaanapali Beach Hotel, Whalers Village/Westin Maui, Maui Marriott/Hyatt Regency, Papakea, Kaanapali Shores, Embassy Suites, Maui Kai, and Mahana. In the Wailea Resort area a complimentary shuttle runs every 15-20 minutes between 6:30 am and 10:30 pm making stops at the tennis center, golf course, shopping centers, the Stouffer Wailea Beach Resort and the Maui Inter-Continental Hotel and no doubt will be adding the new resorts as well. Kapalua has a shuttle running between 6:15 am and midnight between the condos and the hotel. Call the front desk to request it.

Most van tours offer pickup at your hotel or condo.

***RENTAL CARS AND TRUCKS:*** It has been said that Maui has more rental cars per mile of road than anywhere in the nation. This is not surprising when you realize that Maui has virtually no mass transit, a population of 85,000 (1990 figures for Island of Maui), and over two million visitors per year. Recently, shuttles have been initiated to help eleviate this problem. A choice of more than 30 car rental companies offer luxury or economy and new or used models. Some are local island operators, others are nation-wide chains, but all are very competitive. The rates may vary between high and low season and the best values are during price wars, or super summer discount specials.

Given the status of public transportation on Maui, a rental car is still the best bet and sometimes only way to get around the island, and for your dollar a very good buy. Prices for compacts range from $13.95 to $20.95/day, mid-size from $20.95 to $34.95, vans from $45.95, and jeeps from $30. The least expensive choice is a late-model compact, with stick shift and no air conditioning. Often these cars are only 2 - 3 years old and in very good condition. Also available from specialty car rental agencies are a variety of luxury cars. A Porsche or Mercedes will run $200 plus per day. Currently there are no companies which rent camping equipment. Vans are available from a number of agencies, but camping in them is not allowed.

Many of the rental companies have booths near the main terminal building at the Kahului Airport. There is also a large courtesy phone board in the main terminal (not at the United terminal). This free phone is for those rental agencies not having an airport booth, or for regular shuttle service, so that you can call for a pick up. A pay phone is available in the United terminal. A few agencies will take your flight information when your car reservation is made and will meet you and your luggage at the airport with your car.

The policies of all the rental car agencies are basically the same. Most require a minimum age of 21 to 25 and a maximum age of 70. All feature unlimited mileage with you buying the gas ($1.36 - $1.40 per gallon). Be sure to fill up before you return your car, the rental companies charge about $2.25 per gallon to do it for you. A few require a deposit or major credit card to hold your reservation. Insurance is an option you may wish, which can run an additional $5 - $10 a day. A few agencies will require insurance for those under age 25. Add to the rental price a 4% sales tax.

Most of the car rental agencies strongly encourage you to take the additional insurance coverage. Hawaii is a no-fault state and without the insurance, you are required to take care of all the damages before leaving the island. We suggest you check with your own insurance company before you leave to verify exactly what your policy covers.

A few of Maui's roadways are rough and rugged. The rental agencies recommend that cars not traverse these areas (shown on the map they distribute) and that if these roads are attempted, you are responsible for any damage.

As a starting point, we would suggest you call Tropical Rent-A-Car as one possible choice. They have competitive prices and we have always been pleased with their service.

## RENTAL CAR LISTING:

ADVENTURES
UNLIMITED
877-6626

ALAMO RENT A CAR
1-800-327-9633
Kahului 877-3466
Lahaina 661-7187

ANDRES RENT A CAR
Kahului 877-5378

ARTHUR'S
LIMOUSINE SERVICE
1-800-345-4667
Lahaina 661-5466

ATLAS U DRIVE
1-800-367-5238
Kahului 687-7208

AVIS
1-800-331-1212
Kahului 871-7575
Kaanapali 661-4588

BUDGET
1-800-527-0700
Kaanapali 661-8721
Wailea 879-9150
Kahului 871-8811

CHARTON U DRIVE
Kahului 877-7836
Kaanapali 661-3489

DOLLAR RENT A CAR
1-800-367-7006
Kahului 877-6526
Kaanapali 667-2651
Kapalua 669-7400
Interisland
1-800-342-7398

HERTZ
1-800-654-8200
Kahului 877-5167
Interisland
1-800-654-3131

HONOKOWAI
HOLIDAZE
669-8787

ISLAND AUTO
LEASING
Kahului 877-0031

KIHEI HOLIDAZE
Kihei 879-1905

KIHEI RENT A CAR
Kihei 879-7257

KLUNKERS USED CARS
Kahului 877-3197

MAUI RENT A JEEP
Kahului 877-6626

NATIONAL
1-800-227-7368
Kahului 877-5347
Kaanapali 667-9737

PARADISE RENT
A USED CAR
Kihei 879-8788

PAYLESS CAR RENTAL
1-800-345-5230
Kahului 877-5600

PRACTICAL
USED CAR RENTAL
1-800-367-5238
Kahului 871-2860

RAINBOW
Lahaina 661-8734

ROBERTS
Kahului 871-6226

SEARS RENT A CAR
contracts with Budget
1-800-527-0700
Kahului 877-7764
Lahaina 661-3546

SUNSHINE
RENT A CAR
1-800-367-2977
Kahului 871-6222
Lahaina 661-5646
Interisland
1-800-522-8440

SURF RENT A CAR
(flat beds, pick-ups)
Wailuku 244-5544

THRIFTY
1-800-367-2277
Kahului 871-7596
Kaanapali    667-9541
Interisland
1-800-342-1540

TRANS MAUI
RENT A CAR
(Jeep rentals available)
1-800-367-5228
Kahului 877-5222

TROPICAL
RENT A CAR
1-800-367-5140
Kahului 877-0002
Kaanapali    661-0061
Interisland
1-800-352-3923

UNITED CAR RENTAL
(Truck rentals)
Kahului 871-7328
Lahaina 667-2688

UPTOWN SERVICE
Wailuku 244-0869
or 242-7896

VIP CAR RENTAL
1-800-367-6080
Kahului 877-2054

WORD OF MOUTH
RENT A CAR
Kahului 877-2436

# SHOPPING

ᴛo give you an idea of what to expect at the supermarket, here are some grocery store prices. Bread $1.05 - $2.49, chicken $1.09 /lb., hamburger $1.39 - $2.99 /lb., mayonnaise $1.69 - 2.15, Starkist Tuna $1.33, disposable diapers 12-24 count size $4.99 - $6.99, 32 oz. ketchup $2.49, 2% milk $1.72 half gallon.

The three major grocery stores in Kahului are Foodland, Safeway, and Star Market. In Lahaina you can choose between Foodland or Nagasako at the Lahaina Shopping Center or the new Safeway at the Cannery Shopping Center. In Kihei the major markets are Foodland, Azeka's, and Star Market. These larger stores offer the same variety as your hometown store and the prices are better than at the small grocery outlets. In Hana there is, of course, Hasegawa as well as the Hana Ranch Store.

A few of the more unusual or specialty markets include: *Paradise Fruits* ★ in Kihei (open 24 hours) offers an open-air market with a variety of fresh fruits and vegetables which are also available for shipment home. *Azeka's Market* (879-0611) at Azeka's Place in Kihei can provide you with those world famous Azeka (Kalbi) ribs for cooking. *The Farmers' Market* is a group of people who bring produce down from the Kula area. They set up roadside shopping, and you can't find it fresher. Their locations seem to change each time we visit. Just look for the green sandwich board signs that are set up roadside, or look for a brochure in the shopping center information racks.

*Take Home Maui* (661-8067) is located just off Front Street in Lahaina and offers a selection of fruits and vegetables for shipment home. *Fresh Island Fish* (244-9633 or 242-6532) is located near the dock at Maalaea Harbor, open Mon. - Sat. 10 - 5. They offer a wonderful selection of fresh island fish and a cafe open 10 am - 8 am (see restaurants). A newcomer is the *Nagasako Fish Market* on Lower Main Street in Wailuku. They have what may be the most diverse selection of fresh seafood from reef fish to live clams and crabs.

Local grocery shopping is a little more adventuresome. The largest local stores are in Wailuku and Kahului. In addition to the regular food staples, they have wonderful selections of local foods such as marinated seafoods and often have deli sections which feature local favorites and plate lunches. *Takamiya's* at 359 N. Market St. in Wailuku has a huge deli section with perhaps more than 50 cooked foods and salads as well as very fresh meats. *Ah Fook's* at the older Kahului Mall has a smaller deli section with plate lunches running about $3. *Ooka's* is the largest of the three. The packed parking lot and crowded aisles prove its popularity! Its prices are the best, and they run local coupons in mailings to island residents which add to the savings. Lots of selections in sundry items, as well as seafood and meats.

# ANNUAL MAUI EVENTS

**FEBRUARY**
- Marine/Art Expo runs two months at the Maui Inter-Continental Hotel

**MARCH**
- Annual Maui Marathon from Wailuku to Lahaina, sponsored by the Valley Isle Road Runners
- Annual Kukini Run along the Kahakuloa Valley Trail
- The 26th is Prince Kuhio Day, a state holiday.
- LPGA Women's Kemper Open at the Kaanapali Golf Course
- Maui Onion Festival - Whalers Village Shopping Center

**MAY**
- May Lei Day celebration in Wailea (check with the Inter-Continental Hotel for their events)
- Seabury Hall in Makawao sponsors their annual craft fair the Saturday prior to Mother's Day

**JUNE**
- Obon Season (late June through August) - Bon Odori festivals are held at the many Buddhist temples around the island. They are announced in the local newspapers and the public is invited.
- King Kamehameha Day Celebration
- Maui Upcountry Fair

**JULY**
- Annual 4th of July Rodeo & Parade in Makawao
- Canoe races at Hookipa State Park
- Maui Jaycees Carnival at Kahului Fairground
- Annual Sausa Cup races in Lahaina, sponsored by the Lahaina Yacht Club
- Victoria to Maui Yacht Race
- Kapalua Wine Symposium
- Keiki Fishing Tournament at Kaanapali

**AUGUST**
- Kapalua Music Festival (a week of Hawaiian & classical music)
- Run to the Sun Marathon, a grueling trek from sea level up to the 10,000 foot level of Haleakala Crater
- The 21st is Admissions Day, a state holiday.

**SEPTEMBER**
- Maui County Rodeo in Makawao
- Aloha Week Festival
- Labor Day Fishing Tournament
- Wailea Speed Crossing, a windsurfing regatta across the seven mile stretch of Pacific to Molokini and back. Sponsored by the Maui Inter-Continental

## OCTOBER
- Maui County Fair at the Kahului Fairgrounds
- Open Pro-Am Golf Championship
- Parade and Halloween festivities in Lahaina

## NOVEMBER
- Na Mele O'Maui Festival at Lahaina & Kaanapali Beach Resorts
- Kapalua International Championship of Golf
- Queen Kaahumanu Festival at the Maui High School
- Sand Castle contest, check for current beach location in Kihei, usually held Thanksgiving weekend
- Thanksgiving weekend Santa arrives at Kaahumanu Mall
- Thanksgiving, La Hoomaikai, luau celebration at Maui Inter-Continental

## DECEMBER
- Kapalua/Betsy Nagelsen Pro-Am Tennis Invitational
- Christmas House at Hui Noeau, near Makawao, is a non-profit organization featuring pottery, wreaths, and other artwork.

For the exact dates of many of these events, write to the Hawaii Visitors Bureau, 2270 Kalakaua Avenue #801, Honolulu, Hawaii 96815, and request the Hawaii Special Events Calendar. The calendar also gives non-annual information and the contact person for each event. Check the local papers for dates of additional events.

# WEATHER

When thinking of Hawaii, and especially Maui, one visualizes bright sunny days cooled by refreshing trade winds, and this is the weather at least 300 days a year. What about the other 65 days? Most aren't really bad - just not perfect. Although there are only two seasons, summer and winter, temperatures remain quite constant. Following are the average daily highs and lows for each month and the general weather conditions.

| | | | | | |
|---|---|---|---|---|---|
| January | 80/64 | May | 84/67 | September | 87/70 |
| Feb. | 79/64 | June | 86/69 | October | 86/69 |
| March | 80/64 | July | 86/70 | November | 83/68 |
| April | 82/66 | Aug. | 87/71 | December | 80/66 |

*Winter:* Mid October thru April, 70 - 80 degree days, 60 - 70 degree nights. Tradewinds are more erratic, vigorous to none. Kona winds are more frequent causing wide-spread cloudiness, rain showers, mugginess and even an occasional thunderstorm. 11 hours of daylight.

*Summer:* May thru mid October, 80 degree days, 70 - 80 degree nights. Tradewinds are more consistent keeping the temperatures tolerable, however, when the trades stop, the weather becomes hot and sticky. Kona winds are less frequent. 13 hours of daylight.

Summer type wear is suitable all year round. However, a warm sweater or light-weight jacket is a good idea for evenings and trips such as to Haleakala.

If you are interested in the types of weather you may encounter, or are confused by some of the terms you hear, read on. For further reference consult *Weather in Hawaiian Waters*, by Paul Haraguchi, 99 pages, available at island bookstores.

**TRADE WINDS:** Trade winds are an almost constant wind blowing from the northeast through the east and are caused by the Pacific anti-cyclone, a high pressure area. This high pressure area is well developed and remains semi-stationary in the summer causing the trades to remain steady over 90% of the time. Interruptions are much more frequent in the winter when they blow only 40 to 60% of the time. The major resort areas of South and West Maui are situated in the lee of the West Maui Mountains and Haleakala respectively. Here they are sheltered from the trades and the tremendous amount of rain (400 plus inches per year) they bring to the mountains.

**KONA WINDS:** The Kona Wind is a stormy, rain-bearing wind blowing from the southwest, or basically from the opposite direction of the trades. It brings high and rough surf to the resort side of the island - great for surfing and boogie-boarding, bad for snorkeling. These conditions are caused by low pressure areas northwest of the islands. Kona winds strong enough to cause property damage have occurred only twice since 1970. Lighter nondamaging Kona winds are much more common, occurring usually two to five times almost every winter (November thru April).

**KONA WEATHER:** Windless, hot and humid weather is referred to as Kona weather. The interruption of the normal trade wind pattern brings this on. The trades are replaced by light and variable winds and, although this may occur any time of the year, it is most noticeable during the summer when the weather is generally hotter and more humid, with fewer localized breezes.

**KONA LOW:** A Kona low is a slow-moving, meandering, extensive low pressure area which forms near the islands. This causes continuous rain with thunderstorms over an extensive area and lasts for several days. November through May is the most usual time for these to occur.

**HURRICANES:** Hawaii is not free of hurricanes. However, most of the threatening tropical cyclones have weakened before reaching the islands, or have passed harmlessly to the west. Their effects are usually minimal causing only high surf on the eastern and southern shores of some of the islands. At least 21 hurricanes or tropical storms have passed within 300 miles of the islands in the last 33 years, but most did little or no damage. Only Hurricane Dot of 1959 and Hurricane Iwa of 1982 caused extensive damage. In both cases, the island of Kaua'i was hit hardest, with lesser damage to southeast O'ahu and very little damage to Maui.

**TSUNAMI:** A tsunami is an ocean wave produced by an undersea earthquake, volcanic eruption, or land slide. Tsunamis are usually generated along the coasts of South America, the Aleutian Islands, the Kamchatka Peninsula, or Japan and travel through the ocean at 400 to 500 miles an hour. It takes at least 4 1/2 hours for a tsunami to reach the Hawaiian Islands. A 24-hour Tsunami Warning System has been established in Hawaii since 1946. When the possibility exists of a tsunami reaching Hawaiian waters, the public will be informed by the sound of the attention alert signal sirens. This particular signal is a steady one minute siren, followed by one minute of silence, repeating as long as necessary. Immediately turn on a TV or radio; all stations will carry CIV-Alert emergency information and instructions with the arrival time of the first waves. Do not take chances - false alarms are not issued. Move quickly out of low lying coastal areas that are subject to possible inundation.

The warning sirens are tested throughout the state on the first working Monday of every month at 11 am. The test lasts only a few minutes and CIV-Alert announces on all stations that the test is underway. Since 1813, there have been 112 tsunamis observed in Hawaii with only 16 causing significant damage.

Tsunamis may also be generated by local volcanic earthquakes. In the last 100 years there have been only six, with the last one November 29, 1975, affecting the southeast coast of the island of Hawaii. The Hawaiian Civil Defense has placed earthquake sensors on all the islands and, if a violent local earthquake occurs, an urgent tsunami warning will be broadcast and the tsunami sirens will sound. A locally generated tsunami will reach the other islands very quickly, therefore, there may not be time for an attention alert signal to sound. Any violent earthquake that causes you to fall or hold onto something to prevent falling is an urgent warning, and you should immediately evacuate beaches and coastal low-lying areas.

For additional information on warnings and procedures in the event of a hurricane, tsunami, earthquake or flash flood, read the civil defense section located in the forward section of the Maui phone book.

***TIDES:*** The average tidal range is about two feet. Tide tables are available daily in the Maui News or by calling the marine weather number, 877-3477.

***SUNRISE AND SUNSET:*** In Hawaii, day length and the altitude of the noon sun above the horizon do not vary as much throughout the year as at the temperate regions because of the island's low latitude within the sub-tropics. The longest day is 13 hours 26 minutes (sunrise 5:53 am, sunset 7:18 pm) at the end of June, and the shortest day is 10 hours 50 minutes (sunrise 7:09 am and sunset 6:01 pm at the end of December). Daylight for outdoor activities without artificial lighting lasts about 45 minutes past sunset.

# Where to Stay—
# What to See

## *INTRODUCTION*

Maui has more than 16,000 hotel rooms and condominium units in vacation rental programs, with the bulk of the accommodations located in two areas. These are West Maui, a 10-mile stretch between Lahaina and Kapalua, and the South shore of East Maui, which is also about ten miles of coastline between Maalaea and Makena. On the northern side, there are four properties near the Kahului Airport, several complexes in Hana and one in upcountry. This chapter contains a list of essentially all of the condominiums that are in rental programs, as well as the island's hotels. Bed and Breakfast homes are sprinkled around the island.

***HOW TO USE THIS CHAPTER:*** For ease in locating information, the properties are first indexed alphabetically following this introduction. In both South and West Maui, the condominiums have been divided into groups that are geographically distinct and are laid out (sequentially) as you would approach them arriving from the Kahului area. These areas also seem to offer similar price ranges, building style, and beachfronts. At the beginning of each section is a description of the area, sights to see, shopping information, best bets and a sequential listing of the complexes. For each complex, we have listed the local address and/or P.O. Box and the local and toll-free phone numbers. Often times the management at the property does reservations, other times not. In many cases there are a variety of rental agents handling units in addition to the on-site management and we have listed an assortment of these. We suggest that when you determine which condo you are interested in that you call all of the agents. There are also several agents which book just about all the properties on Maui and we have not listed these under each property. Two of these are *Maui 800* and *Gentle Island Holidays* (phone numbers listed under rental agents). Prices may vary and so will availability. While one agent may tell you they have no vacancy, another will have several. The prices we have listed are generally the lowest available (although some agents may offer lower rates but with the reduction of certain services such as maid service on check in only - that means your room is clean when you arrive - rather than daily maid service). At the end of the accommodations chapter is an alphabetical listing of rental agents and the properties they handle. Prices can vary, sometimes greatly, from one agent to another, so we suggest again that you contact them all. Prices are listed to aid your selection and, while these were the most current available at press time, they are subject to change without notice. As island vacationers ourselves, we found it important to include this feature rather than just giving you broad categories such as budget or expensive. After all, one person's "expensive" may be "budget" to someone else!

43

For the sake of space, we have made use of several abbreviations. The size of the condominiums are identified as studio (S BR), one bedroom (1 BR), two bedroom (2 BR) and three bedroom (3 BR). The numbers in parenthesis refers to the number of people that can occupy the unit for the price listed and that there are enough beds for a maximum number of people to occupy this unit. The description will tell you how much it will be for additional persons over two, i.e. each additional person $6/night. Some facilities consider an infant as an extra person, others will allow children free up to a specified age. The abbreviations o.f., g.v., and o.v. refer to oceanfront, gardenview and oceanview units. The prices are listed with a slash dividing them. The first price listed is the high season rate, the second price is the low season rate. A few have a flat yearly rate so there will be only be a single price.

All listings are condominiums unless specified as a (Hotel). Condos are abundant, and the prices and facilities they offer can be quite varied. We have tried to indicate our own personal preferences by the use of a ★. We felt these were the best buys or special in some way. However, it is impossible for us to view all the units within a complex, and since condominiums are privately owned, each unit can vary in its furnishings and its condition.

***WHERE TO STAY:*** As for choosing the area of the island in which to stay, we offer these suggestions. The Lahaina and Kaanapali areas offer the visitor the hub of the island's activities, but accommodations are a little more costly. The beaches are especially good at Kaanapali. The values and choice of condos are more extensive a little beyond Kaanapali in Honokowai, Kahana (Lower Honoapiilani Hwy. area) and further at Napili. However, there are fewer restaurants here and slightly cooler temperatures. Some of the condominiums in this area, while very adequate, may be a little overdue for redecorating. While many complexes are on nice beaches, many are also on rocky shores. Kapalua offers high class and high price condominium and hotel accommodations. Maalaea and Kihei are a half-hour drive from Lahaina and offer some attractive condo units at excellent prices and, although few are located on a beach, there are plenty of easily accessible public beach parks. Many Maui vacationers feel that Kihei offers better weather in the winter months. There are plenty of restaurants here and an even broader selection by driving the short distance to Wailuku. The Wailea and Makena areas are just beyond Kihei and are beautifully developed resort areas. Wailea is experiencing a recent boom in development with a number of plush resorts and condominiums in varying stages of construction. The beaches in this area are excellent for a variety of water activities, however, this area is significantly more expensive than the neighboring Kihei area. We suggest reading the introductory section to each area for additional information.

***HOW TO SAVE MONEY:*** Maui has two "price" seasons. High or "in" season and low or "off" season. Low season is generally considered to be April 15 to December 1, and the rates are discounted at some places as much as 30%. Different resorts and condominiums may vary these dates by as much as two weeks and a few resorts are going to a flat, year round rate. Ironically, some of the best weather is during the fall when temperatures are cooler than summer and there is less rain than the spring months. (See GENERAL INFORMATION - weather for year round temperatures).

For longer than one week, a condo unit with a kitchen can result in significant savings on your food bill. While this will give you more space than a hotel room and at a lower price, you may give up some resort amenities (shops, restaurants, maid service, etc.) There are several large chain grocery stores around the island with fairly competitive prices, although most things at the store will run slightly higher than on the mainland. (See GENERAL INFORMATION - Shopping.)

Money can be saved by using the following tips when choosing a place to settle. First, it is less expensive to stay during the off or low season. Second, there are some areas that are much less expensive. Although Kahului has some motel units, we can't recommend this area as a place to headquarter your stay. The weather is wetter in winter, hotter in summer, generally windier than the other side of the island, and there are few good beaches. Two recently renovated old hotels in Wailuku now offer serviceable, basic and affordable accommodations for the budget minded, and they should especially appeal to the windsurfing community with nearby Hookipa Beach. There are some good deals in the Maalaea and Kihei areas, and the northern area above Lahaina has some older complexes that are reasonably good values. Third, some condo type units without kitchens are less expensive, but you must weigh the cost savings versus doing your own cooking. Fourth, there are some pleasant condo units either across the road from the beach or on a rocky, less attractive beach. This can represent a tremendous savings, and there are always good beaches a short walk or drive away. Fifth, hotel rooms or condos with garden or mountain views are less costly than oceanview or ocean-front rooms. We find the mountain view, especially in Kaanapali, to be, in fact, superior. Not only are the mountains gorgeous, but your room does not get the full day sunlight and stays cooler.

There is a growing trend to offer only limited maid service in the condominiums, perhaps only on check out or once a week. Additional maid service is usually available for an extra charge. Rooms without telephones or color televisions usually have lower prices, and a few condominiums do not have pools. A few words of caution, condominium units within one complex can differ greatly and, if a phone is important to you, ask! More complexes are adding phone service to their rooms, however, there are still some that have only a courtesy phone or a pay phone at the office. Some may also add 50 - 75 cents per in-room local call, others have no extra charge. Some units have washers and dryers in the rooms, while others do not.

Travel agents will be able to book your stay in the Maui hotels and also in most condominiums. If you prefer to make your own reservation, we have listed the various contacts for each condominium and endeavored to quote the best price generally available. Rates vary between rental agents, so check all those listed for a particular condominium. We have indicated toll free 800 numbers for the U.S. when available. For additional Canadian toll free numbers check the rental agent list at the end of this chapter. Look for an 808 area code preceding the non-toll free numbers. You might also check the classified ads in your local newspaper for owners offering their units, which may be a better bargain.

Although prices can jump (and have done so in recent years), most go up only 5-10% per year. Prices listed do not include sales tax.

*GENERAL POLICIES:* Condominium complexes require a deposit, usually equivalent to one or two nights stay, to secure your reservation and insure your room rate from price increases. Some charge higher deposits during winter or over Christmas holidays. Generally a 30 day notice of cancellation is needed to receive a full refund. Most require payment in full either 30 days prior to arrival or upon arrival, and many do not accept credit cards. The usual minimum condo stay is 3 nights with some requiring one week in winter. Christmas holidays may have steeper restrictions with minimum stays as long as two weeks, payments 90 days in advance and heavy cancellation penalties. It is not uncommon to book as much as two years in advance for the Christmas season. ALL CONDOMINIUMS HAVE KITCHENS, T.V.'S. AND POOLS UNLESS OTHERWISE SPECIFIED.

Monthly and oftentimes weekly discounts are available. Room rates quoted are generally for two. Additional persons run $8 - $10 per night per person with the exception of the high class resorts and hotels where it may run as much as $25 to $35 extra. Many complexes can arrange for crib rentals. (See GENERAL INFORMATION - Traveling with Children). We have tried to give the lowest rates generally available, which might not be through the hotel or condo office, so check with the offices as well as the rental agents. When contacting condominium complexes by mail, be sure to address your correspondence to the attention of the manager. The managers of several complexes do not handle any reservations and we have indicated to whom you should address reservation requests. If two addresses are given, use the P.O. Box or R.R. rather than street address.

# BED AND BREAKFAST

An alternative to condominiums and hotels are the Bed and Breakfast organizations. They offer homes around the island, and some very reasonable rates. *Bed & Breakfast Hawaii* is among the best known. To become a member and receive their directory, which also includes the other islands, contact: Bed and Breakfast Hawaii, Directory of Homes, Box 449, Kapaa, Hawaii 96746. Another organization, *Bed and Breakfast Maui Style* can be reached at P.O. Box 886, Kihei, Maui 96753 (808-879-7865) or (808-879-2352). *Go Native Hawaii* also features bed and breakfast vacations, contact them at P.O. Box 13115, Lansing, MI 48901, phone (517-349-9598).

# PRIVATE RESIDENCES

### BELLO REALTY/MAUI BEACH HOMES
P.O. Box 1776, Kihei, Maui, HI 96753. (808-879-3328). 1-800-541-3060. Condos and homes rented by the day, week or month. Specializing in the Kihei area.

### ELITE HOLIDAYS UNLIMITED
P.O. Box 10817, Lahaina, Maui, HI 96761. 1-800-448-9222 U.S., 1-800-448-9223 Canada, (808-667-5527). Condos, family homes and luxury estates available for weekly and monthly rentals on Maui and other islands. Condos include The Whaler, Polo Beach, Kaanapali Plantation and Kapalua Bay and Golf Villas. Condo and car packages available.

### HANA BAY VACATION RENTALS
Stan Collins offers eight homes in the Hana area. Contact Hana Bay Vacation Rentals, P.O. Box 318, Hana, Maui, HI 96713. (808-248-7727)

### HAWAIIAN APARTMENT LEASING ENTERPRISE
479 Ocean Ave., Suite B., Laguna Beach, CA 92651. 1-800-472-8449 California, 1-800-854-8843 U.S. except California, 1-800-824-8968. 150 plus homes and 90 condominium properties on all islands.

### HAWAIIAN ISLAND RESERVATIONS
P.O. Box 1863, Kailua, Hawaii 96734. Home rentals on a weekly basis.

### KIHEI MAUI VACATIONS
1-800-542-6284 US, 1-800-423-8733 ext. 4000 Canada (808-879-7581). In addition to condos they offer homes and cottages in the Kihei, Wailea and Makena areas.

### MAUI and ALL ISLANDS
P.O. Box 1089, Aldergrove, BC V0X 1A0. 1-800-663-6962 from B. C. and Alberta. (604) 533-4190. Approximately 150 homes rented weekly, bi-weekly and monthly on Kaua'i and Maui.

### MAUI CONDO AND HOME RENTAL
P.O. Box 1840, Kihei, Maui, Hi 96753. 1-800-822-3309 U.S., 1-800-648-3301 Canada, (808-879-5445). Homes and condos rented daily, weekly and monthly.

### MAUI LUXURY VACATIONS
1-800-222-6313 or (808-661-4840). Fully furnished luxury 2, 3 and 4 bedroom homes with pool, jacuzzi, BBQ's and more. Choose tropical settings with beachfront, golf course and ocean views. Personal bilingual and Hawaiian concierge service. Honeymoon packages available.

### VACATION LOCATIONS, HAWAII
1-800-522-2757 or (808-874-0077). Rent homes on Maui or neighbor islands. Don't want to cook or clean? Select a home with daily maid service and a cook.

### VILLAS OF HAWAII
4218 Waialae Ave., Suite 203, Honolulu, HI 96816. 1-800-522-3030 U.S. and Canada. Beachfront homes on Maui, Kaua'i, O'ahu and the Big Island. Maid service, chefs, limos arranged on request.

### WINDSURFING WEST, LTD.
P.O. Box 330104, Kahului, Maui, HI 96733. 1-800-367-8047 ext. 170 U.S. (808-572-5601). Private homes and cottages available for vacation rental.

## LONG-TERM STAYS

Almost all condo complexes and rental agents offer the long term visitor moderate to substantial discounts for stays of one month or more. Private homes can also be booked through the agents listed above.

# CONDOMINIUM & HOTEL INDEX

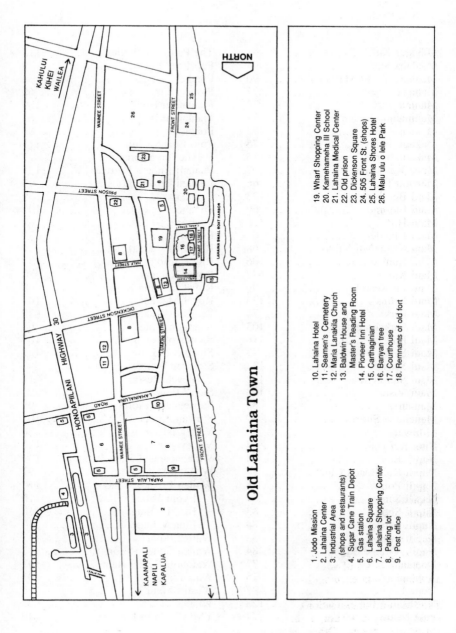

# Old Lahaina Town

1. Jodo Mission
2. Lahaina Center
3. Industrial Area
   (shops and restaurants)
4. Sugar Cane Train Depot
5. Gas station
6. Lahaina Square
7. Lahaina Shopping Center
8. Parking lot
9. Post office
10. Lahaina Hotel
11. Seamen's Cemetery
12. Maria Lanakila Church
13. Baldwin House and
    Master's Reading Room
14. Pioneer Inn Hotel
15. Carthaginian
16. Banyan tree
17. Courthouse
18. Remnants of old fort
19. Wharf Shopping Center
20. Kamehameha III School
21. Lahaina Medical Center
22. Old prison
23. Dickenson Square
24. 505 Front St. (shops)
25. Lahaina Shores Hotel
26. Malu ulu o lele Park

# LAHAINA

## INTRODUCTION

As you leave the Kahului area on Hwy. 38, you plunge immediately into miles of sugar cane. The rugged and deeply carved valleys of the West Maui mountains are on the right, and on the left is the dormant volcano, Haleakala. Its broad base and seemingly gentle slopes belie its 11,000 foot height, and no hint of its enormous moon-like crater is discernible from below. On a clear day the mountains are so distinct and sharp edged they appear to have been cut out with giant scissors. The drive across the isthmus ends quickly as you pass Maalaea Harbor where the gently swaying sugar cane gives way to rugged sea cliffs and panoramic Pacific vistas. Across the bay is the South Maui coastline and in the distance the islands of Kaho'olawe and Lanai. Construction of this road was to accommodate the new resort developments at Kaanapali that began in the 1960's. Traffic must have been far different on the old road which is still visible in places along the craggy cliffside. The tunnel, built in 1951, is the only one on Maui. Just beyond it are enormous metal chain blankets hanging along the rocky cliffs above the road. Termed a protective measure by some and an eyesore by others, they were installed in 1987.

As you descend from the cliffs the first glimpse of the tropical and undeveloped West Maui coastline is always a thrill. Stretching as far as the eye can see are sugar cane fields hugging the lower slopes of the mountains and a series of narrow, white sand beaches lined by kiawe trees and coconut palms. For several miles the constant stream of traffic is the only clue to the populated areas ahead. The first sign of civilization is Olowalu, a mere hamlet along the roadside and an unusual location for one of the island's best restaurants, Chez Paul. Expect to see a great deal more development in Olowalu. This will soon be Maui's hottest new visitor attraction and research center. The *Maui Ocean Center* will be a marine center with a series of exhibits on aquatic and natural science. Scheduled opening is 1991. Public beaches continue to line the highway and the unobstructed view of the ocean may reward the observant with a whale sighting during the December to April humpback season.

A few homes to the left and the monolithic smoke stack of the Pioneer Mill announce your arrival to Lahaina, the now bustling tourist center of Maui. It has maintained the aura of more than a century ago when it was the whaling capitol of the world. Located about a 45 minute drive from the Kahului Airport (depending on traffic), this coastal port is noted for its Front Street, which is a several block strip of shops and restaurants along the waterfront. The Lahaina Harbor is filled with boats of varying shapes and sizes, eager to take the visitor afloat for a variety of sea excursions.

The oldest accommodation on the island, Pioneer Inn, is located here. Still popular among many a visitor, it offers a nostalgic and rustic atmosphere, and very reasonable prices. Other accommodations include a luxuriously expensive condominium complex and two charming new country inns. Although several complexes are located oceanfront, the beaches in Lahaina are fronted by a close-in reef which prohibits swimming. Only Puamana has a beach suitable for swimming. If you want to be in the midst of the action on Maui, you might want to investigate staying in this area.

**BEST BETS: *Puamana*** - A nice residential type area of two-plex and four-plex units, some oceanfront. *Lahaina Shores* - A moderately priced colonial style high rise within walking distance of Lahaina shops. *Plantation Inn* and *Lahaina Hotel* are two tastefully done new bed and breakfast hotels with all the elegance of bygone days.

## WHAT TO DO AND SEE

There is much to see and do in busy Lahaina town. The word Lahaina means "merciless sun," and it does tend to become quite warm, especially in the afternoon with little relief from the tropical trade winds. Parking can be somewhat irksome. Several all day lots are located near the corner of Wainee and Dickenson (only a couple blocks off Front Street) and charge $5 for all day. One nearer to Front Street charges $7 per day. The inexpensive lots fill up early in the day. The Lahaina Shopping Center has a three hour (free) parking area, but it is always very crowded. The new Lahaina Center, across the street from the Lahaina Shopping Center has pay parking, validated with purchase from one of the stores. If you don't mind a short walk, parking is available across the road from the Lahaina Shores Village. (See the Lahaina map for locations of other parking areas). On-street parking is very limited and if you are fortunate enough to find a spot, many are only for one hour. BEWARE, the police here are quite prompt and efficient at towing.

Now that you have arrived, let's get started. Historical memorabilia abounds in Lahaina. The historical landmarks have all been identified by numbered markers. A free pamphlet is available at the historical sites in Lahaina (Baldwin House, Carthaginian, Masters Reading Room, and Wo Hing Temple) for your own self-guided tour. You may be able to purchase admission at a package price. Check at the Wo Hing Temple, Carthaginian or Baldwin House for more information on this option. The following is a brief discussion of the most interesting sites.

*The Banyan Tree* is very easy to spot at the south end of Lahaina adjacent to Pioneer Inn on Front Street. Planted on April 24, 1873 by Sheriff William Owen Smith, it was to commemorate the 50th anniversary of Lahaina's first Protestant Christian Mission.

The stone ruins of *the old fort* can be found harborside near the Banyan Tree. The fort was constructed in the 1830's to protect the missionaries homes from the whaling ships and the occasional cannon ball that would be shot off when the sailors were aroused. The fort was later torn down and the coral blocks reused elsewhere. A few blocks have been excavated and the corner of the fort was rebuilt as a landmark in 1964. On the corner near the Pioneer Inn is a plaque marking the site of the 1987 Lahaina Reunion Time Capsule, which contains newspapers, photos and other memorabilia.

**Pioneer Inn**, built in 1901, is the distinguished green and white structure just north of the Banyan Tree. It was a haven for inter-island travelers during the early days of the 20th century. Having survived the dry years of prohibition, it added a new wing in 1966 along with a center garden and pool area. Two restaurants operate here (one of our favorite haunts for breakfast) and accommodations are available in the original and the newer structure. (See RESTAURANTS - Lahaina and WHERE TO STAY - Lahaina for additional information).

**The Lahaina Courthouse** was built in 1859, at a cost of $7,000, from wood and stone taken from the palace of Kamehameha II. You'll find it near the Lahaina Harbor. The Lahaina Restoration and Preservation Foundation anticipates in the future to begin renovations to restore and convert it into a museum featuring Lahaina's plantation era, and the importance of the reign of Kamehameha III on the Hawaiian Islands. The first floor would also house an information center.

In front of Pioneer Inn is the **Lahaina Harbor**. You can stroll down and see the boats and visit stalls where a wide variety of water sports and tours can be arranged. (See RECREATION AND TOURS) **The Carthaginian,** anchored just outside the harbor, is a replica of a 19th century square rigger, typical of the ships that brought the first missionaries and whalers to these shores. The ship features video movies and recorded songs of humpback whales and an authentic 19th century whale boat. All items are on display below deck. It is open for public inspection daily from 9 to 4:30, with admission $3 for adults and children free.

Whale watching is always an exciting pastime in Lahaina. The whales usually arrive in November and December to breed and calve in the warm waters off Maui for several months.There is also a number to call to report any sightings you make. WHALE WATCH HOTLINE at 879-6530. Numerous whale watching excursions are available. (See RECREATION AND TOURS.)

Adjacent to the Carthaginian is the oldest Pacific lighthouse. "It was on this site in 1840 that King Kamehameha III ordered a nine foot wooden tower built as an aid to navigation for the whaling ships. It was equipped with whale oil lamps kept burning at night by a Hawaiian caretaker who was paid $20 a year." In 1866 it increased to 26 feet in size and was again rebuilt in 1905. The present structure

CARTHAGINIAN

of concrete was dedicated in 1916. (Information from an engraved plaque placed on the lighthouse by the Lahaina Restoration Foundation.)

***The Hauola Stone*** or Healing Rock can be found near the Lahaina Harbor. Look for the cluster of rocks marked with a visitors bureau warrior sign. The rock, resembling a chair, was believed to have healing properties which could be obtained by merely sitting in it with feet dangling in the surf. Here you will also find remnants of the ***Brick Palace*** of Kamehameha the Great. Vandals destroyed the display which once showed examples of the original mud bricks.

***The Baldwin House*** is across Front Street from Pioneer Inn. Built during 1834-1835, it housed the Reverend Dwight Baldwin and his family from 1837 to 1871. Tours of the home, furnished as it was in days gone by, are given every 15 minutes between the hours of 9:30 and 4:30. Adults are $2, no charge for children. The empty lot adjacent was once the home of Reverend William Richards, and a target of attack by cannon balls from angry sailors during the heyday of whaling. On the other side of the Baldwin Home is the Master's Reading Room. Built in 1833, it is the oldest structure on Maui. Its original purpose was to provide a place of leisure for visiting sea captains. It is not open to the public at this time.

***The Old Prison*** (Hale Paahao) on Prison Street just off Wainee, is only a short trek from Front Street. Upon entry you'll notice the large gate house which The Lahaina Restoration Foundation reconstructed to its original state in 1988. Nearby is a 60 year old Royal Palm and in the courtyard an enormous 150 year old breadfruit tree. The cell block was built in 1852 to house the unruly sailors from the whaling vessels and to replace the old fort. (It was reconstructed in 1959.) In 1854 coral walls (the blocks taken from the old fort) were constructed. The jail was used until the 1920's when it was relocated to the basement of the Lahaina Court House next to the Harbor. While you're at Hale Paahao be sure to say hello to the jail's only tenant, George. He is a wax replica of a sailor who is reported to have had a few too many brews at Uncle Henry's Front Street Beer House in the 1850's, then missed his ship's curfew and was tossed into jail by Sheriff William O. Smith. George will briefly converse with you by means of a taped recording. The grounds are open to the public daily, no charge.

BALDWIN HOUSE

Construction of the **Waiola Church** began in 1828 on what was then called the Wainee Church. Made of stone and large enough to accommodate 3,000 people, the church unfortunately did not survive the destructive forces of nature and man. The current structure dates from only 1953. In the neighboring cemetery you will find tombs of several notable members of Hawaiian royalty, including Queen Keopuolani, wife of Kamehameha the Great and mother of Kamehameha II and III. The church is located on Wainee and Shaw Streets. The **Maria Lanakila Church** is on the corner of Wainee and Dickenson. Built in 1928, it is a replica of the 1858 church. Next door is the Seamen's Cemetery.

**Hawaiian Experience Omni Theatre** ★ at 824 Front Street occupies what was once the site of the old Queen's Theatre. The seating for 150 persons is such that everyone gets an unobstructed view of the 180 degree screen which curves up and to the sides of the auditorium. The history of the islands is narrated as the viewer is thrilled to a bird's eye view of the remote Hawaiian leeward islands of Tern, Nihoa and Necker. Travel through the jungles and volcanoes of the major islands as well as the underwater world of the Pacific. The adults in our group found the show realistic enough to cause an occasional "seasick" sensation (especially the bike ride down Haleakala), but the kids were riveted and motionless for the 40 minute show. The film, "Hawaii: Island of the Gods," is an informative as well as entertaining show and the air-conditioned comfort is a pleasant break from the warm sidewalk shopping in Lahaina. The show is offered hourly from 10 am - 10 pm. $5.95 adults, children ages 4 - 12 $3.95, under 4 are free. Phone 661-8314.

**The Wo Hing Temple** on Front Street opened following restoration in late 1984. Built in 1912, it now houses a museum which features the influence of the Chinese population on Maui. Hours are 9 am - 9 pm with a $1 admission donation. The adjacent cook house has become a miniature theatre which features movies filmed by Thomas Edison during his trips to Hawaii in 1898 and 1906.

A small, but interesting **Whaling Museum** is located in the Crazy Shirts shop on Front Street. No admission is charged.

The newest shopping complex in West Maui is the **Lahaina Center,** not to be confused with the adjoining and older, more established Lahaina Shopping Center. This new center has kept the low level, pioneer type architecture and has a paid parking area. Behind in their scheduled opening, little information has been released about the new tenants. Hilo Hatties is the first to open with a 17,000 square foot location. Hatties is famous around the islands for its aloha wear. Another reported soon-to-be tenant is the Hard Rock Cafe. No doubt there will be more art galleries and plenty of places to pick up tee-shirts or postcards.

Follow Front Street towards Kaanapali to find **The Seamen's Hospital.** This structure was once a hideaway for King Kamehameha III and a gaming house for sailors of Old Lahaina. Now it houses one of the Lahaina Printsellers Galleries.

**Hale Pa'i** is on the campus of Lahainaluna school. Founded in 1831, Lahainaluna is the oldest school and printing press west of the Rockies. You will find it located just outside of Lahaina at the top of Lahainaluna Road. Open weekdays 10 - 4. No charge. Donations welcomed.

**The Lahaina Jodo Mission** is located on the Kaanapali side of Lahaina, on Ala Moana Street near the Mala Wharf. The great Buddha commemorated the 100th anniversary of the Japanese immigration to the islands which was celebrated at the mission in 1968. The grounds are open to the public, but not the buildings. The public is welcome to attend their summer O'Bon festivals, usually in late June and early July. Check the papers for dates and times.

## WHERE TO SHOP

Shopping is a prime fascination in Lahaina and it is such a major business that it breeds volatility. Shops change frequently, sometimes seemingly overnight, with a definite "trendiness" to their merchandise. It was a few years back that visitors could view artisans creating scrimshaw in numerous stores. The next few years saw the transformation to T-shirt stores. There still are plenty, but the clothing stores are diminishing in numbers. The theme now is art, art, art, with galleries springing up on every corner. It's a wonderful opportunity to view the fine work of the many local artists with no admission charge! Original oils, watercolors, acrylics, carvings and pottery are on display, as well as fine quality lithographs.

Here are some shops that are unusual or favorites of ours:

**Lahaina Galleries**, 17 Lahainaluna Rd., and 728 Front Street, plus galleries in Kaanapali and Kapalua. Begun in 1976, their art falls in the $500 - $30,000 (and up) range. The works by local artists are the most popular.

New galleries are opening constantly. Some of the larger ones are the Dolphin Gallery, Sunset Galleries (two locations in Lahaina), The Dyansen Gallery, The Larry Dotson Gallery, and The Hansen Gallery. A number of "retired" movie and television stars have turned artist and you'll see the work of Tony Curtis, Red Skelton, Anthony Quinn and Buddy Epsen. Originals, numbered lithographs and poster prints by popular Hawaiian artists Peggy Hopper, Diana Hansen and others can be found in many shops as well. The best representation of local artists may be found at the Lahaina Gallery, The Village Gallery and in the Old Jail in the courthouse basement, which is operated by the Lahaina Arts Society. (Our recent survey of the Lahaina area yellow pages listed 49 galleries!)

JODO MISSION

**The Lahaina Print Sellers** has their newest location at 704 Front Street, with others at Whaler's Village in Kaanapali, Wailea Shopping Center and at the Seamen's Hospital in Lahaina. They specialize in framed antique maps and old prints from around the world. It's an interesting place to browse. The items differ at each location. (Whaler's Village 667-7617 or 661-3579 at the Wharf.)

Friday night in Lahaina is ART NIGHT! Participating galleries feature a special event between 6 and 9 pm that might include guest artists and refreshments.

**Claire the Ring Lady** is at 858-4 Front Street (667-9288). Claire, who learned her craft in Florence, can take your stone, (or choose one of hundreds of hers) and make it up to your specifications.

You will find the rebirth of the aloha shirt at **Kula Bay**, Lahaina Market Place, on Front Street near Lahainaluna Road (667-5852). They've taken original patterns from the 1930's, and 40's, updated and subdued the colors, and then recreated them on comfortable all cotton fabric. Stop by and Erik will be delighted to show you their 100% cotton, made in the U.S.A., slacks, shorts and shirts. Many of the shirts have coconut buttons! They are also available in solid colors.

**Island Sandals** is tucked away in a niche of the Wharf Shopping Center near the postal center at 658 Front Street, Space #125, Lahaina, Maui, Hi 96761, (661-5110). Michael Mahnensmith is the proprietor and creator of custom-made sandals. He learned his craft in Santa Monica from David Webb who was making sandals for the Greek and Roman movies of the late 50's and early 60's. He rediscovered his sandal design from the sandals used 3,000 years ago by the desert warriors of Ethiopia. He developed the idea while living in Catalina in the 1960's and copyrighted it in 1978. The shoes are all leather, which is porous and keeps the feet cool and dry, with the exception of a non-skid synthetic heel. They feature a single strap which laces around the big toe, then over and under the foot, and around the heel, providing comfort and good arch support. As the sandal breaks in, the strap stretches and you simply adjust the entire strap to maintain proper fit (which makes them feel more like a shoe than a sandal). They are clever and functional. His sandals have been copied by others, but never duplicated. So beware of other sandals which appear the same, but don't offer the fit, comfort or function of Michael's! The charge is $ 85 for the right shoe and the left shoe is free. Charges may be slightly higher for men's sandals over size 13. Anyone who gets shoes from Island Sandals becomes an agent and is authorized to trace foot prints of others. Commissions are automatic when your sales reach the "high range." (However, you must like coconuts and bananas.) Michael stresses the importance of good footwear while on Maui, so stop in upon your arrival, or they can be ordered by sending a tracing of your foot and big toe (or by having an "authorized agent" do so) along with $85 to Island Sandals. Michael also can assist with leather repair of your shoes, purses, bags, or suitcases.

**Seegerpeople** at the Wharf has an interesting photographic twist. Located in the lower level corner, they have on display hundreds of samples of their work. After a photographic sitting that lasts about half an hour with a dozen poses, selections are made. The prints are first adhered to a heavy plastic board, then to white plastic, after which the photographs are cut out closely around the head and body.

This results in miniature people that can be creatively arranged and mounted on stands. They're not cheap, but they sure are fun. We enjoyed just looking at all their samples! A sitting, three mounted poses, and the stand runs about $100.

Shell Hansson describes his shop as a fossil remainder. According to Hansson, the *Mad Hatter*, located on the street level of the Wharf Center, is the only place in the world still producing custom-made straw hats. While he specializes in straw, you can find just about any kind of hat or cap and in any price range. At any one time there are 4,000 - 5,000 hats on display with 20,000 - 30,000 hats in stock. The Mad Hatter is one of the original Wharf tenants, opening his shop in 1973. Notable purchasers include Red Skelton, Vincent Price, Charlton Heston and Buddy Epsen, to name only a few. Stop in and try on his motto, "If you have the head - I have the hat."

For the collector of just about anything the *Coral Tree* at the Wharf Shopping Center needs to be penciled in on your Lahaina itinerary. The buyer, Pat Adams, has collected items from around the world and they fill shelves from the floor to ceiling of this small shop. (She adds jokingly that the smallness of the shop is because in Lahaina space is rented by the square inch per minute.) Turtles, frogs, trunk-up elephants (trunk-down elephants are bad luck) and cats are the most popular collectibles. However, if pigs or dogs or zebra are your hobby then you'll find them here! You won't find things made out of just coral either, there is black jet from England, amber from the Dominican Republic, turquoise pieces made by the Zumi Indians or bone and alabaster from Indonesia. The collector of jewelry will find this an intriguing stop as well.

A three screen movie theater was added in 1989 to the second level of the *Wharf Shopping Center* and has a seating capacity of 330 people (current special matinee prices are only $1.99). The *Fun Factory* is located in the lower level with video games and prize oriented games. The shopping center has also introduced "Aloha Friday" - a free show each Friday at 6:30 pm.

*Dickenson Square* (On Dickenson St. off Front St.) bears a strong resemblance to Pioneer Inn. A clothing shop, several small shops, a quick stop grocery, Lahaina Coolers restaurant and a new Nautilus center are located here, however, the main attraction is its historic architecture.

*505 Front Street* is a short walk past the Banyan Tree. Originally developed to be a shopping center, then unsuccesfully converted into condominiums, it has now been restored into busy shops and restaurants. The Old Lahaina Luau is held on the beachfront and restaurants include Bettino's, Sam's and the Old Lahaina Cafe.

*Dan's Green House* at 133 Prison Street (661-8412) has a variety of beautiful tropical birds for sale as well as an array of plants for shipping home. Their specialty is the Fuku-Bonsai "Lava Rock" plants. These bonsai are well packaged to tolerate the trip home.

An area slightly removed from Lahaina's Front Street is termed the industrial area. Follow Honoapiilani Road and turn by the Pizza Hut. *The Sugar Cane Train* main depot is here. This nostalgic railroad will transport you between Kaanapali and Lahaina. One way for adults is $4.25, and children $2.00, round trip for adults runs $6.50 and children $3.25. "Babes in arms are free." Special package options include train trips combined with lunch in Lahaina, a visit to the Baldwin House and the Carthaginian, or a trip in the glass bottom boat, the Lin Wai. Make your plans early as space is limited and sometimes the return trips are booked. The red double decker bus stops at the Wharf Shopping Center and in front of Pioneer Inn to transport you to the Sugar Cane Depot. (See Land tours for additional information.)

Also in the industrial area, *The Bakery* is a personal favorite for some really fine pastries and breads. *MGM, Maui Gold Manufacturing* (661-8981) not only does standard repairs, but designs outstanding jewelry pieces. They can design something to your specifications, or choose a piece from one of their many photograph books. A limited number of pieces are ready made for sale as well. *J.R.'s Music Shop* (661-0801), on the back side of The Bakery building, has a large selection of Hawaiian tapes and records as well as just about any other type of music.

Just on the Kaanapali side of Lahaina, a drive of less than a mile, is *The Cannery Shopping Center* which opened in 1987. The original structure, built in 1920, was used as a pineapple cannery until its closure in 1963, and this new facility was built to resemble its predecessor. It's easy to spot as you leave Lahaina heading for Kaanapali. A large parking area makes for convenient access. This enclosed air conditioned mall is anchored by Safeway and Longs Drug Store. Within the mall are several fast food eateries, Marie Callendar's restaurant, Walden's bookstore, Sir Wilfred's coffee house, jewelry, clothing and sporting goods stores. The surf board display at *Hobie Sports* is very interesting.

BE FOREWARNED!!! If you have the time, do a lot of window shopping before you buy. Prices can vary significantly on some items from one store to another.

## ACCOMMODATIONS - LAHAINA

| | |
|---|---|
| Puamana | Plantation Inn |
| Lahaina Shores | Lahaina Hotel |
| Pioneer Inn | Lahaina Roads |
| Maui Islander | Puunoa |

### PUAMANA ★
P.O. Box 515, Lahaina, Maui, HI 96761. (808-667-2551) 1-800-367-5630. Agents: RSVP 1-800-663-1118. 228 units in a series of duplexes and four-plexes in a garden setting. This large oceanside complex resembles a residential community much more than a vacation resort. The variation in price reflects location in the complex, oceanfront to gardenview. Wkly/mnthly discounts. Limited maid service. *1 BR (4) $ 95-180, 2 BR (6) $155-240, 3 BR (6) $260-290     3-night minimum.*

### LAHAINA SHORES ★
475 Front Street, Lahaina, Maui, HI 96761. (808-661-4835) Agents: Classic Resorts 1-800-628-6699, Rainbow Reservations 1-800-367-6092, Kumulani 1-800-367-2954, Kaanapali Vacation Rentals 1-800-367-8008.

200 oceanfront units in this 7-story building of Victorian style offer air conditioning, lanais, full kitchens, daily maid service, and laundry facilities on each floor. The beach here is fair and the water calm due to offshore reefs, but shallow with coral. Lahaina town is only a short walk away, plus this complex neighbors the Lahaina Shores Village which offers several restaurants and a small grocery store.

| | | | |
|---|---|---|---|
| *SBR* | *mtn.v.-o.v.* | *$ 89- 99 /$ 99-110* | *Extra persons, $10/day* |
| *1BR* | *o.v.-o.f.* | *120-140 / 130-155* | *Children under 12, no charge* |
| *Penthouse* | *mtn.v.-o.f.* | *155-175 / 165-185* | |

## PIONEER INN  (Hotel)

658 Wharf St., P.O. Box 243, Lahaina, Maui, HI 96764. (808-661-3636) Agent: Maui 800 1-800-367-5224. If you want rustic, this is it, and the prices can't be beat. The original building was constructed in 1901 as accommodations for inter-island travelers. Pioneer Inn has been a Lahaina landmark ever since. Many of the units in the old building have shared baths and the furnishings are spartan. Don't let the "new" in new wing give you ideas of grandeur. This wing was added in 1966 and is only a little more modern, with each having a private bath, air conditioning, and lanais. Here you are in the hub of activity in Lahaina and sounds of the music downstairs will lull you to sleep.

*Original building: shared bath single $21, 2 $24; private bath single $27, 2 $30*
*New wing: Single $42, double $45, superior $55.*

A guest in 1901 would have been required to adhere to the following bizarre "house rules:" "You must pay you rent in advance. You must not let you room go one day back. Women is not allow in you room. If you wet or burn you bed you going out. You are not allow to gamble in you room. You are not allow to give you bed to you freand. If you freand stay overnight you must see the mgr. You must leave you room at 11 am so the women can clean you room. Only on Sunday you can sleep all day. You are not allow in the down stears in the seating room or in the dinering room or in the kitchen when you are drunk. You are not allow to drink on the front porch. You must use a shirt when you come to the seating room. If you cant keep this rules please dont take the room."

PIONEER INN

## MAUI ISLANDER HOTEL

660 Wainee St., Lahaina, Maui, HI 96761. (808-667-9766) 1-800-367-5226. Agents: RSVP 1-800-663-1118. 372 rooms include hotel rooms w/ refrigerators. Studio and 1 BR suites w/ kitchens. Located in the heart of Lahaina town, less than a 5 minute walk to the sea wall, yet far enough away to be peaceful. The back of the building borders the Honoapiilani Hwy., so there may be more traffic noise in those units. Daily maid service, air conditioning, laundry facilities, tennis courts, pool. *Room-no kitchen (2) $93/81, Studio (4) $105/93, 1 BR (4) $117/105*

## PLANTATION INN ★

174 Lahainaluna Rd., Lahaina, Maui, HI 96761 (808-667-9225) 1-800-433-6815. It's wonderful to see this kind of development in Lahaina. This new building has all the charm of an old inn, while all the benefits of modernization. Filled with antiques, hardwood floors, and stained glass, they also offer air conditioning, refrigerators and even VCR's. Located a block from the ocean in the heart of Lahaina, it also has a 12 foot deep tiled pool, and a spa. Developed by the owners of Central Pacific Divers, they of course offer diving packages. An added bonus is the outstanding Gerard's Restaurant, which provides guest breakfasts and the option for dinner as well.
*Deluxe double with breakfast $95, front double with breakfast $109*
*Suite with breakfast $129     Dinner available at Gerard's with 40% discount*

## LAHAINA HOTEL ★ (Hotel)

127 Lahainaluna Rd., Lahaina, Maui, HI 96761. (808-661-0577) 1-800-669-3444. Rick Ralston, who also owns Crazy Shirts, has undertaken renovations at this ideally situated location and the transformation is dramatic. Gone are the $25 a night "rustic" units. The fully air conditioned hotel will have 13 rooms for single or double occupancy only.

The hotel has been restored exactly as if it were sent into a time warp between 1860 and 1900. No details have been overlooked from the authentic antiques to the ceiling moldings. All the furnishings have come from Rick Ralston's personal collection so each room is different. The headboard/footboards are intricately carved as are the highboy dressers. Each room is unique with lush wallpaper in deep greens, burgundy, blues and golds and offers a small, but adequate private bathroom. Ten of the rooms are standard and three are larger parlour suites. Each has its own lanai complete with rocking chairs. Manager Ken Eisley emphasizes that this is a service oriented hotel with 24 hour desk service. Adjacent is the new David Paul's Lahaina Grill.
*$110 - $170 with continental breakfast included.*

## LAHAINA ROADS

1403 Front St., Lahaina, Maui, HI 96761. (808-661-3166) 1-800-624-8203. 42 oceanview units, covered parking and elevator to upper levels. Microwaves, washer/dryer, cable TV. Maid service available for extra charge. A very unpretentious, non-resort looking property. Additional person $10/night, 7-night/3-night minimum, two nights deposit, deposit forfeited if unit not re-rented. Weekly and monthly discounts. NO CREDIT CARDS.
*1 BR (2,max 4) $85-100 / 60-75, 2 BR (2,max 6) $115/100,  PH (4) $200/250*

**PUUNOA**
45 Kai Pali Place, Lahaina, Maui, HI 96761. Agents: Classic Resorts (808-667-1400) 1-800-642-MAUI. Amenities include full size swimming pool, jacuzzi, his and hers sauna, and paddle tennis courts. Units include laundry rooms, lanais, master bath with jacuzzi, full bar and daily maid service. These luxury units are located on Puunoa Beach in a residential area just north of Lahaina. Beautiful and spacious air-conditioned units, convenient to restaurants and shops. The beachfront has a coral reef which makes for calm conditions for children, but swimming or snorkeling are poor due to the shallowness and coral. 3-night deposit.
*2 BR 2 bath o.f. (4) $525/495, 3 BR 4 bath with loft (8) $660-$725/$550-$600*

# KAANAPALI

## INTRODUCTION

The drive through Lahaina is quick (unless it's rush hour). All that is really visible are a couple gas stations, the old mill, a few nondescript commercial buildings, and a Pizza Hut. Old Lahaina and the waterfront cannot be seen as they are a couple of large blocks off to the left. The large shopping center on the left is the Cannery, described above. As you leave Lahaina, the vista opens with a view of the Hyatt Regency and the beginning of the Kaanapali Beach Resort a mile off in the distance. The resort is beautifully framed by the West Maui mountains on the right, the peaks of Molokai appearing to be another part of Maui in the background, the island of Lanai off to the left, and of course, the ocean. The name Kaanapali means "rolling cliffs" or "land divided by cliffs" and refers to the wide, open ridges that stretch up behind the resort toward Pu'u Kukui, West Maui's highest peak. The beaches and plush resorts here are what many come to Hawaii to find.

Kaanapali is an Amfac Development that began in the early 1960's with the first hotels, the Royal Lahaina and the Sheraton, opening in late 1962 and early 1963 respectively. The Kaanapali Resort, 500 acres along three miles of prime beachfront, is reputed to be the first large-scale planned resort in the world. There are six beachfront hotels and seven condominiums which total more than 5,000 rooms and units, two golf courses, 37 tennis courts, and a shopping village.

Now that the Kaanapali Airport has been closed, there are another 700 acres available for development. Construction has been awaiting the road improvements in the Kaanapali to Lahaina area. The resort boasts the most convention space of any of the neighboring islands, with the Marriott, Westin Maui and the Hyatt Regency being popular locations. All the hotels are located beachfront, although some of the condos are situated above the beach in the golf course area. All are priced in the luxury range. The wide avenues and the spaciousness of the resort's lush green and manicured grounds are most impressive. No on-street parking and careful planning has successfully given this resort a feeling of spaciousness. Nestled between a pristine white sand beach and scenic golf courses with a mountain range beyond, this may be the ideal spot for your vacation.

This may be paradise, but traffic congestion between Kaanapali and Lahaina may have reminded you more of L.A. in the past few years. Non-synchronized traffic lights, roads designed for 20 years ago, and greatly increased traffic, caused the three mile transit through Lahaina to Kaanapali (or Kaanapali to Lahaina) to consume over an hour during the afternoon rush (most other times there was only light traffic). Of deep concern to the government, residents and business interests alike, this situation was eased considerably with the recent completion of all four lanes from Kaanapali to Lahaina. The major bottleneck is now at the first Kaanapali entrance where the four lanes end. Getting past this point in either direction can be difficult. Hopefully, the four lanes will extend up to at least Napili or Kahana in the near future.

**BEST BETS: Hyatt Regency Maui** - An elegant and exotic setting with a wonderful selection of great restaurants. **Marriott** - Beautiful grounds with a nice pool area and attractively decorated rooms. **Westin Maui** - A gorgeous resort and a pool aficionados paradise. **Kaanapali Alii** - One of only three condominiums that are oceanfront. Luxurious, expensive and spacious. (Our choice to purchase a unit with future lottery winnings!). **Royal Lahaina Resort** - A beautiful property on sandy Kaanapali Beach. **The Whaler** condominiums on the heart of Honokaoo Beach adjacent to the Whaler's Shopping Center.

## WHAT TO DO AND SEE

**The Hyatt Regency** and **The Westin Maui** must be put at the top of everyone's list of things to see. Few hotels can boast that they need their own wildlife manager, but upon entry you'll see why they do. Without spoiling the surprises too much, just envision the Hyatt with palm trees growing through the lobby, peacocks strolling by, and parrots perched amid extraordinary pieces of oriental art. The lagoon and black swans are spectacular. The pool area occupies two acres and features two swim-through waterfalls and a cavern in the middle with a swim up bar. A swinging bridge is suspended over one of the two pools and a water slide offers added thrills for hotel guests. The newest project at Kaanapali is the Westin Maui. To appreciate this property, a little background may be necessary. The Maui Surf was the original hotel with the single curved building and a large expanse of lush green lawn and two pools. The transformation has been extraordinary. The pool areas are unsurpassed, with five swimming pools on various levels fed by waterfalls and connected by two slides. There are exotic birds afloat on the lagoons which greet you upon your arrival and glide gracefully by two of the hotel's restaurants. The oriental art collection surpasses even the Hyatt's. Both resorts feature glamorous shopping arcades, with prices to match of course. Both developments were designed by the remarkable, champion hotel builder of Hawaii, Chris Hemmeter.

## WHERE TO SHOP

**Whaler's Village Shopping Center** is located in the heart of Kaanapali. Some part or other of this center always seems to be under construction or renovation. It offers several small shops for grocery items, as well as a bounty of jewelry and clothing shops, and restaurants. A multi-level parking structure is adjacent to the

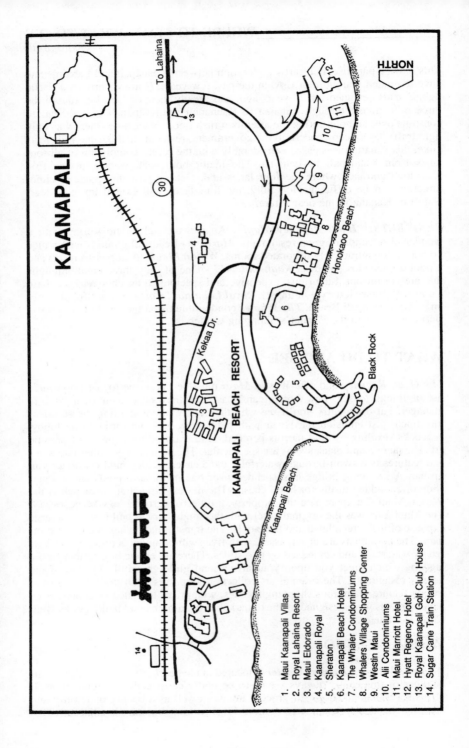

# KAANAPALI

**NORTH**

To Lahaina

(30)

Kekaa Dr.

**KAANAPALI BEACH RESORT**

Kaanapali Beach

Honokaoo Beach

Black Rock

1. Maui Kaanapali Villas
2. Royal Lahaina Resort
3. Maui Eldorado
4. Kaanapali Royal
5. Sheraton
6. Kaanapali Beach Hotel
7. The Whaler Condominiums
8. Whalers Village Shopping Center
9. Westin Maui
10. Alii Condominiums
11. Maui Marriott Hotel
12. Hyatt Regency Hotel
13. Royal Kaanapali Golf Club House
14. Sugar Cane Train Station

mall and parking is $1 for the first two hours or fraction thereof, and 50 cents for each additional half hour, with a $10 maximum charge. Restaurants can provide validation. The Whaling Museum on the upper level is free. (Donations are welcomed.) Historian Conrad Justel gives lectures on topics from scrimshaw to the life of a sailor. Call for times at 661-5992. Private group lectures are also a possibility. Near the front of the complex is a complete whale skeleton displayed along with models and information on many different whales. Restaurants include The Rusty Harpoon, Leilani's, El Crab Catcher, Cafe Kaanapali, Ricco's, and Chico's. The major store is Liberty House, along with a diverse array of shops including The Sharper Image, and an assortment of jewelry, clothing and novelty shops. Walden's has a very good bookstore here with an excellent selection of Hawaiian literature. The mall is a pleasant place for an evening stroll and shop browsing, before or after dinner, followed by a seaside walk back to your accommodations on the paved beachfront sidewalk.

## ACCOMMODATIONS - KAANAPALI

| | |
|---|---|
| Hyatt Regency | Royal Lahaina Resort |
| Marriott | Maui Kaanapali Villas |
| Kaanapali Alii | Kaanapali Plantation |
| Westin Maui | International Colony Club |
| The Whaler | Maui Eldorado |
| Kaanapali Beach Hotel | Kaanapali Royal |
| Sheraton | |

**HYATT REGENCY ★ (Hotel)**
200 Nohea Kai Drive, Lahaina, Maui, HI 96761. (808-661-1234) 1-800-228-9000. This magnificent complex is located on 18 beachfront acres and offers 815 rooms and suites. The beach here is beautiful, but has a steep dropoff. Adjoining Honokaoo Beach Park has a gentler slope into deeper water. The pool area is an impressive feature, covering two acres and resembling a contemporary adventure that Robinson Crusoe could only have dreamt. The pool is divided by a large cavern that can be reached on either side by swimming beneath a waterfall. Once inside there is a swim up bar! Guests can choose to walk into the cavern. One side of the pool is spanned by a large swinging rope bridge. Stand on the bridge and enjoy guests being whisked into the water by the long waterslide.

Penquins, jewel-toned koi, parrots, peacocks and flamingos around the grounds require full time game keepers. The lobby is a blend of beautiful pieces of oriental art and paths that lead to the grounds. Originally there had been a huge tree, the focal point of the lobby, however, it quite unexpectedly toppled over one night. Fortunately no one was injured. Part of the base remains and has become a popular parrot walk. The birds are apparently so at home here that one exotic pair surprisingly gave birth, a rarity for this species in captivity! Non-guests should definitely visit the Hyatt for a self guided tour of the grounds, the art, the elegant shops and for an opportunity to enjoy one of this resort's fine restaurants. It's worth coming here just to look around! Restaurants include Swan Court, Lahaina Provision Company, The Pavilion and Spats II.
*Terrace rooms $195, golf/mtn.v. $235, o.f. $280-310*
*Suites: Golf $275, Ocean $475, Deluxe $775, Regency $1,100, Presidential $2,000*

The Regency Club consists of certain floors that feature special services, including continental breakfast, evening cocktails and appetizers, and a complimentary health club membership. Room rates are based on single/double occupancy (for additional persons 13 years or older $25 charge per night/Regency Club level $45 charge per night.) 3 adults or 2 adults 2 children maximum per room.
*Regency Club (mtn.v.) $340, (o.f.) $375*

### MARRIOTT ★ (Hotel)
100 Nohea Kai Drive, Lahaina, Maui, HI 96761. (800-228-9290) 1-808-667-1200. This 720 room complex has a large, open lobby in the middle featuring an array of fine shops. Although not as exotic as its neighbor the Hyatt, this is still a very attractive, upscale property. The pool area is large and a keiki (children's) wading pool is a welcome addition for families. On site restaurants are Nikko's Steak House, Lokelani's, Moana Terrace and the Kau Kau Bar.
*Mtn./o.v. $195-230, o.v. $250, Deluxe o.v. $285*

### KAANAPALI ALII ★
50 Nohea Kai Dr., Lahaina, Maui, HI 96761. 1-800-367-6090. Agents: Classic Resorts 1-800-642-MAUI, Condo Resorts 1-800-854-3823, Whaler's Realty 1-800-367-5632, Kaanapali Vacations 1-800-367-8008, Hawaiian Apt. Leasing 1-800-854-8843, 1-800-472-8449 CA.

This is the newest condo development in Kaanapali. All 264 units are very spacious and beautifully furnished, with air conditioning, microwaves, washer/dryer, and daily maid service. Other amentities include security entrances and covered parking. The 1-bedroom units have a den, which actually makes them equivalent to a 2-bedroom. Three lighted tennis courts, pool (also a children's pool), and exercise room. No restaurants on the property, but shops and restaurants are within easy walking distance. A very elegant, high-class and quiet property with a very cordial staff and concierge department. They charge for local phone calls from room, as do most hotels.
*1 BR (2) g.v. $205/180, o.v. $255/225          Extra person $15/night*
*2 BR (4) g.v.  245/230, o.v.  315/280, o.f. $395/375     3-night min., cribs $10*

WATER LILY

**WESTIN MAUI ★ (Hotel)**
2365 Kaanapali Parkway, Lahaina, Maui, Hi 96761. (808-667-2525). Westin
Central Reservations 1-800-228-3000. Under the direction of Chris Hemmeter,
champion hotel builder in Hawaii, this new resort offers 762 deluxe rooms,
including 28 suites. The Westin Maui has an ocean tower of 11 stories with 556
guest rooms and a beach tower with 206 guest rooms and suites. Guest rooms are
provided for those with disabilities as well as non-smoking floors. The rooms
have been designed in comfortable hues of muted peach and beige. The top two
floors of the new tower, house the Royal Beach Club, which offers guests
complimentary continental breakfast buffet, afternoon cocktails, evening cocktails
and hors d'oeuvres, and a guest relations coordinator. Complimentary shuttle
service to the Royal Lahaina Tennis Ranch, the largest tennis facility on Kaanapali
with 11 tennis courts and 6 courts lit for night play. Complimentary shuttle
service is also available to the Royal Kaanapali Golf Courses. Conference and
banquet facilities are available as well as an array of gift, art, and fashion shops.
The focal point of this resort is the 55,000 square foot aquatic playground,
complete with meandering streams, 15 - 20 foot waterfalls, and a 25,000 square
foot pool area featuring five free-form pools, two waterslides and a swim-up
jacuzzi hidden away in a grotto. The pool areas are spacious and well arranged.
Eight restaurants and lounges overlook the ocean, waterfalls and pools. The hotel
exercise room includes complete exercise and weight rooms, with sauna and
whirlpool. Tour the grounds with a guide to learn more about the Westin's family
of birds and their tropical surroundings. This resort's 2.5 million art collection
could put a museum to shame and each piece was carefully selected and placed
personally by Chris Hemmeter. Numerous nooks with comfortable chairs and art
work provide intimate conversation areas. Parents may appreciate the resort's
Camp Keiki which offers activities and care for children during vacation periods.

Rates are based on single or double occupancy. Third person add $25, to Royal
Beach Club add $45 (maximum 3 persons to a room). Family Plan offers no extra
charge for children 18 or under sharing the same room as parents. A 25% discount
is available for additional rooms occupied. Complimentary valet parking.
*Standard $185, garden view $230, golf/mtn. view $250, oceanview $265-$320*
*Royal Beach Club $375, Suites $500-$1,600*

**THE WHALER ★**
2481 Kaanapali Parkway, Lahaina, Maui, Hi 96761. (808-661-4861) Managed by
Village Resorts 1-800-367-7052. Agents: Condo Resorts 1-800-854-3823, Maui
Network 1-800-367-5221, Whaler's Realty 1-800-367-5632, Kaanapali Vacations
1-800-367-8008, Hawaiian Apt. Leasing 1-800-854-8843 (U.S. except California),
1-800-472-8449 Calif., RSVP 1-800-663-1118.

Choice location in the heart of Kaanapali next to the Whaler's Village Shopping
Center and on an excellent beach front. A large pool area is beachfront and they
provide an excellent children's program during the summer. Underground parking.
$200 deposit, 3-night minimum, balance on check-in. 2-week refund notice.
*S BR 1 bath (2) o.v. $180/165, g.v. $160/145      Cribs $10/night*
*1 BR 1 bath (4) o.v.  230/220, g.v.  195/185      Rollaway beds $12/night*
*1 BR 2 bath (4) o.v.  240/230, g.v.  205/195, o.f. $290/275*
*2 BR 2 bath (6) o.v.  335/320, o.f.  405/380*

## KAANAPALI BEACH HOTEL  (Hotel)

2525 Kaanapali Pkwy, Lahaina, Maui, HI 96761 (808-661-0011) 1-800-367-5170. 430 units located on Kaanapali beach near Black Rock and Whaler's Village Shops. Air conditioning. Tennis available at nearby Royal Lahaina. Try their whale-shaped swimming pool! Restaurants on site include the Kaanapali Beach Hotel Koffee Shop and the Tiki Terrace Restaurant. A great location, but not a "posh" resort. Some freshening up would do wonders for this comfortable hotel, but then it would also be reflected in the prices!

A good value and this is Maui's most Hawaiian hotel where the staff are actually instructed in Hawaiiana. Be sure and ask about the free deluxe compact rental car, automatic with air, which may be available during certain times of the year. Extra persons $20/night. Children under 17 free when sharing room with parents using existing bedding. *Std. $135, courtyard $145, partial o.v. $155, o.v.$165, o.f. $175. Suites available on request.*

## SHERATON HOTEL  (Hotel)

2605 Kaanapali Parkway. (808-661-0031) 1-800-325-3535. This 494 unit hotel winds around the side of Black Rock and was one of the first completed in Kaanapali. There are also six-unit cottages on the grounds. The hotel is continuing updating and renovations of the rooms. The beachfront here is excellent for snorkeling and everyone comes here to enjoy the nearly tame fish. The only problem is finding a place to park if you're not staying here. The complex features a variety of shops. Restaurants include The Discovery Room and Black Rock Terrace. They continue to feature the dramatic cliff dive nightly. *Std $195, g.v. $205, partial o.v. $235, o.v. $255, o.f. cottages $300 O.v. suites $400-$850     Extra adults $25/night*

## ROYAL LAHAINA RESORT ★

2780 Kekaa Drive, Lahaina, Maui, HI 96761. (808-661-3611) 1-800-621-2151. 542 units located on excellent Kaanapali Beach just north of Black Rock. All cottage suites have kitchens and are situated around the lush, spacious grounds. A mini-mall is conveniently located on the property. Sailing Activities Center, 11 tennis courts, three swimming pools. Restaurants on the property include Royal Ocean Terrace, which features a very good Sunday Brunch, Moby Dick's and Chopsticks Restaurant. Nightly luaus in the luau gardens. *Std. $150, superior $175, dlx. $200, dlx. o.f. $225, garden cottage $175 Oceanfront cottage $225, 1 & 2 BR suites $375 - $1,000*

## MAUI KAANAPALI VILLAS

2805 Honoapiilani Hwy., Lahaina, Maui, 96761 (808-667-7791). Agents: Aston 1-800-367-5124, Whaler's Realty 1-800-367-5632, Kumulani 1-800-367-2954, All About Hawaii 1-800-336-8687, Hawaiian Apt. Leasing 1-800-854-8843 (U.S. except Calif.), 1-800-472-8449 Calif., Kaanapali Vacation Rentals 1-800-367-8008, 1-800-423-8733 ext. 515 Canada, RSVP 1-800-663-1118.

Located on fabulous, sandy Kaanapali Beach, this was once a part of the Royal Lahaina Resort, and before that the Hilton, prior to being converted into condominiums. *Room with refrigerator $159/129, studio with kitchen $149-169/$179-199, 1 BR with kitchen $175-205/$205-235     Extra person $10*

## KAANAPALI PLANTATION

150 Puukolii Rd., (PO Box 845) Lahaina, Maui, HI 96761 (808-661-4446). No rental units available at this time from on-site management. Agents: Condominium Connection 1-800-423-2976, Maui (800) 1-800-367-5224, Elite Holidays 1-800-448-9222. 62-unit one, two and three bedroom units in a garden setting situated on a hillside with golf course and ocean views. Washer/dryers. No air conditioning, only ceiling fans. *$90 - $130.*

## INTERNATIONAL COLONY CLUB

2750 Kalapu Dr., Lahaina, Maui, HI 96761 (808-661-4070) 1-800-526-6284. 44 private cottages on 10 lush acres. Full kitchens, lanais, most have washer/dryers, but coin-op laundry also on premises. Maid service. Two heated swimming pools. This property is located across Honoapiilani Highway and up the road from the Kaanapali resorts. It is a bit of a walk to the beach. 4-day minimum April - Dec. 15, 7 day minimum Dec. 16 - March. Deposit equal to 2 nights plus 9.43% tax is refundable with 2 week notice prior to check-in, some restrictions for high season. NO CREDIT CARDS.

*1 BR (2) $95, 2 BR (2) $110, 3 BR (3) $130    Extra persons over age 6, $10/nite*

## MAUI ELDORADO

2661 Kekaa Drive, Lahaina, Maui, Hi 96761. (808-661-0021) 1-800-367-2967, Canada 1-800-663-1118. Agents: Maui 800 1-800-367-5224, Hawaiian Apt. Leasing 1-800-472-8449 CA, 1-800-854-8843 U.S. except CA, Kaanapali Vacations 1-800-367-8008, 1-800-423-8733 ext. 515 Canada.

204 air conditioned units located on golf course. Private lanais with free HBO and Disney cable TV. Daily maid service. Three pools. Cabana on nearby beachfront. Variation in rates reflect location.

| | | |
|---|---|---|
| *S BR (1-2) $124-139 / $109-119* | *Extra persons $10.* | |
| *1 BR (1-4)   149-169 /   123-134* | *5-day minimum, weekly/monthly discounts.* | |
| *2 BR (1-6)   195-219 /   183-205* | *Rollaways and cribs $10/day* | |

## KAANAPALI ROYAL ★

2560 Kekaa Dr., Lahaina, Maui, HI 96761. (808-667-7200) Agents: Resorts Pacifica (808-661-7133) or 1-800-367-5637. Hawaiiana Resorts 1-800-367-7040, Whaler's Realty 1-800-367-5632, Hawaiian Apt. Leasing 1-800-472-8449 CA, 1-800-854-8843 U.S. except CA, RSVP 1-800-663-1118.

These very spacious condos, 1,600 - 2,000 sq. ft., offer air conditioning and lanais and are situated on the Kaanapali golf course overlooking the Kaanapali resort and Pacific Ocean. Daily maid service. Washer/dryers. Note that while all units have two bedrooms, they may be rented as a one bedroom based on space availability. One bedroom reservations may be wait listed outside of 30 days of arrival. No minimum stay except over Christmas holiday. One night deposit.

*1 BR (2,max 4) garden or golf view  $160/135, o.v. or dlxe golf view  $175/150*
*2 BR (2,max 6) garden or golf view   205/180, o.v. or dlxe golf view   190/165*

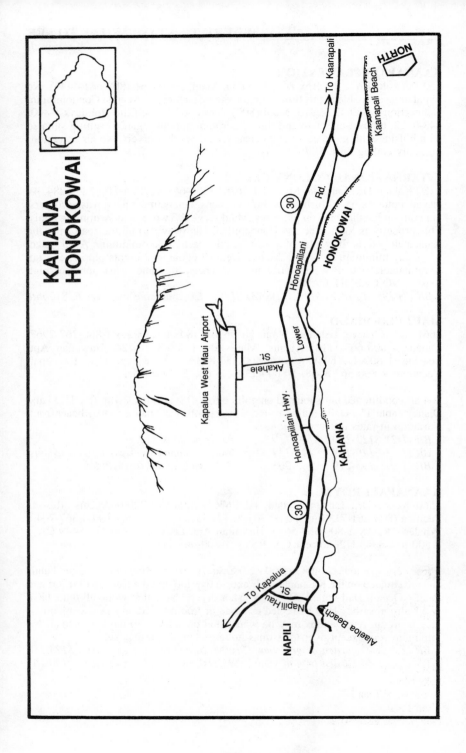

KAHANA
HONOKOWAI

NORTH

To Kaanapali

Kaanapali Beach

30

Honoapiilani Rd.

HONOKOWAI

Lower Akahele St.

Kapalua West Maui Airport

Honoapiilani Hwy.

30

KAHANA

To Kapalua

Napili Hau St.

Alaeloa Beach

NAPILI

# HONOKOWAI

## INTRODUCTION

As you leave the Kaanapali Resort there is a stretch of yet undeveloped beachfront on the left still planted with sugar cane. This was the site of the old Kaanapali Airport. Resorts will be stretched along this beach within the next few years. Ahead, four large condo complexes signal the beginning of Honokowai, which stretches north along Lower Honoapiilani Highway. Accommodations are a mix of high and low rise, some new, however, most are older. The beachfront is narrow and many complexes have retaining walls. A close-in reef fronts the beach and comes into shore at Papakea and at Honokowai Park. Between the reef and beach is generally shallow water unsuitable for swimming or other water activities. The only wide beach and break in the reef for swimming and snorkeling is at the Kaanapali Shores and Embassy Suites. In late 1987, several condominiums made a major investment in saving the beachfront by building a seawall beneath the sand to prevent winter erosion. A number of the condominiums are perched on rocky bluffs with no sandy beach. Many people return year after year to this quiet area, away from the bustle of Lahaina and Kaanapali and where prices are in the moderate range. A couple of small grocery stores are nearby. For dining out there is the Beach Club restaurant at the Kaanapali Shores and The Embassy Suites offers three restaurant choices. Just a short walk up the road is one of our Best Bets for great food. Fat Boy's features enormous, juicy hamburgers and great affordable plate lunches. Feed the family and have change left over.

The condominiums are individually owned for the most part, and the quality and care of each (or lack of) is reflected by the owner. Perhaps it is the shape of the sloping ridges of Pu'u Kukui that cause this area to be slightly cooler and cloudier with more frequent rain showers in the afternoon than at neighboring Kaanapali.

*BEST BETS: **Kaanapali Shores** - A high-rise surrounded by lovely grounds on the best beach in the area. **Papakea** - A low rise complex with attractive grounds and pool. **Embassy Suites Hotel** - A mix between a condo and a hotel, spacious rooms and breakfast is included, a good sandy beach.*

## ACCOMMODATIONS - HONOKOWAI

| | | |
|---|---|---|
| Mahana Resort | Hale Kai | Hale Ono Loa |
| Maui Kai | Pikake | Lokelani |
| Embassy Suites Hotel | Hale Maui | Hale Mahina |
| Kaanapali Shores |   Apt. Hotel |   Beach Resort |
| Papakea | Nohonani | Hoyochi Nikko |
| Maui Sands | Kulankane | Kuleana |
| Paki Maui | Makani Sands | Polynesian Shores |
| Honokowai East | Kaleialoha | Mahinahina Beach |
| Maui Park | Hale Royale | Mahina Surf |
| Honokowai Rsrt. Apts. | Hono Koa | Noelani |

71

## MAHANA
110 Kaanapali Shores Place, Lahaina, Maui, HI 96761. (808-661-8751) Agents: Aston 1-800-367-5124, Whaler's Realty 1-800-367-5632, Kumulani 1-800-367-2954, RSVP 1-800-663-1118, Hawaiian Apt. Leasing 1-800-472-8449 CA, 1-800-854-8843 U.S. except California.

All units oceanfront. Two twelve-story towers with two tennis courts, heated pool, central air conditioning, saunas, elevators, small pool area. Located on narrow beachfront with offshore coral reef precluding swimming and snorkeling. A better swimming area is 100 yards up the beach. This is a rather drab looking condominium on the outside, the pool sits out on a slab of cement without much in the way of atmosphere. They have been experiencing trouble with beach erosion as have the other properties along this stretch of Honokowai.
*S BR 1 bath (1-2) $159/139        3-night minimum*
*1 BR 1 bath (1-4)   185/165*
*2 BR 2 bath (1-6)   230/210*

## MAUI KAI
106 Kaanapali Shores Pl., Lahaina, Maui, Hi 96761. (808-661-0002) 1-800-367-5635. Agents: Blue Sky Tours 1-800-678-2787, Condo Network 1-800-321-2525, Paradise Resorts 1-800-367-2644, Maui 800 1-800-367-5224, Kumulani 1-800-367-2954.

A single ten-story building with 79 units. 2-night deposit, 2-night minimum. Units offer central air conditioning, private lanais, full equipped kitchens. Property amenities include swimming pool, jacuzzi, laundry facilities, free parking. 5th person in room $10 per night. No charge children under 2. Built in 1970, it is one of the older properties. Weekly/monthly discounts.
*Studio (1-3) $110/80,  1 BR (1-4) $130/97,  2 BR (1-4) $180/140*

## EMBASSY SUITES RESORT ★
104 Kaanapali Shores Place, Lahaina, Maui, HI  96761. (808-661-2000). 1-800-462-6284 U.S., 1-800-458-5848 Canada. On 7 1/2 acres this pink pyramid structure with a three-story blue waterfall cascading down the side can't be missed. A new concept in resorts on Maui, it blends the best of condo and resort living together. The pool area is large and tropical with plenty of room for lounge chairs. For the past few years this beach has suffered from erosion problems, but seems to have somewhat stablized since the neighboring Kaanapali Shores conducted some extensive erosion control measures. The lobby is open air. (When it opened they discovered there was a little too much breeze and glass walls were installed to keep the wind from whistling in too briskly!) Glass enclosed elevators whisk you up with a view!

Each one bedroom suite is a spacious 840 sq. ft, two bedroom suites are 1,200 sq. ft. Each features lanais with ocean or terrace views. Master bedrooms are equipped with a remote control 29" television and an large adjoining master bath with soaking tub. The living room, decorated in comfortable hues of blue and beige contains a massive 35" television, stereo receiver, VCR player and cassette player. Living rooms have a sofa that makes into a double bed. A dining area with a small kitchenette is equipped with a microwave, small refrigerator and sink.

Toaster ovens, hot plates and ironing equipment available upon request. One phone in the living room and another in the bedroom have two lines which connect to a personal answering machine for your own recorded message. Their two presidential suites are 2,100 sq. ft. and offer two bedrooms, two full baths and a larger kitchen. One features an Oriental theme, the other is decorated with a contemporary California flare. Made-to-order breakfasts are served in their Oihana restaurant. The Maui Rose offers more formal dining and the Lokahi Terrace features a prime rib buffet.

*1 BR terrace view $195, mt. view $225, o.v. $245, o.f. $295*
*2 BR mt. view $390, o.v. $450, Presidential Suite $1,000*

## KAANAPALI SHORES ★
100 Kaanapali Shores Place, Lahaina, Maui, Hi 96761, (808-667-2211). Agents: Aston 1-800-367-5124, Condo Resorts 1-800-854-3823, Whaler's Realty 1-800-367-5632, Kumulani 1-800-367-2954, RSVP Reservations 1-800-663-1118.

463 units, all offer telephones, free tennis, daily maid service, and air conditioning. Nicely landscaped grounds and a wide beach with an area of coral reef cleared for swimming and snorkeling. This and the Embassy Suites are the only resorts on north Kaanapali Beach that offers a good swimming area. Putting green, jacuzzi and the Beach Club Restaurant are all located in the pool area.

*S BR (1-4) g.v. $159/129, o.v. $169/139*       *7-night minimum over*
*1 BR (1-4) g.v.  189/159, o.v.  199/169*       *the christmas holidays*
*2 BR (1-6) g.v.  219/189, o.v.  239/209, o.f. $310/280*

## PAPAKEA ★
3543 L. Honoapiilani, Lahaina, Maui, HI 96761. (808-669-4848) 1-800-367-5637. Agents: Whaler's Realty 1-800-367-5632, Condo Resorts 1-800-854-3823, Rainbow Reservations 1-800-367-6092, RSVP 1-800-663-1118, Resorts Pacifica 1-800-367-5637, Hawaiian Apt. Leasing 1-800-472-8449 CA, 1-800-854-8843 U.S. except CA.

364 units in five four-story buildings. Two pools, two jacuzzi's, two saunas, tennis courts, putting green, washer/dryers, and BBQ area. A seawall was installed in an effort to prevent further beach erosion. The shallow water is great for children due to a protective reef 10-30 yards offshore, but poor for swimming or snorkeling. A better beach is down in front of the Kaanapali Shores. One of the nicer grounds for a condominium complex with lush landscaping and pool areas. A comfortable, and quiet property that we recommend especially for families. No smoking units available. Crib or roll-away $6/day. Christmas holiday 14-day minimum with no refunds after October 1.

*S BR (2) partial o.v. $104, o.f. $117*    *Wkly/monthly discounts*
*1 BR (4) partial o.v.  119, o.f.  149*    *2-nite deposit, 7-dy refund notice*
*2 BR (6) partial o.v.  159, o.f.  189*    *Add $20/nite between 12/20 - 1/2*

## MAUI SANDS
3559 L. Honoapiilani, Lahaina, Maui, HI 96761. (808-669-4811) 1-800-367-5037. Agent: Rainbow Reservations 1-800-367-6092. All 76 units have air conditioning. Limited maid service. A very friendly atmosphere where old friends have been gathering each year since it was built in the mid-sixties. All units have kitchens,

a large central laundry facility is available and a large pool area also has barbecues. Large boulders line the beach. It is a good family facility.

*1 BR (2,max 4) std. $65,   g.v. $ 85-95,   o.v. $130     Extra persons $9/nite*
*2 BR (2,max 6) std.  80,   g.v.  125,    o.v.  145     15% monthly discounts*

## PAKI MAUI
3615 L. Honoapiilani, Lahaina, Maui, HI 96761. (808-669-8235) Agents: Aston 1-800-367-5124, Condo Resorts 1-800-854-3823, Kumulani 1-800-367-2954, All About Hawaii 1-800-336-8687, RSVP 1-800-663-1118. This complex surrounds a garden and waterfall. No air conditioning.

*S BR (1-2) o.f.  $139/119              2-nite deposit*
*1 BR (1-4) g.v.  149/129,  o.f. $169/149   cribs $2/nite*
*2 BR (1-6) o.f.   185-195 / $165-175       children under 2 free*

## HONOKOWAI EAST
3660 L. Honoapiilani Hwy., Lahaina, Maui, HI 96761 (808-669-8355) 51 units, mostly studios, in a 4-story building. Long term property.

## MAUI PARK
3626 L. Honoapiilani Hwy., Lahaina, Maui, HI 96761. (808-669-6222) AGENTS: Aston 1-800-922-7866, RSVP Reservations 1-800-663-1118, Maui Condominiums 1-800-663-6962. Located across the road from a Honokowai Beach Park which lacks a sandy shoreline. A quiet area of West Maui with nearby grocery store. All units have complete kitchen. Coin operated laundry facility. Originally built as residential apartments they offer phones, complimentary morning coffee and daily maid service. Because of its original intention, this property does resemble a residental area more than a vacation resort. *S BR (1-2) $89/79, 1 BR (1-4) $109/99*

## HONOKOWAI PALMS RESORT
3666 L. Honoapiilani, Lahaina, Maui, HI 96761. (808-669-6130) 1-800-843-1633. 30 units across road from Honokowai Beachfront Park. Built of cement blocks this property lacks a great deal of ambience as a vacation retreat. Perhaps for the real budget conscious it would be suitable, but it is a very basic, functional complex. Deposit $200 or 25% of first 28 days. Cancellations must be made 30 days prior to arrival. NO CREDIT CARDS.

*1 BR (2,max 4) non-o.v. $60,  o.v. $65    Extra person $6/nite*
*2 BR (2,max 6) non-o.v.  65               Wkly/monthly discounts*

## HALE KAI
3691 L. Honoapiilani Hwy., Lahaina, Maui, HI 96761. (808-669-6333) 40 units in a two-story building. The units do have lanais, kitchens, and a pool, but the beach is somewhat rocky. A simple and quiet property. 3-night minimum except Christmas. $250 refundable with 45 day notice. NO CREDIT CARDS. Weekly and monthly discounts. *1 BR (2) $80,  2 BR (4) $110-115*

## PIKAKE
3701 L. Honoapiilani, Lahaina, Maui, HI 96761. (808-669-6086) 1-800-446-3054. A low-rise, two-story, Polynesian style building with only twelve apartments completed in 1966. Private lanais open to the green lawn or balconies, with a beach protected by sea wall. Central laundry area. Light housekeeping provided

after two week's stay. 3-night deposit, 3-night minimum, Extra persons $10/night. NO CREDIT CARDS. *1 BR (2,max 4) $75,  2 BR (4,max 6) $108*

## HALE MAUI APARTMENT HOTEL
P.O. Box 516, Lahaina, Maui, HI 96761. (808-669-6312). Limited maid service. Coin-operated washer/dryer. BBQ. Weekly and monthly discounts. 3-day minimum, 7-day during Christmas. Extra persons $8/night. *1 BR (2, max 5) $60-80*

## NOHONANI
3723 L. Honoapiilani, Lahaina, Maui, HI 96761. (808-669-8208) 1-800-822-7368. Office open 9 am - 5 pm Mon. - Sat. Two 4-story buildings containing 22 two-bedroom units and 5 one-bedroom units. All units are oceanfront. Complex has large pool, telephones, and is one block to grocery store. Maid service on checkout. Extra persons $15/night. $200 deposit with 60-day refund notice, 4-day minimum stay. Weekly/monthly discounts. NO CREDIT CARDS.
*1 BR (1-2) $102/83,  2 BR (1-4) $126/101*

## KULAKANE
3741 L. Honoapiilani (P.O. Box 5238), Lahaina, Maui, HI 96761. (808-669-6119) 1-800-367-6088. 42 oceanfront units with fully equipped kitchen, laundry facilities on premise. Lanais overlook ocean but no sandy beach. $10 extra person. 3 night minimum low season, 5 night high season. $150 deposit. 10% monthly discounts.
*1 BR 1 bath (1-2) $90,  2 BR 2 bath (1-4) $135*

## MAKANI SANDS
3765 L. Honoapiilani Hwy., Lahaina, Maui, HI 96761. (808-669-8223). Thirty units in a four-story building. Dishwashers, washer/dryers, elevator. Oceanfront with small sandy beach. Weekly maid service. Deposits vary, weekly/monthly discounts, 3-night minimum, extra persons $10/night. NO CREDIT CARDS.
*1 BR (2) $80,  2 BR (4) $111,  3 BR (6) $135,  Penthouse (6) $160*

## KALEIALOHA
3785 L. Honoapiilani, Lahaina, Maui, HI 96761. (808-669-8197) 1-800-222-8688. 67 units in a 4-story building. 3-night minimum. Deposit equal to three nights stay, $7.50 extra persons over age 2. Refundable if cancelled 45 days prior to arrival. Washer/dryers. Discounts (10-15%) for low season.
*Studio (1-2) mtn.v. $65,  1 BR (1-4) o.v. standard $75,  deluxe $85*

## HALE ROYALE
3788 L. Honoapiilani, Lahaina, Maui, HI 96761. (808-669-5230). No rental units available at this time.

## HONO KOA
3801 L. Honoapiilani, Lahaina, Maui, HI 96761. (808-669-0979) 1-800-225-7215. Agent: Hawaiian Island Resorts 1-800-367-7042, RSVP 1-800-663-1118. 28 units in one four-story building. Washer/dryer, dishwasher, microwave, BBQ. Nice pool area with jacuzzi. No minimum stay. Advance deposit of one or two nights, refunded with 72 day cancellation notice except high season. $10 extra person, children 6 and under free. Limited maid service.
*2 BR 2 bath (1-4, max 6) g.v. 130/110,  o.v. $150/135,  o.f. $175/160*

## HALE ONO LOA ★
3823 L. Honoapiilani, Lahaina, Maui, HI 96761. (808-669-6362) AGENTS: Real
Hawaii 1-800-367-5108 U.S., Maui Accommodations 1-800-252-MAUI (U.S.) 67
oceanfront and oceanview units. Maid service extra charge. Beachfront is rocky.
The units we toured were roomy and nicely furnished with spacious lanais. The
grounds and pool area were pleasant and well groomed. A good choice for a quiet
retreat. Grocery store nearby.
*1 BR 1 bath (4) g.v. $100 (560/wk), o.v. $110 (640), partial o.v. $105 (600)*
*2 BR 2 bath (6) g.v.  150 (840/wk), o.v.  160 (920), o.f. $170 (1,000/wk)*

## LOKELANI
3833 L. Honoapiilani, Lahaina, Maui, HI 96761. (808-669-8110) 1-800-367-2976.
Three 3-story 12 unit buildings with beachfront or oceanviews. The 1-BR units
are on beach level with lanai, 2-BR units are townhouses with bedrooms upstairs
and lanais on both levels. Units feature washer/dryers and dishwashers. Weekly
discount, 3-night minimum low season, 7-night high season, extra persons $7.50,
$10 cancellation fee. Three night deposit, balance due two weeks prior to arrival.
NO CREDIT CARDS. *1 BR (1-2) $80/70, 2 BR (townhouses) (1-4) $105/95*

## HALE MAHINA BEACH RESORT
3875 L. Honoapiilani, Lahaina, Maui, HI 96761. (808-669-8441) 1-800-367-8047
ext. 441. AGENTS: Kaanapali Vacation Rentals 1-800-367-8008. 52 units in two,
four-story buildings and one two-story building featuring lanais, ceiling fans,
microwaves, washer/dryer. BBQ area, jacuzzi. Extra persons $10/night. 3-day
minimum, deposit within two weeks of reservations, balance on arrival.
*1 BR (1-2) $100/90, 2 BR (1-4) $135/120*

## HOYOCHI NIKKO
3901 L. Honoapiilani, Lahaina, Maui, HI 96761. (808-669-8343) Agents: Rainbow
Reservations 1-800-357-6092, Kihei Maui Vacations 1-800-542-6284. 18 one-
bedroom oceanview units (on a rocky beachfront) in two-story building bearing
an oriental motif. Underground parking, "Long Boy" twin beds, some with
queens, half size washer/dryers in units. Maid service on check-out. $100-250
deposit with 30-day refund notice low season, 60-day high season. Prepayment
required. NO CREDIT CARDS. *1 BR ($80-90) with extra persons $10/night.*

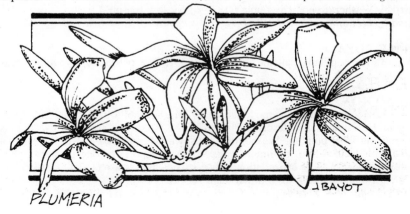

PLUMERIA

## KULEANA

3959 L. Honoapiilani, Lahaina, Maui, HI 96761. (808-669-8080) U.S. Mainland 1-800-367-5633, Canada 1-800-237-8256, RSVP 1-800-663-1118. 118 1-bedroom units with queen size sofa bed in living room. Large pool with plenty of lounge chair room and tennis court. A short walk to sandy beaches. Weekly/monthly discounts. 3 night minimum stay. Extra persons $5. Children under 2 free. Cribs $4/night, rollaways $6. 3-night deposit refundable with 14 day notice.
*1 BR o.v. $85/80, o.f. 95/90, 2 BR $140/125*

## POLYNESIAN SHORES

3975 L. Honoapiilani, Lahaina, Maui, HI 96761. (808-669-6065) 1-800-433-6284, from Canada call collect. 52 units on a rocky shore but nice grounds with deck overlooking the ocean.

*1 BR 1 bath (1-2) $ 95/ 80*  *$200 deposit, 3-day minimum*
*2 BR 2 bath (1-4) 135/120*  *10% disount for stays of*
*3 BR 3 bath (1-6) 160/145*  *29 days or more*

## MAHINAHINA BEACH

4007 L. Honoapiilani, Lahaina, Maui, HI 96761. Units only through owners.

## MAHINA SURF

4057 L. Honoapiilani, Lahaina, Maui, HI 96761 (808-669-6068) 1-800-367-6086. 56 one-bedroom and one-bedroom with loft units. Dishwashers, maid service available at hourly charge. Located on rocky shore, the nearest sandy beach is a short drive to Kahana. Large lawn area around pool offers plenty of room for lounging. $150 deposit, 4-week refund notice.

*1 BR 1 bath $ 90/80 (2,max 4)* *Extra persons $8/nite including children*
*2 BR 1 bath  105/95,  2 bath $110/100* *Wkly/monthly discounts*

## NOELANI

4095 L. Honoapiilani, Lahaina, Maui, HI 96761. (808-669-8375) 1-800-367-6030. Agent: Maui 800 1-800-367-5224, Condominium Connection 1-800-423-2976. 50 oceanfront units in one 4-story building and two 2-story structures. Kitchens with dishwashers and washer/dryers only in 1, 2, and 3-bedroom units. Three bedroom units feature a sunken livingroom as do the two bedroom units on the third floor. Complex has two pools and maid service mid-week. Located on a rocky shore, nearest sandy beach is short drive to Kahana. Weekly/monthly discounts.

*S BR 1 bath (1-2) $ 77,  1 BR 1 bath (1-2) $ 97  3-nite deposit*
*2 BR 2 bath (1-2) 120,  3 BR 3 bath (1-6) 145  2-wk refund notice*

# KAHANA

## INTRODUCTION

To the north of Honokowai, and about seven miles north of Lahaina is a prominent island of high rise condos with a handful of two-story complexes strung along the coast in its lee. This is Kahana. The beach adjacent to the high rises is fairly wide, but tapers off quickly after this point. Several of the larger complexes offer very nice grounds and spacious living quarters with more resort type activities than in Honokowai. The prices are lower than Kaanapali, but higher than Honokowai. Several restaurants and quick-stop groceries are located nearby. Over the past year or so there has been a continuing problem with a green algae along the shoreline from Honokowai to Kapalua. The major problem, though, appears to be in the Kahana area where shoreline conditions, ocean currents and wind bring the algae into shore, frequently rendering the beaches here less than desirable places to play. The algae tends to accumulate in large blankets under water at the other beaches, like Napili and Kapalua, but doesn't tend to come onto shore. The most recent reports indicate that it is an algae bloom resulting from the runoff of chemicals such as nitrates. The state officials are, as yet, unable to pinpoint the source and indicate that it could be a combination of factors. Monitoring stations will be established in the waters around West Maui to track the situation. The county may find it necessary to institute controls on the amount of fertilizers used in the area.

***BEST BETS: Sands of Kahana*** - Spacious units on a nice white sand beach. ***Kahana Sunset*** - Low rise condos surrounding a secluded cove and beach.

## WHERE TO SHOP

There are a few shops in the lower level of the Kahana Manor. A site adjacent to this has been proposed for development. Everything from a bowling alley to a movie theater has been mentioned as possible inclusions in this center. No doubt, however, there will be more shops!

## ACCOMMODATIONS - KAHANA

Kahana Beach Hotel      Pohailani Maui
Kahana Villa            Kahana Reef
Sands of Kahana         Kahana Outrigger
Valley Isle Resort      Kahana Village
Royal Kahana            Kahana Sunset
Hololani

### KAHANA BEACH HOTEL
4221 L. Honoapiilani, Lahaina, Maui, HI 96761. (808-669-8611) Agent: Pleasant Hawaiian Holidays (package tours) 1-800-242-9244. All units offer oceanview. The studios sleep up to four and have kitchenettes. The 1 BR units have kitchens, 2 lanais, living room with queen-size sofa bed, bedroom with 2 queen beds, 2 full-size baths, dressing room, and will accommodate 7. Coin-op laundry on premises. Nice, white sandy beach fronting complex.

### KAHANA VILLA
4242 L. Honoapiilani, Lahaina, Maui, HI 96761. (808-669-5613) Agents: Colony Resorts 1-800-367-6046, Condo Resorts 1-800-854-3823, Hawaiian Apt. Leasing 1-800-854-8843 (1-800-472-8449 CA), Kumulani 1-800-367-2954 RSVP 1-800-663-1118. Across the road from the beach. Units have microwaves, washer/dryers, telephones. Daily maid service. Sauna, tennis courts, store, restaurants.
*1 BR 1 bath g.v. $105/ 90,   o.v. $125/105, o.v. sup. $145/125*
*2 BR 1 bath g.v.  145/125,   o.v.  165/145, o.v. sup.  175/155*

## SANDS OF KAHANA ★

4299 L. Honoapiilani, Lahaina, Maui, HI 96761. (808-669-0400) Agents: Colony Resorts 1-800-367-5599, All About Hawaii 1-800-336-8687, Hawaiian Apt. Leasing 1-800-854-8843 (1-800-472-8449 CA), Kaanapali Vacations 1-800-367-8008, RSVP 1-800-663-1118.

96 units on Kahana Beach. Underground parking. If you're looking to be a little away from the hustle of Lahaina/Kaanapali, with quarters large enough for a big family, and luxuries such as microwaves and full-size washer/dryers then this may be just what you seek. Located on a sandy beachfront and only a couple miles from Kaanapali, it is also less than a mile from the West Maui Airport. Four 8-story buildings surround a central restaurant (The Kahana Terrace) and a dual pool area. Sands of Kahana is family oriented from the size of their rooms to their children's playground and summer programs. Spacious one, two and three bedroom units have enormous kitchens and beautifully appointed living rooms. Plenty of room to spread out! Molokai is beautifully framed in the large picture windows of the oceanview units, or select among the slightly less expensive garden view units. Although an annoying green algae has been invading the beaches lately, which pretty well prohibited swimming, there was plenty of fun in the sun. The beachside volleyball court was filled each afternoon, and the large three foot deep children's pool was popular, as was another larger and deeper pool with jacuzzi, three tennis courts and a putting green.

Most complexes restrict the use of snorkel gear or flotation equipment in the pool, however, here it is allowed to the delight of the children. A small children's play area offers diversion while parents make use of several garden area charcoal barbecues. The summer children's program is very reasonable and invites children between the ages of 5 and 12 to participate in activities. Across the street is a small grocery store and several restaurants are within walking distance.

*1 BR 1 bath o.v. $209/179,  o.f. $229/199, mtn.v. $189/159*    *Extra persons $10*
*2 BR 1 bath o.v.  245/215,  o.f.  270-300 / 240-270*            *Under age 18 free*
*3 BR 2 bath o.v.  310/280,  o.f.  340/310*                      *in existing space*

BOUGAINVILLEA                                                    J. BAYOT

79

# WHERE TO STAY - WHAT TO SEE
*Kahana*

## VALLEY ISLE RESORT
4327 L. Honoapiilani, Lahaina, Maui, HI 96761. (808-669-5511) Agent: Rainbow Reservations 1-800-367-6092. Partial air conditioning. Telephones. Located on Kahana Beach. On site restaurant and grocery store. Prices are for stays of 1 to 3 days. Substantial discounts for 5 or more days (i.e. 5 day stay, low season studio is $65). Payment in full 30 days prior to arrival. Weekly maid service. Add $20/day for daily maid service.

*S BR 1 bath (2) o.f.   $128/118            Extra persons $10, under 3 free*
*1 BR 1 bath (2) o.v.   128-133 / 118-123,  o.f. $138-148 / 128-138*
*2 BR 2 bath (4) o.v.   170/160,  o.f. $190/180*

## ROYAL KAHANA
4365 L. Honoapiilani, Lahaina, Maui, HI 96761. (808-669-5911) Agents: Resort Mgmt. 1-800-524-3405, Kumulani 1-800-367-2954. High-rise complex with 236 oceanview units on Kahana Beach. Underground parking and air conditioning. Daily maid service. A nice pool area with sauna. Tennis courts. Units have full kitchens and microwaves. Near grocery stores, restaurants and shops.

*1 BR 1 bath (1-2) o.v.  $120/ 94, Studio (1-3) o.v.  $ 92/ 78      2-nite deposit*
*2 BR 2 bath (1-6) o.v.  146/118,  o.f. $176/152            15-day refund notice*

## HOLOLANI
4401 L. Honoapiilani, Lahaina, Maui, HI 96761. (808-669-8021) 1-800-367-5032, 1-800-423-8733 ext. 318 Canada. 27 oceanfront units on sandy, reef protected beach. Covered parking. Grocery store. 7-day/3-day minimum. $250 deposit, full payment 60/30 days prior to arrival. Children under 5 free. NO CREDIT CARDS. Extra persons $10/night. *1 and 2 BR 2 bath (2,max 6) $125-135 / $110-115*

## POHAILANI
4435 L. Honoapiilani, Lahaina, Maui, HI 96761. (808-669-6125)
No short term rentals

## KAHANA REEF ★
4471 L. Honoapiilani, Lahaina, Maui, HI 96761. (808-669-6491. Agent: Condo Resorts 1-800-854-3823, Kumulani 1-800-367-2954. 88 well-kept units. Limited number of oceanfront studios available. Laundry facilities on premises. 15% monthly discounts. Maid service daily except Sunday. NO CREDIT CARDS. Room and car packages available. $200 deposit, extra persons $8/night.
*1 BR 1 bath (2,max 5) o.v. $95/85*

## KAHANA OUTRIGGER
4521 L. Honoapiilani, Lahaina, Maui, HI 96761. (808-669-5544) 1-800-852-4262. Agents: Rainbow Reservations 1-800-367-6092. Twelve spacious 3-bedroom oceanview condo suites in low-rise complex on a narrow sandy beachfront. Units have microwaves and washer/dryers and are appointed with lots of Italian tile. On a recent inspection we noticed that these beautiful units have not been well maintained. *3 BR 2 bath (6) $200/180,  3 BR 3 bath (6) $226/190*

## KAHANA VILLAGE ★
4531 L. Honoapiilani, Lahaina, Maui, HI 96761. (808-669-5111) 1-800-824-3065. Agents: Kumulani 1-800-367-2954, RSVP 1-800-663-1118.

Attractive townhouse units. Second level units are 1,200 sq.ft.; ground level 3-bedroom units have 1,700 sq.ft. with a wet bar, sunken tub in master bath, Jenn-aire ranges, microwaves, lanais, and washer/dryers. Pool area jacuzzi. Nice but narrow beach offering good swimming. 5-day minimum. Bi-weekly maid service. NO CREDIT CARDS. $300 deposit, balance due prior to arrival. Monthly discounts. Additional person $20.
*2 BR o.v. $175/140, o.f. $200/170, 3 BR o.v. $225/185, o.f. $250/210*

## KAHANA SUNSET ★

P.O. Box 10219, Lahaina, Maui, HI 96761. (808-668-8011) 1-800-367-8047 ext. 354. Agents: Parker Pacific 1-800-426-0494, Maui 800 1-800-367-5224, Hawaiian Leasing 1-800-472-8449 CA, 1-800-854-8843 U.S., RSVP 1-800-663-1118.

Ninety units on a beautiful and secluded white sand beach. Units have very large lanais, telephones, and washer/dryers. Each unit has its own lanai, but they adjoin one another, adding to the friendly atmosphere of this complex. One of the very few resorts with a heated pool and heated children's pool, BBQ. You can drive up right to your door on most of the two bedroom units making unloading easy (and with a family heavy into suitcases that can be a real back saver.) Extra persons $6/night including infants, 10% monthly discounts.
*1 BR 1 bath (2) o.v. $125, 2 BR 2 bath (2) o.v. $155, o.f. $195*

# NAPILI

## INTRODUCTION

This area's focal point is the beautiful Napili Bay with good swimming, snorkeling and boogie boarding, and it even has tide pools for children to explore. The condominium units here are low-rise, with prices mostly in the moderate range, and are clustered tightly around the bay. A number are located right on the beach, others a short walk away. The quality of the units vary considerably, but generally a better location on the bay and better facilities demands a higher price. The complexes are small, most under 50 units, and all but one has a pool. A small grocery store or two is within walking distance and, depending on your location, a couple of restaurants could be reached on foot. A new shopping center is scheduled for development here soon.

***BEST BETS: Napili Sunset*** - Centered right on the edge of Napili Bay, rooms are well kept. ***Napili Kai Beach Club*** - A quiet facility on the edge of Napili Bay. Large grounds and a restaurant are on site. Resort activities are offered.

## ACCOMMODATIONS - NAPILI

| | |
|---|---|
| Honokeana Cove | Napili Bay |
| Napili Sands | Napili Sunset |
| Coconut Inn | Hale Napili |
| Napili Point | Napili Village Suites |
| Napili Shores | Mauian |
| Napili Surf | Napili Kai Beach Club |

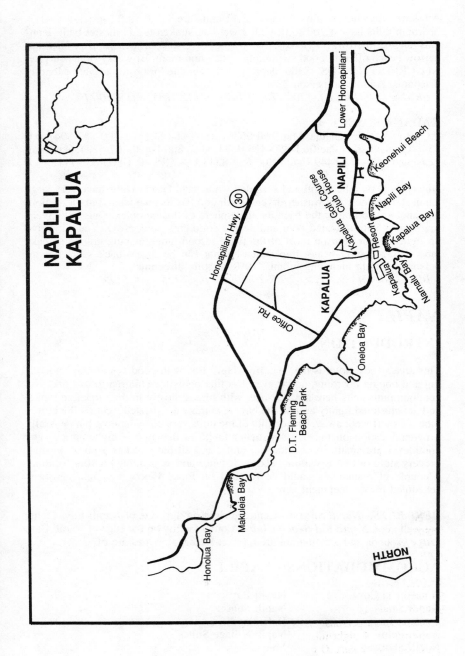

NAPLILI
KAPALUA

Honoapiilani Hwy. (30)

Lower Honoapiilani

Keonehui Beach

Kapalua Golf Course
Club House

NAPILI

Napili Bay

Resort

Kapalua Bay

KAPALUA

Kapalua

Namalu Bay

Office Rd.

Oneloa Bay

D.T. Fleming
Beach Park

Makuleia Bay

Honolua Bay

NORTH

## HONOKEANA COVE
5255 L. Honoapiilani, Lahaina, Maui, HI 96761. (808-669-6441) 1-800-237-4948. Thirty-eight oceanview units on Honokeana Cove near Napili Bay. Direct-dial phones. Attractive grounds. Maid service weekly only if staying two or more weeks. 5-night minimum, 5-night depoist. NO CREDIT CARDS.
*1 BR 1 bath (2,max 2) $ 85, 2 BR 2 bath (4,max 4) $125     Extra persons $9/nite*
*3 BR 2 bath (6,max 6)   143, Townhouse (4) $140          Wkly/monthly discounts*

## NAPILI SANDS
Hui Rd. "F", Lahaina, Maui, HI 96761. 44 studios and 88 one-bedroom units in 11 two-story buildings. One block to small grocery store and walking distance to bus stop and beach. Long term stays only.

## COCONUT INN
Hui Rd. "F", P.O. Box 10517, Napili, Maui, HI 96761. (808-669-5712) 1-800-367-8006. Agents: Maui 800 1-800-367-5224, Condo Resorts 1-800-854-3823. Forty units in two-story retreat about 1/4 mile above Napili Bay. No oceanviews. Pool, spa, daily maid service. Attractive tropical grounds. Complimentary continental breakfast by pool. Extra persons $10/night.
*Studio (2) $69-74,  1 BR (2) $79-89,  loft (2) $89-99*

## NAPILI POINT
5295 L. Honoapiilani, Lahaina, Maui, HI 96761. (808-669-5611)  Agents: Aston 1-800-367-5124, Condo Resorts 1-800-854-3823, Kumulani 1-800-367-2954, Hawaiian Apt. Leasing 1-800-854-8843 (1-800-472-8449 CA), RSVP 1-800-663-1118. Located on rocky beach, but next door to beautiful Napili Bay. Units have washer/dryer, direct dial phones, daily maid service. Two pools.
*1 BR 1 bath (2,max 4) o.v. $159/129,  o.f. $174-184/144-164*
*2 BR 2 bath (2,max 6) o.v.  185/155,  o.f.  205-215/175-185*

## NAPILI SHORES
5315 L. Honoapiilani, Lahaina, Maui, HI 96761. (808-669-8061)  Agents: Colony Resorts 1-800-367-6046, Rainbow Reservations 1-800-367-6092, Condo Resorts 1-800-854-3823, Hawaiian Apt. Leasing 1-800-854-8843 (1-800-472-8449 CA), RSVP 1-800-663-1118.

152 units on Napili Bay. Rooms offer lanais and the one-bedroom units have dishwashers. Laundry facilities on premises as well as two pools, hot tub, croquet, and BBQ area. Restaurant, cocktail lounge and grocery store on property. Extra persons $15/night, no minimum stay, daily maid service.
*S BR (2,max 3) g.v. $130/115,  o.v. $145/130,  o.f. $155/140*
*1 BR (2,max 4) g.v.  155/140,  o.v.  170/155*

## NAPILI SURF
50 Napili Place, Lahaina, Maui, HI 96761. 1-800-541-0638. 53 units on Napili Bay. Two pools, BBQ, shuffleboard, lanais, daily maid service, and laundry facilities. Extra persons $15/night, 10% monthly discount, $300 deposit, 14-day refund notice, 4-night minimum except 10-day during Christmas, NO CREDIT CARDS. *SBR (2,max 3) g.v. $78,  o.v. $82,  1 BR (2,max 5) $120*

## NAPILI BAY
33 Hui Drive, Lahaina Maui, HI 96761. (808-669-6044). Agents: Maui 800 1-800-367-5224, Hawaiian Apt. Leasing 1-800-854-8843 (1-800-472-8449 CA), Kumulani 1-800-367-2954, RSVP 1-800-663-1118.

This older complex on Napili Bay is neat, clean and affordably priced. Studio apartments offer 1 queen & 2 single beds, lanais, kitchens, daily maid service. Coin-op laundromat with public phones. Extra persons $8/night, children under 12 free. 3-night minimum, 2-night deposit, 7-day refund notice, weekly and monthly discounts. *Studio (2,max 4) on ocean $95/85, off ocean $80/70*

## NAPILI SUNSET ★
46 Hui Rd., Lahaina, Maui, HI 96761. (808-669-8083) 1-800-447-9229. Forty-one units located on Napili Bay. Daily maid service.These units have great oceanviews and are well maintained. A very friendly atmosphere. Deposits vary $262 - $656. 15 day notice for full refund. 10% monthly discount. 3-day minimum. *Studio (2) g.v. $80, 1 BR 1 bath (2) o.f. $138, 2 BR 2 bath (4) o.f. $200*

## HALE NAPILI
65 Hui Rd., Napili, Maui, HI 96761. (808-669-6184) 1-800-245-2266. 18 units oceanfront on Napili Bay. Lanais. Daily maid service. Laundry facilities on property. No pool. 3-night minimum. Extra persons $8/night, $150 deposit, 7-day refund notice, monthly discounts, NO CREDIT CARDS. *Studio (2) g.v. $70, o.f. $90, 1 BR (2) o.f. $110*

## NAPILI VILLAGE SUITES
5425 Honoapiilani, Lahaina, Maui, HI 96761. (808-669-6228) 1-800-336-2185. All rooms have king or queen size beds, daily maid service. Free laundry facilities on premises. Located a short walk from Napili Bay. Extra persons $6/night, 3-night deposit. *Studio (2) $85/75*

## MAUIAN
5441 Honoapiilani, Lahaina, Maui, Hi 96761. (808-669-6205) 1-800-367-5034. Studio apartments on Napili Bay. Kitchen plus microwave, one queen and two twin day beds. BBQ area. Two public phones on property, one courtesy reservations phone. Television only in recreation center. Units cleaned every third day with trash removal and fresh towels daily. 3-day minimum, 3-night deposit, 14-day refund notice. 5% two week, 12% monthly discounts. Rental crib $8/night. Extra person $8/night. *Studio garden $90/75, o.v. $95/80, o.f. $115/100*

## NAPILI KAI BEACH CLUB ★
5900 Honoapiilani, Lahaina, Maui, HI 96761. (808-669-6271) 1-800-367-5030, Canada 1-800-263-8183. Agents: Maui 800 1-800-367-5224, Condo Resorts 1-800-854-3823. Units feature lanais, kitchenettes, telephones, and washer/dryer facilities. Complimentary tennis equipment, beach equipment, croquet, putters, and snorkel gear. Daily coffee and tea party in Beach Club. Sea House Restaurant located on grounds. Two tennis courts, 4 pools, very large jacuzzi. The grounds are extensive and the area very quiet. A relaxed and friendly atmosphere, a great beach, and a wide variety of activities may tempt you to spend most of your time enjoying this very personable and complete resort. Approximately 30 new luxury

units with an oriental motif have been added in what used to be the large lawn area. 2-night deposit. 14-day refund notice. NO CREDIT CARDS.

| | | |
|---|---|---|
| *S BR (2) std. o.v. $145, dlx. o.v. $175, dlx. o.f. $195* | *Extra persons $15/nite* |
| *1 BR (2) std. o.v.   180, dlx. o.v.   200, dlx. o.f.   235* | *Children under 3 free* |
| *2 BR (4) std. o.v.   195, dlx. o.v.   310, dlx. o.f.   385* | *Cribs $15* |

# KAPALUA

## INTRODUCTION

Colin Cameron chose 750 acres of his family's 23,000 acre pineapple plantation for the development of this up-scale resort. The result is the Kapalua Bay Hotel and Villas which opened in 1979. The mood reflected here is serene with their philosophy being quality of food and service in a resort setting offering the ultimate in privacy and luxury living. The grounds are spacious with manicured lawns and an oasis of waterfalls and gardens. Recent renovations have meant a new butterfly-shaped pool located nearer the beachfront. More than 400 condominium units are located in the Ridge, Golf and Bay Villas, of which about 125 are available for rent. There are two 18-hole championship golf courses, a tennis garden, a shopping area with a myriad of boutiques and a deli/restaurant.

There are several excellent restaurants to choose from ranging from The Pool Terrace Restaurant and Bar located beachfront to the posh and elegant Bay Club Restaurant located on a scenic promontory of the bay.

Kapalua Bay is a small cove of pristine white sand, nestled at the edge of a coconut palm grove. The protected bay offers good snorkeling and a safe swimming area for all ages.

***BEST BETS: The Kapalua Bay Hotel & Villas*** - This resort offers quiet elegance, top service, great food with all the amenities. Any of the condominium units in this area would be excellent, however, they are not all located within easy walking distance of the beach. A shuttle service is available. The condominiums at Kapalua offer spacious living and complete kitchen facilities.

## WHAT TO SEE AND DO

Kapalua, "arms embracing the sea," is the most north-western development on Maui. The logo for Kapalua is the butterfly, and with a close look you can see the body of the butterfly is a pineapple. One might enjoy a stop at the elegant Kapalua Bay Hotel. The lobby bar is ideally situated for an evening refreshment, music, and sunset viewing. The resort has a small shopping mall located just outside the hotel. The resort grounds are attractive and spacious. The road beyond Kapalua is paved and in excellent condition, and offers some magnificent shoreline views. However, it finally turns to a rough dirt road and rental car agencies are not responsible should you choose this route. (A wash out has closed this section of the road for some time, check with the county to see if it is accessible.) After several miles the road returns to pavement before you arrive at Wailuku. Slaughterhouse Beach is only a couple of miles beyond Kapalua and you may find it interesting to watch the body surfers challenge the winter waves. Just beyond is Honolua Bay where winter swells make excellent board surfing conditions. A good viewing point is along the roadside on the cliffs beyond the bay. Continuing on, you may notice small piles of rocks. This is graffiti Maui style. They began appearing a few years ago and these mini-monuments have been sprouting up ever since. There are some wonderful hiking areas here as well. One terrain resembles a moonscape, while another is windswept peninsula with a symbolic rock circle formation. See What to Do hiking for more details.

## WHERE TO SHOP

The Kapalua shopping area was recently renovated, now offering an entirely covered mall, is a showcase of treasures. Here you will find *The Kapalua Shop* (669-4172) where everything from men's and women's resort wear to glassware display the Kapalua butterfly logo. The newest shop, *Kapalua Kids* features fashions for infants through boys and girls size 7 (669-0033). *The Market Cafe* (669-4888) has fresh pastries, wines, gourmet items to go, or enjoy breakfast, lunch, or dinner at their restaurant.

## ACCOMMODATIONS - KAPALUA

Kapalua Bay Hotel & Villas
Ritz-Carlton
Kapalua Bay and Golf Villas
The Ridge
Ironwoods

### KAPALUA BAY HOTEL & VILLAS ★

One Bay Drive, Kapalua, Maui, HI 96761. (808-669-0244) 1-800-367-8000. 194 hotel rooms plus villa condominiums. Rooms have service bars and refrigerators, no kitchens. All rooms have air conditioning with decor in warm neutral shades of taupe, rose and muted terra cotta. Five-star restaurants include The Bay Club and The Plantation Veranda. A cafe is available in the Kapalua Shops, located adjacent to the hotel. Lovely grounds, excellent beach and breathtaking bay. The resort includes The Bay Course and The Village Course, two 18-hole championship golf courses designed by Arnold Palmer, (a third course is being developed) and a tennis garden with 10-plexipave courts with 4 lighted for night play. The new pool is butterfly-shaped, located near the ocean. The expanse of lawn gives way to lush tropical foliage, waterfalls, pools and gardens. This is elegance on a more sophisticated scale than the glitter and glitz of the Kaanapali resorts. Children 14 or younger free if sharing room with parent. Extra persons $35 high season, $25 low season. Cribs available at no charge. 3-night's deposit high season, one low season. 14-day refund notice. Modified American Plan is available at $65 per person.
*g.v. $215-250/$185-205, o.v. $315-340/$270-$295, o.f. $360-385*
*Parlor Suite $430-630/500-680, 1 BR suite $750-950, 2 BR suite $1,050-1,250*

## THE RITZ-CARLTON, KAPALUA

Discovery of some ancient burial sites temporarily halted construction during 1989 on the newest Kapalua resort, the Ritz-Carlton. The most recent projection for completion is for early 1992. The 550 room oceanfront resort will be comprised of 314 Kings, 170 doubles, 63 executive suites and 3 Ritz-Carlton suites. Room amenities will include plush terry bathrobes, twice daily maid service and marble bathrooms with telephone. The three resort restaurants will include The Cafe and Terrace, serving breakfast, lunch, dinner and Sunday brunch in a casual indoor/outdoor atmosphere. The Dining Room, seating 80, will offer continental cuisine, nightly entertainment in an elegant environ, and The Grill and Lounge will serve cocktails, lunch and dinner in an oceanfront club-like atmosphere. Three lounges and numerous meeting rooms will also be available.

## KAPALUA BAY AND GOLF VILLAS ★

Agents: Kapalua Hotel 1-800-367-8000, Condo Resorts 1-800-854-3823, Whaler's Realty (golf villas only) 1-800-367-5632, Kaanapali Vacations 1-800-367-8008, Hawaiian Apt. Leasing 1-800-854-8843 (1-800-472-8449 CA). There are over 400 units in the golf, bay and ridge villages. Units include kitchens, washer/dryers, and daily maid service. Each are beautiful appointed. Several pools and tennis courts. Extra persons $35/25 per night. 3-night deposit high season, 2-night low season.
*1 BR (4, max 6) fairway v. $275/250,   o.v. $315/285,   o.f. $375/335*
*2 BR (4, max 6) fairway v.  375/330,   o.v.  415/375,   o.f.  475/435*

## THE RIDGE ★

Agents: Ridge Rentals (808-667-2851) 1-800-367-8047 ext. 133 U.S., 1-800-423-8733 ext. 133 Canada, Kapalua Hotel 1-800-367-8000, Kumulani 1-800-367-2954, Kaanapali Vacations 1-800-367-8008. Part of the Kapalua condominiums, these are also well appointed but slightly less expensive. With its location above the hotel in the golf course area it is quite a walk to the beach, however, a free on-call shuttle is available for transport to the hotel and beach. 5-day minimum, $100 deposit, 14-day notice of cancellation between 12/15 and 4/15, other dates 48 hours. Maid service only on check in. NO CREDIT CARDS.
*1 BR 2 bath o.v. $165/90, 2 BR 2 bath o.v. $240/135   (prices-Ridge Rentals)*

## IRONWOODS

Beautiful and very expensive oceanview homes. Currently no rentals available.

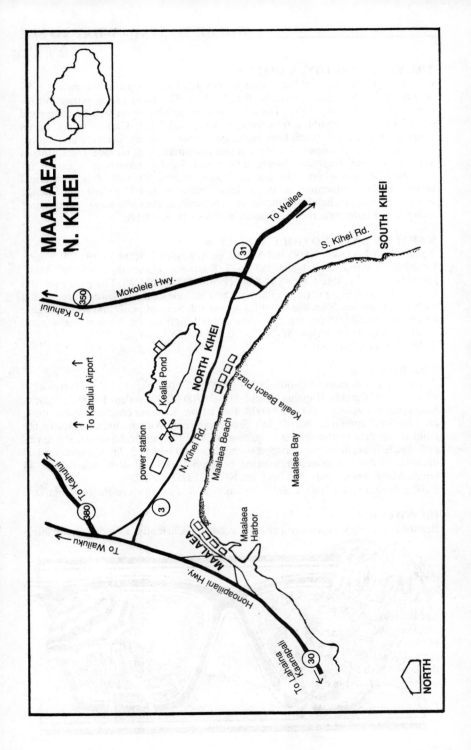

MAALAEA
N. KIHEI

To Wailea

SOUTH KIHEI

S. Kihei Rd.

Mokolele Hwy.

350

To Kahului

Kealia Pond

NORTH KIHEI

Kealia Beach Plaza

To Kahului Airport

Maalaea Beach

power station

Maalaea Bay

To Kahului

N. Kihei Rd.

380

3

To Wailuku

MAALAEA

Maalaea Harbor

Honoapiilani Hwy.

To Lahaina
Kaanapali

30

NORTH

# MAALAEA

## INTRODUCTION

Maalaea to many is just a signpost enroute to Kaanapali, or a harbor for the departure of a tour boat. However, Maalaea (which means "area of red dirt") is the most affordable and centrally located area of the island. A short 10 minutes from Kahului, 30 minutes from Lahaina and 15 minutes from Wailea makes it easy to see all of the island while headquartered here. You can hop into the car for a beach trip in either direction. Even better is the mere six-mile jaunt to Kahului/Wailuku for some of Maui's best and most affordable eateries. This quiet and relaxing area is a popular living area for local residents. Seven of the ten condominium complexes are located on a sea wall on or near the harbor of Maalaea, while the other three are on one end of the three mile long Maalaea Bay beach. The two end complexes, Maalaea Mermaid and the Maalaea Yacht Marina are actually within the harbor.

The ocean and beach conditions are best just past the last condo, the Makani A Kai. There is less turbidity providing fair snorkeling at times, good swimming and even two small swimming areas protected by a reef. These are found on either side of the small rock jetty with the old pipe. This length of beach is owned by the government and is undeveloped, providing an excellent opportunity for beach walkers who can saunter all the way down to Kihei. The condominium complexes are small and low-rise with moderate prices and no resort activities. The vistas from many of the lanais are magnificent, with a view of the harbor activity and the entire eastern coastline from Kihei to Makena, including majestic Haleakala, as well as Molokini, Kaaholawe and Lanai. The view is especially pleasing at night and absolutely stunning when a full moon shimmers its light across the bay and through the palm trees with the lights of Upcountry, Kihei and Wailea as a backdrop. No other part of the island offers such a tranquil setting. Another plus are the almost constant trade winds which provide non-air conditioned cooling as opposed to the sometimes scorching stillness of the Lahaina area. Summer in Maalaea is time for surfing. Summer swells coming into the bay reportedly create the fastest right-breaking rideable waves in the world and are sometimes referred to as "the freight train." The local kids are out riding from dawn to dusk. Winter brings calmer seas with fair snorkeling over the offshore reef. The calm conditions, undisturbed by parasailing and jet-skiing, also entices the Humpback whales into the shallow waters close to shore.

Local eating options are limited. One nice restaurant, The Waterfront, at the Milowai condominiums is open for dinner. In addition to sandwiches at the Maalaea Mermaid market, a limited number of snacks are available at the Maalaea store. Casual seafood lunch and dinners are served Monday through Saturday at the Island Fish Market. Buzz's Wharf, with their new renovations, sits at the end of the Harbor and is open for lunch, dinner and cocktails.

Under development is the new Maalaea Fishing Village. Plans call for a shopping center about as large as the Lahaina Cannery. The two-story design is reported to incorporate a local Hawaiian theme with lots of small shops and vendors selling fresh produce and island crafts. Five or six restaurants are being figured into the preliminary reports. There is also mention that the Maui Historical Society is reportedly considering a museum and museum shop. It will be located across from the harbor near the old fishing shrine.

All in all, Maalaea is a quiet and convenient choice. It is not for those seeking the hub of activity or convenient fine dining, but for the independent and quiet travelelr. Undecided about where to stay? Maybe you should try Maalaea!

**BEST BETS: *Kana'i A Nalu*** - Attractive complex with all two-bedroom units on a sandy beachfront, affordably priced. *Lauloa* - Well designed units oceanfront on the seawall. *Makani A Kai* - Located on a sandy beachfront, two bedroom units are townhouse style.

## WHAT TO DO AND SEE

The Maalaea Harbor area is a scenic port from which a number of boats depart for snorkeling, fishing and whale watching. Also in this area is Buzz's Wharf Restaurant, open for dinner. Fresh Island Fish Company, a seafood market, open Monday thru Saturday 10 - 5.

## ACCOMMODATIONS - MAALAEA

| | |
|---|---|
| Maalaea Mermaid | Island Sands |
| Maalaea Yacht Marina | Maalaea Banyans |
| Milowai | Kana'I A Nalu |
| Maalaea Kai | Hono Kai |
| Lauloa | Makana A Kai |

### MAALAEA MERMAID
No rental information available. Located within the Maalaea Seawall. Small market on the ground floor.

### MAALAEA YACHT MARINA
Hauoli St., Maalaea, Maui, HI 96793. (808-244-7012). Agents: Maalaea Bay Rentals 1-800-367-6084, Kihei Maui Vacations 1-800-542-6284, Hawaiian Apt. Leasing 1-800-854-8843 (1-800-472-8449 CA). All units are oceanfront, a beach is nearby. The units we viewed were pleasant with a wonderful view of the boats from most units, and the added plus of having security elevators and stairways. Many of the units have no air conditioning and laundry facilities are located in a laundry room on each floor. A postage stamp size grassy area in front and a small, but adequate pool. *1 BR (2) $80/$70,  2 BR (4) $90-$100/$80*

### MILOWAI
Hauoli St., Maalaea, Maui 96793. Agents: Kihei Maui Vacations 1-800-542-6284, Maalaea Bay Rentals 1-800-367-6084, Hawaiian Apt. Leasing 1-800-854-8843 (1-800-472-8449 CA). One of the larger complexes in Maalaea with a restaurant on location, The Waterfront. This complex has a large pool area, with a BBQ along

the seawall. The corner units are a very roomy 1,200 square feet with windows off the master bedrooms. Depending on condo location in the building, the views are of the Maalaea harbor or the open ocean. The one bedroom units have a lanai off the living room and a bedroom in the back. Washer/dryer. Weekly/monthly discounts. *1 BR $90/70, 2 BR $110/80*

## MAALAEA KAI
Hauoli St., Maalaea, Maui, HI 96793. (244-7012). Agents: Maalaea Bay Rentals 1-800-367-6084, Kihei Maui Vacations 1-800-542-6284. 70 oceanfront units. Laundry facilities, putting green, BBQ, and elevator to upper levels. Located on the harbor wall, the rooms were standard and quite satisfactory. Some do not have washer and dryers in the rooms. There is a pool area and large pleasant grounds in front along the harbor wall. A few blocks walk down to a sandy beach. Monthly discounts. *1 BR (2) $80/70, 2 BR (4) $100/80*

## LAULOA ★
Hauoli Street,(RR 1 Box 383) Maalaea, Maui, HI 96793. (808-242-6575). Agent: Larry Jessup 1-800-635-4540. Forty-seven 2-bedroom, 2-bath units of 1,100 sq.ft. One of the Lauloa's best features is their floor plan. The living room and master bedroom are on the front of the building with a long connecting lanai and sliding glass patio doors which offer unobstructed ocean views. A sliding shoji screen separates the living room from the bedroom.

Each morning from the bed you have but only to open your eyes to see the palm trees swaying and a panoramic ocean view. The second bedroom is in the back of the unit. These two bedroom units are spacious (only two bedroom units are currently available in the rental program) and are in fair to good condition (depending on the owner). Each has a washer/dryer in the unit. The pool area and grounds are along the seawall. With a stairway in the sea wall, there often are local fishermen throwing nets and lines into the ocean. $200 deposit. Maid service extra charge. Monthly discounts. NO CREDIT CARDS. *2 BR, 2 bath (4) 1-6 nights $120/95, 7-29 nights $110/85*

GINGER

*Maalaea*

## ISLAND SANDS
Hauoli St., (RR 1, Box 391) Wailuku, Maui, HI 96793. Agents: Rainbow Rentals 1-800-451-5366. Eighty-four units in a 6-story building. This is one of Maalaea's larger complexes. Maui shaped pool, a grassy lawn area, BBQ. Many of these units also have a lanai off the master bedroom, however the lanais have a concrete piece in the middle of each railing which somewhat limits the view while sitting or laying in bed. Washer/dryers, and air conditioning. Elevators. Extra person $7.50/night. Weekly and monthly discounts. 4-night minimum, $200 deposit with 15-day refund notice. Children under 3 free.
*Studio (2) $60/50, 1 BR 1 Bath (2) $75/65, 2 BR 2 Bath (4) $90/75*

## MAALAEA BANYANS
Hauoli St., (RR 1 Box 384) Maalaea, Maui, HI 96793. (808-242-5668). Agents: Maalaea Bay Rentals 1-800-367-6084, Oihana 1-800-367-5234, Maui 800 1-800-367-5224.

Seventy-six oceanview units with lanai and washer/dryer. Weekly/monthly discounts. Oceanfront on rocky shore, short walk to beach. Pool area, jacuzzi, BBQ's. Extra persons $10/night. 7-night minimum. NO CREDIT CARDS.
*1 BR 1 bath (2,max 4) $80/70, 2 BR 2 bath (4,max 6) $90/80*

## KANA'I A NALU ★
Hauoli Street, Maalaea. Agents: Maalaea Bay Rentals 1-800-367-6084, Oihana 1-800-367-5234. 80 units with washer/dryers in four buildings with elevators. No maid service. This is the first of the three condominiums along a sandy beach-front. Its name means "parting of the sea, surf or wave." The complex is V-shaped with a pool area in the middle. Nicely landscaped grounds, a decent beach and only a short walk along the beach to the best swimming and playing area. Overall one of the best values in the Maalaea area. One of the places we keep coming back to! 5-day minimum, $200 deposit, 30-day refund notice. Weekly discounts. *2 BR (4) o.v. $110/80, o.f. $130/100      Extra persons $7.50/night*

## HONO KAI
Hauoli St., Maalaea. (244-7012) Agent: Maalaea Bay Rentals 1-800-367-6084. Forty-six units located on the beach. Choice of garden view, oceanview or oceanfront. Laundry facilities, BBQ, pool. This is one of three properties managed by Maalaea Bay Rentals and according to the agent this property is the "chevrolet" model. This complex is on the beach and bears attention for the budget conscious traveler, but don't expect any frills.
*1 BR $75-$85/$60-70, 2 BR $88-98/$77-87, 3 BR $103/93*

## MAKANI A KAI
Hauoli St., Maalaea. Agent: Maalaea Bay Rentals 1-800-367-6084. These deluxe units are also on the beach and offer a choice of oceanfront or oceanview. Laundry room on property, pool, BBQ. This is the last property along the beach in Maalaea. Beyond this is a long stretch of sandy beach along undeveloped state land and about a four mile jaunt down to North Kihei. Great for you beach walkers! The two bedroom units are townhouse style. Unfortunately, these do not have in-room washer/dryers. In our opinion, a must for the traveling family!
*1 BR $80-105/70-80, 2 BR $110-130/100-130      5-day minimum stay*

# NORTH KIHEI

## INTRODUCTION

North Kihei is 15 minutes from the Kahului Airport and located at the entrance to South Kihei. The condominiums here stretch along a gentle sloping white sand beach. The small Kealia Shopping Center is located between the Kihei Sands and Nani Kai Hale. Another small shopping area is found at the Sugar Beach Condominiums. Several snack shop restaurants can be found along Kihei Road in this area. A little to the south down Kihei Road are additional restaurants, grocery stores and large shopping areas. Along with Maalaea, this is one of our favorite places to stay because of the good units, central, but quiet location, nice beach, cooling breezes and certainly some of the island's best vacation buys.

***BEST BETS: Kealia** and **Maalaea Surf**.*

## ACCOMMODATIONS - NORTH KIHEI

| | |
|---|---|
| Kealia | Kihei Kai |
| Sugar Beach | Maalaea Surf |
| Kihei Sands | Kihei Beach Resort |
| Nani Kai Hale | |

### KEALIA ★
191 N. Kihei Rd., Kihei, Maui, HI 96753. (808-879-9159) 1-800-367-5222. Fifty-one air conditioned units with lanais, washer/dryers, and dishwashers. Maid service on request. The one bedroom units are a little small, but overall a good value. Well maintained and quiet resort with a wonderful sandy beach. Shops nearby. Extra person $10. 10% monthly discount. $125 deposit, $10 cancellation fee. 100% payment required 30 days prior to arrival. 7-day minimum winter, 4-day in summer. NO CREDIT CARDS. *Studio (2) $80/65, 1 BR (2) $90/75*

### SUGAR BEACH RESORT
145 N. Kihei Rd. Kihei, Maui, HI 96753. (808-879-7765). Agents: Condo Rental HI 1-800-367-5242, The Ching Connection 879-7866, Hawaiian Apt. Leasing 1-800-854-8843 (1-800-472-8449 CA), Maui Condos 1-800-663-6962 Canada, RSVP 1-800-663-1118, Rainbow Rentals 1-800-451-5366.

215 units in several six-story buildings with elevators. Air conditioning. Jacuzzi, putting green, gas BBQ grills. Sandwich shop and quick shop market on location. A fairly large property with a nice pool area and located on an excellent swimming beach. In previous editions we gave this a recommended rating, however, on a recent stay we found our particular unit to be long overdue for some major renovations and down deep cleaning. That is not to say that there aren't a number of excellently cared for units but this is one of many properties that is very popular and heavily rented leaving little time for remodeling and refurbishing. Apartment owners that don't make regular inspections are another part of the problem. This condo and others with similar discrepancies from unit to unit need

management established criteria which would eliminate those units from the rental pool that are not up to certain standards. Extra persons $7.50/night, weekly discounts. *1 BR  g.v. $95/75, o.f. $105/85, 2 BR o.v. $175/130*

## KIHEI SANDS
115 N. Kihei Rd., Kihei, Maui, HI 96753. (808-879-2624) Thirty oceanfront air-conditioned units. Shops and restaurant nearby. 3-day minimum, $100 deposit with $10 cancellation fee. Full payment on arrival. No maid service or room phone. Laundry area. NO CREDIT CARDS. Extra persons $6/night.
*1 BR (2) $70-90/65-75,  2 BR (4) $85-105/70-90*

## NANI KAI HALE ★
73 N. Kihei Rd., Kihei, Maui, HI 96753. (808-879-9120) 1-800-367-6032. Agent: Village Rentals 1-800-367-5634. 46 units in a six-story building. Under building parking, laundry on each floor, elevator. No maid service, no room phones. Patio and BBQ's by beach. Lanais have ocean and mountain views.
Prices based on 7-day minimum/3-day minimum stay.
*S BR w/o kitchen (2) $ 40/30, with kitchen $70/45    $100 deposit*
*1 BR 2 bath (2) o.f.    100/70, o.v. $85/60         10% monthly discount*
*2 BR 2 bath (2) o.v.    120/90    Extra persons $10/12/nite, children under 5 free*

## KIHEI KAI
61 N. Kihei Rd., Kihei, Maui, HI 96753. (808-879-2357) 1-800-367-8047 ext. 248. Twenty-four units in a two-story beachfront building. Recreation area and laundry room. Units have air conditioning or ceiling fans. BBQ. On seven-mile stretch of sandy beach, near windsurfing, grocery stores. A very good value. Minimum 7-days winter, 4-days summer. $100 deposit ($200 deposit for 2 weeks or longer.) Full payment upon arrival. NO CREDIT CARDS. Extra persons $5/night. *1 BR (2,max 4) $70-80/$50-70*

## KIHEI BEACH RESORT
36 S. Kihei Rd., Kihei, Maui, HI 96753. (808-879-2477) 1-800-367-6034. Agents: Maui Network 1-800-367-5221, Maui Condo & Home 1-800-822-4409, Maui Condo 1-800-663-6962 Canada. 54 beachfront units with oceanview, phones. Resort offers central air conditioning, recreation area, elevator, maid service. Extra person $9/night. 3-day deposit with 3-day refund notice low season, 30-day high season. Balance must be prepaid in full. Weekly and monthly discounts. NO CREDIT CARDS. *1 BR (2,max 4) $96/82, 2 BR (4,max 6) $133/113*

## MAALAEA SURF ★
12 S. Kihei Rd., Kihei, Maui, HI 96753. (808-879-1267) 1-800-423-7953. Sixty oceanview units in 8 two-story oceanfront buildings. Units have air conditioning and microwaves. Daily maid service, except Sundays and holidays. Two pools, two tennis courts, shuffleboard. Laundry facilities in each building. Very attractive and quiet low rise complex on a great beach. In this price range, these spacious and attractive units, along with the beautiful grounds, are impressive and hard to beat. Extra persons $8/night. NO CREDIT CARDS. $200-300 deposit. High season balance due 30 days prior to arrival, low season on arrival. 60-day refund notice with $10 cancellation fee during high season, 30-day notice low season.
*1 BR 1 bath (2,max 4) $145/125,  2 BR 2 bath (4,max 6) $200/170*

# SOUTH KIHEI

## INTRODUCTION

South Kihei began its growth after that of West Maui, but unfortunately with no planned system of development. The result is a six-mile stretch of coastline littered with more than 50 properties, nearly all condominiums, with some 2,400 units in rental programs. Few complexes are actually on a good beach, however, many are across Kihei Road from one of the Kamaole Beach Parks. A variety of beautiful beaches are just a few minutes drive away.

This section of East Maui has a much different feel than West Maui or Lahaina. There are no large resorts with exotically landscaped grounds, very few units on prime beachfront, and more competition among the complexes making this area a good value for your vacation dollar. (And a good location for extended stays.)

Kihei always seemed to operate at a quieter and more leisurely pace than that of Kaanapali and Lahaina, but the last couple of years has seen a significant upsurge of development, not of condos, but of shopping complexes. Even parts of South Kihei Rd. have been repaved and regular curbs installed. These changes indicate the increasing tourist activity along with a corresponding loss of Kihei's once laid-back charm. Restaurant selections remain more limited and many vacationers staying in condominiums utilize their kitchens. Most needs can be filled locally at one of several large grocery stores or the growing number of small shopping centers. Kahului, Wailuku, Wailea and Lahaina remain an easy drive for additional shopping and dining out.

***BEST BETS: Laule'a*** - Units are beautifully furnished and well equipped. While not beachfront, they have a pool and jacuzzi and are located on a large grassy park that offers tennis courts. The ocean is on the other side of the park. Better beaches are a short drive away. ***Maui Hill*** - Situated on a hillside across the road from the ocean, some units have excellent ocean views. The three-bedroom units here are roomy and a good value for large families. ***Haleakala Shores*** - Across from Kamaole III Beach Park. ***Mana Kai Maui*** - One of Kihei's larger resorts, and the only one right on a very good beach.

## WHAT TO DO AND SEE

The only historical landmark is a totem pole near the Maui Lu Resort which commemorates the site where Captain Vancouver landed.

## WHERE TO SHOP

Every corner of Kihei is sprouting a new shopping mall. The complexes all seem to have quick markets, video stores and a T-shirt shop. The following are now complete. Several others are in various stages of construction.

Traveling down Kihei Road the first center is **Azeka's.** At the market you can pick up Azeka's famous ribs to cook yourself. The *International House of Pancakes, Island Thai* and *Luigi's* are the restaurants here in addition to a small *Liberty House*, the *Silversword Bookstore*, a *Ben Franklin* and a few small tourist shops.

Ground has been broken for the new $14 million **Azeka Place-Mauka** shopping center, across the street from the Azeka Place center on South Kihei Road. The new 60,000 square foot center will include a *Long's Drug Store* and is scheduled to open in late 1990.

The **Lipoa Shopping Center** is a block down Lipoa Street and offers a medical center and pharmacy, a cycle and sport shop, *Sweet Cream's*, and *Henry's Bar and Grill.*

Just past the Kapulanikai condominiums is the **Kukui Mall**, at 1819 S. Kihei Rd., gets our vote for the most attractive mall. This large complex is done in a Spanish style of architecture with a wide assortment of shops. *Subway Sandwich* shop is a handy stop for lunch enroute to one of Wailea's fine beaches. For a cool treat there's *I Can't Believe It's Yogurt.* A new *Walden's* bookstore is a welcome addition for visitors and residents alike, with a great selection of books on just about any subject, also *Kihei Art Gallery,* and several clothing shops. Add to the list of ethnic eateries the new *Kihei Chicken and Ribs.* Owners Evelio and Carmen Mattos have brought their recipes all the way from Celi Columbia. There is also an all-you-can-eat restaurant called *Bonanza.*

Just beyond is **The Kihei Town Center** which offers a selection of shops including sporting goods, novelty shops, pharmacy, grocery, and clothing. The only restaurant here is *Chuck's*, but two places for a fast bite are *Paradise Fruits* across the street and *McDonald's* on the corner.

The next few shopping areas run almost together. The **Dolphin Plaza,** 2395 S. Kihei Rd., across from Kamaole I Beach is one of the smaller new shopping centers. Here you'll find *The New York Deli, Pizza Fresh, Baskin Robbins Ice Cream,* the *Kihei Bakery* and a video store.

Between the Dolphin Plaza and Rainbow Mall is the **Kamaole Beach Center**, 2411 S. Kihei Rd. *The Sports Page Grill and Bar* is the focal point here with a yogurt place and pizza restaurant rounding out the center.

**The Rainbow Mall** is a small center also located on the mauka side (towards the mountain) of South Kihei Road offering an ice cream shop, *Chris's Smokehouse* restaurant and a limited number of assorted small clothing and souvenir shops.

**Kamaole Shopping Center** is one of the larger new malls and offers several restaurant selections, including *Denny's*, the *Canton Chef* and *Erik's Seafood Broiler, Lappert's Ice Cream* and *Cinnamon Roll Fair.*

The last shopping center in Kihei, across from Kamaole III Beach, is the **Nani Kai Center**. *La Familia, Kihei Prime Rib* and *The Greek Bistro* restaurants make up this complex.

# ACCOMMODATIONS - SOUTH KIHEI

Nona Lani
Kihei Holiday
Wailana Sands
Resort Isana
Pualani
Sunseeker Resort
Maui Lu Resort
Kihei Bay Vista
Kihei Bay Surf
Menehune Shores
Kihei Resort
Koa Lagoon
Koa Resort
Kauhale Makai
Leinaala
Luana Kai
Laule'a
Maui Sunset

Leilani Kai
Kihei Garden
Hale Kai O Kihei
Waiohuli Beach
Kihei Beachfront
Kapulanikai
Island Surf
Kihei Park Shores
Shores of Maui
Punahoa
Kalama Terrace
Beach Club Apts.
Lihi Kai Cottages
Maui Vista
Kamoa Views
Kamaole One
Kamaole Beach
  Royale

Kihei Alii Kai
Royal Mauian
Kamaole Nalu
Hale Pau Hana
Kihei Kai Nani
Kihei Akahi
Haleakala Shores
Maui Parkshore
Kamaole Sands
Hale Kamaole
Maui Kamaole
Maui Hill
Kihei Surfside
Mana Kai Maui
Surf and Sand
Hale Hui Kai

## NONA LANI
455 S. Kihei Rd.,(P.O. Box 655) Kihei, Maui, HI 96753. (808-879-2497). Eight individual cottages with kitchens, queen bed plus 2 day beds, full bath with tub and shower, and lanais. Large grounds, public phone, two BBQ's, and laundry facilities. Located across the road from sandy beach. Weekly rates available. Extra person $10/$7 night. $100 deposit, $50 non-refundable. No personal checks or credit cards. *1 BR (2) $75 (7-night minimum) / $60 (4-night minimum)*

## KIHEI HOLIDAY
483 S. Kihei Rd., Kihei, Maui, HI 96753. (808-879-9228) Agents: Kihei Maui Vacations 1-800-542-6284, Hawaiian Apt. Leasing 1-800-854-8843 (1-800-472-8449 CA), RSVP 1-800-663-1118.

Units are across the street from the beach and have lanais with garden views. Pool area jacuzzi and BBQ's. $100 deposit, full payment 30 days prior to arrival. Maid service on request. NO CREDIT CARDS.
*2 BR (4) $80/65     Extra persons $5/night under age 12, $10 over age 12*

## WAILANA SANDS
25 Wailana Place, Kihei, Maui, HI 96753. (808-879-2026) 1-808-879-3661. 10 units, overlook courtyard and pool area, in a two-story structure. Quiet area on a dead end road. A one block walk to Kihei Beach. 4-day minimum, $100 deposit. NO CREDIT CARDS. *Studio $50/35, 1 BR $60/40, 2 BR $95/65*

## PUALANI TOWNHOUSES
15 Wailana Place, Kihei, Maui, HI. No vacation rental information available.

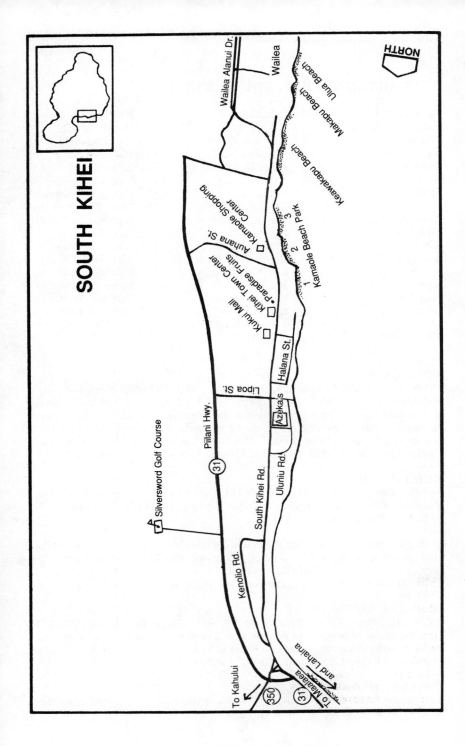

SOUTH KIHEI

## RESORT ISANA
515 S. Kihei Road, Kihei, Maui, HI 96753 (808) 879-7800, 1-800-633-3833. Agent: Hawaiian Pacific Resorts 1-800-367-5004. One of Kihei's newest, these fifty one-bedroom units are decorated in muted beige and blues and are complete down to an electric rice cooker and china dishes. A spacious pool area is the focal point of the central courtyard. *1 BR $130/100, 2BR and 3BR suite available.*

## SUNSEEKER RESORT
551 S. Kihei Rd. (P.O. Box 276) Kihei, Maui, HI 96753. (808-879-1261). Six units including studios with kitchenettes, one and two bedrooms with kitchens. Weekly and monthly discounts available. No room phones, no pool. Across street from beach. Popular area for windsurfing. *SBR $50, 1 BR $60, 2 BR (4) $85*

## MAUI LU RESORT
575 S. Kihei Rd., Kihei, Maui, HI 96753. (808-879-5881) Agent: Aston 1-800-367-5124. 180 units on 26 acres. Cottages have full kitchens, fans, but no air conditioning. Hotel rooms have air conditioning, small refrigerator and hot water maker. Pool is shaped like the island of Maui. Many of the hotel rooms are set back from South Kihei Road and the oceanfront units are not on a sandy beachfront. One of the first resorts in the Kihei area and unusual with its spacious grounds. Recently sold, we hope the new owners undertake some major renovations to upgrade this property. Extra persons $10, under 18 no charge. *Hotel rooms (2) $99-118/$79-$108, 1 & 2 BR cottages (1-3) $118/108*

## KIHEI BAY VISTA
679 S. Kihei Rd., Kihei, Maui, HI 96753. (879-75811), 1-800-367-7040 U.S., 1-800-423-8733 ext. 159 Canada. Agents: Kumulani 1-800-367-2954, Hawaiian Apt. Leasing 1-800-854-8843 (1-800-472-8449 CA), Maui condos 1-800-663-6962 Canada, RSVP 1-800-663-1118. New in 1989 this complex offers pool, spa, jacuzzi, putting green, air conditioning, washer/dryer, lanais and full kitchens. A short walk across the road to the Kamaole I Beach. *1 BR $75 - $105*

## KIHEI BAY SURF
715 S. Kihei Rd. (Manager Apt. 110), Kihei, Maui, HI 96753. (808-879-7650) Agents: Kihei Maui Vac. 1-800-542-6284, Kumulani 1-800-367-2954, Hawaiian Apt. Leasing 1-800-854-8843. 118 studio units in 7 two-story buildings. Pool area jacuzzi, recreation area, gas BBQ's, laundry area, tennis. Across road from Kamaole I Beach. Phones. Weekly discounts. *Studios $60/50*

## MENEHUNE SHORES
760 Kihei Rd., Kihei, Maui, HI 96753. (808-879-1508) Agents: Kihei Kona Rentals 1-808-879-5828, Kihei Maui Vacations 1-800-542-62824, RSVP 1-800-663-1118, Menehune Reservations 1-800-558-9117 U.S. 115 units with dishwashers, washer/dryers and lanais in a 6-story building. Recreation room, roof gardens with whale-watching platform, and shuffleboard. The ocean area in front of this condominium property is the last remnant of one of Maui's early fish ponds. These ponds, where fish were raised and harvested, were created by the early Hawaiians all around the islands. Extra persons $7/night, $150 depoist. Rates quoted are for seven-night stay. Slightly higher for 4-6 day stay. 4-day minimum. *1 BR 1 bath (2) $ 85/70, 2 BR 2 bath (2) $102/ 82, 3 BR 2 bath (5) $140/130*

## KIHEI RESORT
777 S. Kihei Rd., Kihei, Maui, HI 96753. (808-879-7441) 1-800-367-6006. Agents: Village Rentals 1-800-367-5634, RSVP 1-800-663-1118, Maui Condo 1-800-663-6962 Canada, Rainbow Rentals 1-800-451-5366. Sixty-four units, BBQ's, pool area jacuzzi. NO CREDIT CARDS.

*1 BR (2) $ 80/60   7-nite minimum, $100 deposit, 10-day refund notice*
*2 BR (4) $110/85   Extra persons $7/night, 10% monthly discount*

## KOA LAGOON
800 S. Kihei Rd., Kihei, Maui, HI 96753. (808-879-3002) 1-800-367-8030. Agent: Rainbow Rentals 1-800-451-5366. 42 oceanview units in one 6-story building. Bar and ice-maker, washer/dryers. Pool area pavilion, BBQ's. Located on a small sandy beach that is often plagued by seaweed which washes ashore from the offshore coral reef. This stretch of Kihei is very popular with windsurfers. Extra person $15 additional. $30 charged for 5 days or less. $200 deposit, full payment 30 days prior to arrival. 60-day cancellation notice. 14-day minimum stay during Christmas holiday. Supreme units $15 per night more.

*1 BR 1 bath (2,max 4) $120/80,  2 BR 2 bath (4,max 6) $140/110*

## KOA RESORT
811 S. Kihei Rd., Kihei, Maui, HI 96753. (808-879-1161) 1-800-367-8047 ext.407, Canada 1-800-423-8733 ext 407. 54 units (2,030 sq.ft.) located on spacious 5 1/2-acre grounds in 2 five-story buildings across road from beach. Two tennis courts, spa, jacuzzi, putting green. Units have washer/dryers. 5-night minimum stay, $100 deposit. 10% monthly discount. NO CREDIT CARDS.

*1 BR 1 bath (2,max 4) $ 90/ 70   Extra persons $10/day, children under 2 free*
*2 BR 1 bath (4,max 4)  105/ 70,  2 bath (4,max 6) $115/ 95*
*3 BR 2 bath (6,max 8)  140/120,  3 bath (6,max 8)  165/145*

## KAUHALE MAKAI (Village by the Sea)
930-938 S. Kihei Rd. Kihei, Maui, HI 96753. (808-879-8888) Agents: Oihana 1-800-367-5234, Village Rentals 1-800-367-5634, Kihei Maui Vac. 1-800-542-6284, Kumulani 1-800-367-2954, RSVP 1-800-663-1118, Rainbow Rentals 1-800-451-5366.

HUMUHUMUNUKUNUKUAPUAA

169 air-conditioned units in 2 six-floor buildings with phones. Complex features putting green, gas BBQ's, children's pool, sauna, laundry center. The beach here is usually strewn with coral rubble and seaweed. $5 additional person. 5-night minimum. *Studio (2) $55-60/45-50, 1 BR (2) $80-85/$60-65, 2 BR (4) $110/85*

## LUANA KAI ★
940 S. Kihei Rd., Kihei, Maui, HI 96753. (808-879-1268) Agents: Hawaiian Island Resorts 1-800-367-7042, Maui 800 1-800-367-5224, Kihei Maui Vacations 1-800-542-6284, All About Hawaii 1-800-336-8687, Kumulani 1-800-367-2954, Rainbow Rentals 1-800-451-5366.

113 units with washer/dryers are located adjacent to a large oceanfront park with public tennis courts. The beach, however, is almost always covered with coral rubble and seaweed. The grounds are nicely landscaped and include a putting green, BBQ area, pool area sauna and jacuzzi. Towel service mid-week, linen service weekly. Children under 12 free. Extra person $10. 2-night minimum high season. Weekly discounts. 3 BR units may be available from some agents.
*1 BR (2) g.v. $115/80, o.v. $125/95, 2 BR (4) g.v. $130/100, o.v. $145/115*

## LAULE'A MAUI BEACH CLUB ★
980 S Kihei Rd, Kihei, Maui HI 96753. (808-879-5247) Agents: Hawaiiana Resorts 1-800-367-7040, Maui Condo 1-800-663-6962 Canada, RSVP 1-800-663-1118, Hawaiian Apt. Leasing 1-800-854-8843.

Completed in 1984, these 58 units have been tastefully decorated in mauves and blues and have washer/dryer, microwave, phones and maid service every other day. Full prepayment. Only one building has an elevator to the upper floors. Fronting these condos is a public park with 4 tennis courts and a beach that is seasonally strewn with coral rubble. On-site are a nice pool area, separate men's and women's saunas, wet bar area, and hot tub.
*1 BR (4) g.v. $110/ 90, o.v. $125/105   7-nite/2-nite deposit*
*2 BR (6) g.v.   135/115, o.v.   150/130   7-day refund notice*
*3 BR (8) garden $160/140           persons under 18 free*

## LANAKILA
992 S. Kihei Rd., Kihei, Maui, HI 96753. (808-879-5629)
No vacation rentals, long term only.

## LEINAALA
998 S Kihei Rd, Kihei, Maui, HI 96753. (808-879-2235) Agent: Rose Mary Verhulp 1-800-334-3305. 24 one and two bedroom units in a 4-story building. Tennis courts, pool, cable color TV. Oceanview. Monthly discounts. $200 deposit with 30-day refund notice. NO CREDIT CARDS. Extra persons $10 night.
*1 BR (2) $85/65, 2 BR (4) $110/100*

## WAIPUILANI
1002 S. Kihei Rd., Kihei, Maui, HI 96753. (808-879-1465). 42 units in three 3-story buildings. No charge for children under age 3, others $7 per night. Monthly discounts available. $100 deposit with full payment 30 days in advance of arrival. NO CREDIT CARDS. *1 BR (2) $75/65, 2 BR (4) $105/95*

## MAUI SUNSET

1032 S. Kihei, Rd., Kihei, Maui, HI 96753. (808-879-0674) Reservation Assistance 1-800-843-5880. Agents: Kihei Maui Vac. 1-800-542-6284, Kumulani 1-800-367-2954, Hawaiian Apt. Leasing 1-800-854-8843 (1-800-472-8449 CA), RSVP 1-800-663-1118. 225 air conditioned units in 2 multi-story buildings. Tennis courts, pitch and putt golf green, and sauna. Large pool, exercise facility, barbecues. Located on beach park with tennis courts, however, this beach is generally covered with seaweed and coral rubble. Extra persons $7/night.
*1 BR $90-105/80-95, 2 BR $110-125/95-105, 3 BR $150/125*

## LEILANI KAI

1226 Uluniu St.,(P.O. Box 296) Kihei, Maui, HI 96753. (808-879-2606). Eight garden apartments with lanais. $200 deposit. Extra person $7.50. Full payment prior to arrival. 4-day minimum stay. NO CREDIT CARDS.
*Studio (2) $54/39, 1 BR (2) $68/48, 1 BR dlx (4) $79/59, 2 BR (4) $85/65*

## KIHEI GARDEN ESTATES

1299 Uluniu St., Kihei, Maui, HI 96753. (808-879-6123). Agents: Kihei Maui Vacations 1-800-542-6284, RSVP 1-800-663-1118. 84 units in eight 2-story buildings. Jacuzzi, BBQ's. Across road and short walk to beaches. Monthly discounts. $100 deposit, full payment 30 days prior to arrival. NO CREDIT CARDS. *1 BR (2,max 4) $65/50, 2 BR (4,max 6) $95/75*

## HALE KAI O KIHEI

1310 Uluniu Rd., Kihei, Maui, HI 96753. (808-879-2757). 59-oceanfront units with lanais in 3-story building. Sandy beachfront. Shuffleboard, putting green, BBQ's, laundry, recreation area. Maid service on request for extra charge. Extra person $8.50/night. 10% monthly discount. $100 deposit. 60-day/30-day cancellation notice (less $20 handling fee). NO CREDIT CARDS. Daily rates upon request. *1 BR (2) $525 weekly, 2 BR 2 bath (4) $675/495 weekly*

## WAIOHULI BEACH HALE

49 West Lipoa St., Kihei, Maui, HI 96753. (808-879-5396). 52 units in four 2-story buildings. Large pool, BBQ area, shuffleboard, near shopping center. Located on beachfront that is poor for swimming or snorkeling, often covered with coral rubble and seaweed. Spacious park-like lawn area around pool. Extra person $8/night. Weekly discounts. *1 BR (2) $75/65, 2 BR (4) 95/85*

## KIHEI BEACHFRONT RESORT

Located at end of Lipoa St. 8 oceanview 2-bedroom units with washer/dryers, microwaves, dishwashers, and air conditioning in a single 2-story building. Large lawn area fronting units. Lanais on upper level. No elevator. Pool area jacuzzi. *No rental information available at this time.*

## KAPULANIKAI APTS

73 Kapu Place, PO Box 716, Kihei, Maui, HI 96753. (808-879-1607) 12 units are oceanview with private lanais or open terraces. Beachfront is poor for swimming or snorkeling. Grassy lawn area in front. BBQ's, laundry facilities, pay phone on property. *1 BR 1 bath ($65 all seasons)*

## ISLAND SURF
1993 S. Kihei Rd., Kihei, Maui, Hi 96753. Agent: Kumulani 1-800-367-2954. This property is turning many of its units into commercial offices and shops on the first three floors which really detracts from a resort-like setting, although the economy of the place may attract some. Across the road from the Kamaole I Beach Park. $100 deposit. *Hotel rooms $45/$35, 1 BR $70/60, 2 BR $90/75*

## KIHEI PARK SHORES
No rental information available

## SHORES OF MAUI
2075 S. Kihei Rd., Kihei, Maui, HI 96753. (808-879-9140) 1-800-367-8002. 50-unit complex offers BBQ's, tennis courts, and spa. Located across the street from a rocky shoreline and 1 blocks north of Kamaole I Beach Park. $100 deposit, 30-day cancellation notice, full payment 30 days prior to arrival. Extra persons $5/night, monthly discounts. 3-day minimum (Christmas holiday 1-week minimum). Weekly discounts. NO CREDIT CARDS. *1 BR $75/55, 2 BR $100/80*

## PUNAHOA
2142 Iliili Rd., Kihei, Maui, HI 96753. (808-879-2720). 15-oceanview units with large lanais, telephones. No pool. Elevator, laundry facilities, beaches nearby. NO CREDIT CARDS. $200/100 deposit, 60-day refund notice. Extra persons $9/night. 10% discount 14-day low, or 28-day high season stay. 5-day minimum. *Studio (2) $70/52, 1 BR (2,max 4) $93-95 / 66-68, 2 BR (2,max 6) $97/72*

## KALAMA TERRACE
35 Walaka, Kihei, Maui, HI 96753. No rental information available.

## BEACH CLUB APARTMENTS
2173 Iliili Rd., Kihei Maui, HI 96753, (808) 874-6474. One block from beach, washer dryers, kitchens. *1 BR $65-85.*

## LIHI KAI COTTAGES
2121 Iliili Rd., Kihei, Maui, HI 96753. (808-879-2335) 1-800-544-4524. Nine beach cottages. All units are 1 BR 1 bath with kitchen and lanai. Self-service laundromat. Next to Kamaole I Beach. $100 deposit, 3-day minimum, 60/30 day refund notice. *1 BR $59/54*

## MAUI VISTA
2191 S. Kihei Rd., Kihei, Maui, HI 96753. (808-879-7966) 1-800-367-8047 ext.330. Agents: Aston 1-800-922-7866, Oihana 1-800-367-5234, Kihei Maui Vac. 1-800-542-6284, RSVP 1-800-663-1118, Rainbow Rentals 1-800-451-5366.

280 units in three 4-story buildings, across from the beach. Some units have air conditioning, some have washer/dryers. All have kitchens with dishwashers. The 2-bedroom units are fourth floor townhouses. Some oceanview units. 6 tennis courts, 3 pools, BBQ's. 5-night minimum low, 7-night high season. We had some problem with sound carrying from a neighboring unit, but overall, a good value. Extra person $10. One nights deposit. Rollaway $10 additional, $6 cribs. *1 BR (2,max 4) $129/109, 2 BR (2,max 6) $149-159 / 129-139*

### KAMOA VIEWS
Kanani Rd., Kihei, Maui, HI 96753. Agents: Oihana Property 1-800-367-5234.
Secured property, pool. Washer/dryer and dishwashers in units. *1 BR $80*

### KAMAOLE ONE
2230 S. Kihei Rd., Kihei, Maui, HI 96753. (808-879-4811). 2-story building. No
elevators or pool, covered parking. Beachfront. Telephones, washer/dryers,
compactors, air conditioning, ceiling fans and cable TV. Some have microwaves.
NO CREDIT CARDS. *2 BR ground floor $145/125, upper floors $135/115*

### KAMAOLE BEACH ROYALE
2385 S. Kihei Rd.,(P.O. Box 370) Kihei, Maui, HI 96753. (808-879-3131) 1-800-
421-3661. Agent: Maui Beach Homes/Bella Realty 1-800-541-3060. 64 units with
washer/dryers and lanais in a single 7-story building across from Kamaole I
Beach. Recreation area, elevator, roof garden. Extra person $10/$8 per night, 10%
monthly discount, 5-day minimum, $200 deposit, balance due 30 days prior to
arrival. $25 service charge. NO CREDIT CARDS.
*1 BR 1 bath (2) $90/55, 2 BR 2 bath (2) $200/65, 3 BR 3 bath (2) $110/75*

### KIHEI ALII KAI
2387 S. Kihei Rd., Kihei, Maui, HI 96753. (808-879-6770) Agents: Leisure
Properties 1-800-888-6284, Kihei Maui Vacations 1-800-542-6284, Maui Condos
1-800-663-6962, RSVP 1-800-663-1118, Rainbow Rentals 1-800-451-5366.

127 units in four buildings. All units have washer/dryers. No maid service.
Complex features pool, jacuzzi, sauna, two tennis courts, BBQ. Across road and
up street from beach. Nearby restaurants and shops. Extra persons $7/night. $100
deposit, 3 night minimum. Full payment 30 days prior to arrival.
*1 BR (2) $85/60, 2 BR (4) $95-115 / 70-80, 3 BR (6) $115/90*

### ROYAL MAUIAN
2430 S. Kihei Rd., Kihei, Maui, HI 96753. (808-879-1263) 1-800-367-8009.
Agent: Maui 800 1-800-367-5224. 107 units with lanai, washer/dryer, and phone
in a 6-story building. Complex has shuffleboard, carpeted roof garden, and is next
to the pleasant Kamaole II Beach Park. $10 extra person (no charge children
under age 6), 5-night minimum, $200/100 deposit with $25 cancellation fee. 14-
day discount summer, 30-day winter. Maid service every 5 days.
*1 BR (2) $120/92, 2 BR (2) $140/108, 3 BR (4) $195/160*

### KAMAOLE NALU
2450 S. Kihei Rd., Kihei, Maui, HI 96753 (808-879-1006) 1-800-367-8047
ext.435. Agents: Maui Condominiums 1-800-663-6962 Canada. Thirty-six 2-
bedroom, 2-bath units with large lanai, dishwasher, and washer/dryer in a 6-story
building. Located between Kamaole I and II Beach Parks with all units offering
oceanview. Weekly maid service during high season. $12 extra person. 3-day
minimum. NO CREDIT CARDS. $200/100 deposit with $25 cancellation fee.
*2 BR 2 bath (2) o.v. $120/95/75, o.f. $130/100/85*

## HALE PAU HANA

2480 S. Kihei Rd., Kihei, Maui, HI 96753. (808-879-2715) 1-800-367-6036. AGENTS: Condominium Rentals Hawaii 1-800-367-5242. 78 oceanview units in four buildings. Laundry area, elevator. Located on Kamaole II Beach. Limited maid service. NO CREDIT CARDS. Monthly discounts. Extra person $15-20.
*1 BR 1 bath (2,max 4) $100-125/95,  1 BR 2 bath (2, max 4) $105-130/85*

## KIHEI KAI NANI

2495 S. Kihei Rd., Kihei, Maui, HI 96753. (808-879-9088) 1-800-367-8047 ext.332. Agents: Kihei Maui Vacations 1-800-542-6284, Hawaiian Apt. Leasing 1-800-854-8843. 180 one-bedroom units with lanai or balcony in a 2 and 3-story structure. This complex is one of the older ones along Kihei Rd. Laundry room and recreation center. Across from Kamaole II Beach. $7 extra person. 4-night minimum, $100 deposit, balance due 30 days prior to arrival. *1 BR (2) ($80/65)*

## KIHEI AKAHI

2531 S. Kihei Rd., Kihei, Maui, HI 96753. (808-879-1881) Agents: Maui Condo & Home Realty (808-879-5445), Condo Rentals Hawaii (808-879-2778), Maui Condominium Rentals 1-800-367-5242, CANADA 1-800-663-2101, Kihei Maui Vac. 1-800-542-6284, Kumulani 1-800-367-2954, RSVP 1-800-663-1118. 240 units with washer/dryers. 2 pools, tennis court, BBQ's. Across from Kamaole II Beach. $5/nite discount after 6 nites. 10% monthly discount. Extra person $12. 4-day minimum, $125 deposit, full payment 30 days prior to arrival. NO CREDIT CARDS. *Studio (2) $70/60,  1 BR 1 bath (2) $85/70,  2 BR 2 bath (4) $110/$85*

## HALEAKALA SHORES ★

2619 S. Kihei Rd., Kihei, Maui, HI 96753. (808-879-1218) 1-800-367-8047 ext. 119, 1-800-423-8733 ext. 119 Canada. Seventy-six, 2-BR units in two four story buildings. Located across the road from Kamaole III Beach. Maid service available for additional charge. Washer/dryers. Covered parking. They also have a very thoughtful policy where returning guests receive an extra discount. A very good value, especially during summer season! Monthly discounts. 5-day minimum stay low season, 7-days high season. $100 deposit refundable with 30 day cancellation, $25 cancellation fee charged. (Special restrictions during Christmas holiday). NO CREDIT CARDS. $6 extra person. *2 BR (1-4) $105/75*

## MAUI PARKSHORE

2653 S. Kihei Rd., Kihei, Maui, HI 96753. (808-879-1600) AGENT: Condominium Rentals Hawaii 1-800-367-5242. Sixty-four, 2-bedroom, 2-bath oceanview condos with washer/dryers, and lanais in a 4-story building (elevator) across from Kamaole III Beach. Pool area sauna. 10% monthly discount. $100 deposit, payment in full upon arrival. NO CREDIT CARDS. Extra person $10/5 per night. *2 BR 2 bath (4) $100/80*

## KAMAOLE SANDS

2695 S. Kihei Rd., Kihei, Maui, HI 96753. (808-879-0666) Agents: Aston Resorts 1-800-922-7866, About Hawaii 1-800-336-8687, Hawaiian Apt. Leasing 1-800-854-8843 (1-800-472-8449 CA), Kumulani 1-800-367-2954, RSVP 1-800-663-1118, Rainbow Rentals 1-800-451-5366. 440 units in 10 four-story buildings. Includes daily maid service. 4 tennis courts, wading pool, 2 jacuzzi's and BBQ's.

Located on 15 acres across the road from Kamaole III Beach.
*1 BR 2 bath (1-4) $125-180 / 110-165*
*2 BR 2 bath (1-6)    160-200 / 145-185*
*3 BR 3 bath (1-8)    215-240 / 205-215*

## HALE KAMAOLE

2737 S. Kihei Rd., Kihei, Maui, HI 96753. (808-879-2698) 1-800-367-2970.
Agents: Condo Rental Hawaii 1-800-367-5242, Kumulani 1-800-367-2954, RSVP
1-800-663-1118.

188 units in 5 buildings (2 & 3-story, no elevator) across the road from Kamaole
III Beach. Laundry building, BBQ's, two pools, tennis courts. Some units have
washer/dryers. Courtesy phone at office. Limited maid service. $100 deposit per
week, balance due 30-60 days before arrival. 3 night minimum stay. Monthly
discounts. NO CREDIT CARDS. *1 BR $87/67, 2 BR $115/87*

## MAUI KAMAOLE

2777 S. Kihei Rd., Kihei, Maui, HI  96743. (879-7668). AGENTS: Kihei Maui
Vacations 1-800-542-6284, Kumulani 1-800-367-2954, RSVP 1-800-663-1118.

The newest development on South Kihei Rd. is on a bluff overlooking the ocean
and across the street and a short walk down to Kamaole III Beach Park or
Keawakapu Beach. One bedroom units are 1,000 - 1,300 square feet and two
bedroom units are 1,300 to 1,600 square feet. Some have oceanviews. This is a
four phase development that will eventually have 316 residential units on 23 acres.
They are low rise, four-plex buildings grouped into 13 clusters, each named after
Hawaiian flora. Phase One is nearing completion, and Phase Two will include a
pool, spa and pavilion. Phase four will add a second pool and two tennis courts.
*1 BR $70-155, 2 BR $135-169*

## MAUI HILL ★

2881 S. Kihei Rd., Kihei, Maui, HI 96753. (808-879-6321) Agents: Aston Hotels
1-800-367-5124, Maui 800 1-800-367-5224, Kumulani 1-800-367-2954, RSVP 1-
800-663-1118.

140 attractively furnished units with washer/dryers, air conditioning, microwaves,
dishwashers, and large lanais. Daily maid service. There are 12 buildings with a
Spanish flair clustered on a hillside above the Keawakapu Beach area. There is
a moderate walk down and across the road to the beach. Upper units have
oceanviews. The 3-bedroom units are very spacious. Large pool and tennis courts.
1-night deposit. *1 BR (1-4) $165/135, 2 BR (1-6) $185/155, 3 BR (1-8) $105/175*

## KIHEI SURFSIDE

2936 S. Kihei Rd., Kihei, Maui, HI 96753. (808-879-1488) 1-800-367-5240.
Agents: Kihei Maui Vacations 1-800-542-6284.

83 units on rocky shore with tidepools, a short walk to Keawakapu Beach. Large
grassy area and good view.  Maid service every fourth day. Extra persons $6. 3-
night minimum, 3-day deposit, 14-day cancellation notice. 10% monthly discount.
*1 BR 1 bath (2) $120/90, 1  bath (2) $135/100, 2 BR 2 bath (4) $175/140*

## MANA KAI ★
2960 S. Kihei Rd., Kihei, Maui, HI 96753. (808-879-1561) 1-800-525-2025 (FAX 808-874-5042). Agents: Kumulani 1-800-367-2954, Maui Condos 1-800-663-6962.

132 rooms in an 8-story building. The studio units have a room with an adjoining bath. The 1-bedroom units have a kitchen and the 2-bedroom units are actually the hotel unit and a 1-bedroom combined, each having separate entry doors. Complex has laundry facilities on each floor, an oceanfront pool, and a restaurant off the lobby. Daily maid service. The Mana Kai is nestled at the end of Keawakapu Beach, and offers a majestic view of the blue Pacific, the 10,000 foot high Haleakala and Upcountry Maui. It is the only major facility in Kihei on a prime beachfront location. Keawakapu Beach is not only very nice, but generally very under used. 1-night deposit, balance due 30 days in advance of arrival summer season, 60 days winter season. 14-day cancellation notice.
*S BR 1 bath (2) $ 90/ 85, includes breakfast (no kitchen)*
*1 BR 1 bath (2)  152/136*
*2 BR 1 bath (2)  180/162, rates include compact car*

## SURF AND SAND HOTEL
2980 S. Kihei Rd., Kihei, Maui, HI 96753. (808-879-7744) 1-800-367-2958. Hawaiian Pacific Resorts 1-800-367-5004. 100 units with direct-dial phones, air conditioning, and daily maid service in several 2-story buildings. The exterior of these units is an Oriental brown and orange. The room we viewed was very small, and while ocean front, there was only one small window from which to view the surf, and no lanais. There was a small refrigerator, toaster, and hot plate. A shower, but no bathtub. The room appeared very clean, but the furnishings had seen a better day. Located on Keawakapu Beach and next door to the Outrigger Restaurant. *Standard $70/60, Superior $78/68, Deluxe $88/78*

## HALE HUI KAI
2994 S. Kihei Rd., Kihei, Maui, HI 96753. (808-879-1219). Agents: Bellow Realty/Maui Beach Homes 1-800-541-3060. Oceanfront location on Keawakapu Beach. 5-night minimum, $100 deposit, $5/day for over double occupancy.
*2 BR 2 bath oceanview $125/105, side oceanview $105/85*

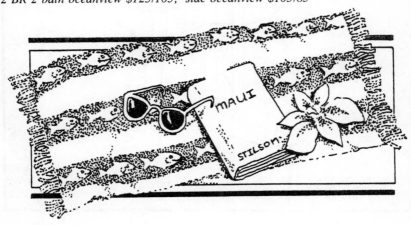

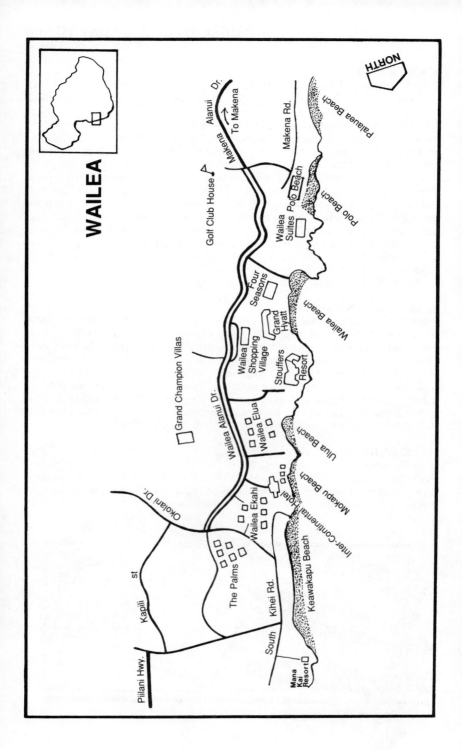

# WAILEA

## INTRODUCTION

Wailea is a well planned and well manicured resort on 1,500 acres just south of Kihei. Developed by Alexander and Baldwin, Wailea encompasses five resort hotels and four condominium complexes which offer high quality accommodations. Included are two 18-hole championship golf courses, a large tennis center and a shopping center. The spacious and uncluttered layout is impressive, as are its series of lovely beaches.

The first two resort hotels were the 550-room Maui Inter-Continental Hotel which opened in 1976, and the 350-room Stouffer Wailea Beach Resort which opened in 1978. During the past year and for the next several years, this area has and will be the hub of development on Maui.

The Four Seasons Hotel which opened in the spring of 1990 will be followed by the neighboring Grand Hyatt Wailea, a 812 room resort with a scheduled opening in mid 1990. Wailea Suites will open on Polo Beach with 450 rooms in 1991 and a private hotel-club, The Diamond Resort recently opened their 72 room property. The Palms at Wailea is a 150 unit condo property that has a projected completion date of 1991.

The Wailea condominium villages are divided into four locations, two are beachfront, while two are adjacent to the golf course. The newest of these villages is the Grand Champion Villas which opened in 1989.

The Polo Beach condominiums are located adjacent to Wailea resorts on Makena Road. Wailea Point offers no vacation rentals.

**BEST BETS: *Maui Inter-Continental Wailea*** - A topnotch resort hotel featuring a tropical flavor with lovely spacious grounds, excellent restaurants, and two great beaches. ***Stouffer Wailea Beach Resort*** - This complex has lush, tropical grounds, and it is fronted by one of the island's finest beaches. ***Wailea Villas*** - Our choice among the four areas would be the Elua Village. These are more expensive, but beach aficionados will love having Ulua Beach at their front door. ***Polo Beach*** - Luxury units in a very secluded location with two small but good beaches. ***Four Seasons Resort*** and ***Grand Hyatt*** - Judging by the advance information, these resorts should be Best Bets when open.

## WHAT TO SEE AND DO

The lovely Wailea beaches are actually well planned and nicely maintained public parks with excellent access, off-street parking and all but one have restrooms and rinse-off showers. Ulua Beach is our personal favorite. The Stouffer Wailea Beach Resort also offers lovely grounds you might want to enjoy.

# WHERE TO SHOP

Wailea Shopping Center is located at the southern end of Wailea. It offers a small pantry market, a mall of shops, and a restaurant. The Stouffer Wailea Beach Resort and the Maui Inter-Continental Wailea both offer shopping areas.

## ACCOMMODATIONS - WAILEA

| | |
|---|---|
| Diamond Resorts | Wailea Suites Resort |
| The Four Season Resort | Wailea Villas |
| Grand Hyatt Wailea | Ekolu |
| Maui Inter-Continental Wailea | Ekahi |
| The Palms at Wailea | Elua |
| Polo Beach Club | Grand Champion |
| Stouffer Wailea Beach Resort | |

### WAILEA VILLAS

3750 Wailea Alanui, Wailea Maui, HI 96753. (808-879-1595) 1-800-367-5246. Agents: Gerry Howell (Elua Condos only) (808-879-4726), Destination Resorts Wailea (808-879-1595) 1-800-367-5246, Kihei Maui Vac. 1-800-542-6284, Village Rentals 1-800-367-5634, Kumulani (only Ekahi condos) 1-800-367-2954, Hawaiian Apt. Leasing 1-800-854-8843.

Some agents may have a limited number of units for slightly better prices than those quoted below. The price range reflects location in the complex. Children under 16 free in parent's room. Extra persons $20/night. 3-night minimum, 3-night deposit, 30-day refund notice, $25 cancellation fee.

| EKOLU VILLAGE | EKAHI VILLAGE | ELUA VILLAGE | | |
|---|---|---|---|---|
| S BR (2) | $130/110 | | | |
| 1 BR (2) $160/135 | 175/145 | garden-o.v. $220/185, | o.f. | $265/230 |
| 2 BR (4)   200/160 | 240/200 | garden-o.v.   265/250, | o.f. | 380/325 |
| 3 BR (6) | | garden-o.v.   360/300, | o.f. | 450/400 |

Elua Village is located on Ulua Beach, one of the best in the area. We would recommend these units. Ekahi Village is on the hillside above the south end of Keawakapu Beach, some units are right above the beach. Ekolu Village is a bit farther away near the tennis center and the Wailea golf course. The newest village is Grand Champion, listing follows.

### GRAND CHAMPION VILLAS

155 Wailea Iki Place, Wailea, HI 96754 Agent: Destination Resorts 1-800-367-5246. The fourth and newest of the Wailea Villas, this is a sportsman's dream, located between Wailea's Blue Golf Course and the "Wimbledon West" Tennis Center. 188 luxury condominium units on 12 lush acres with garden view, golf view or oceanview units. Daily maid service, grocery delivery service, concierge service. Golf, tennis and/or car packages available.
*1 BR (2) $175-190 / 145-160,  2 BR (4) $200-215 / 225-240*

## DIAMOND RESORTS OF JAPAN (Hotel)

(874-0500) This is the first Hawaii resort for this Japanese based company. Diamond resorts also operates 16 mountain resort spas in Japan. Located on 14 acres of oceanfront property in Wailea, the 72 detached villas have been designed after traditional mountain spas. However, members only are allowed to stay. We are told that they anticipate the resort to have a 100% Japanese occupancy. The on-site restaurant as yet has no name and may open to non-members by referral at a later date.

## STOUFFER WAILEA BEACH RESORT ★ (Hotel)

3550 Wailea Alanui, Wailea, Maui, HI 96753. (808-879-4900) 1-800-992-4532.

347 units including 12 suites. This luxury resort covers 15 acres above beautiful Mokapu Beach. Each guest room is 500 sq.ft. and offers a refrigerator, individual air conditioner, a stocked mini-bar, and lanai. The rooms have been recently redecorated in soothing rose, ash and blue tones. An assortment of daily guest activities are available as well as summer and holiday programs for children. The Mokapu Beach Club is a separate beachfront building with 26 units that feature open beamed ceilings and rich koa wood furnishings, plus a small swimming pool. The resort's restaurants are the Maui Onion, a pool-side gazebo; Palm Court, serving international buffets in an open air atmosphere; Raffles', an award-winning gourmet restaurant; and a weekly Hawaiian luau. The Sunset Terrace, located in the lobby area, offers an excellent vantage point for a beautiful sunset and evening entertainment Monday - Friday, 5:30 - 8:30. The beach offers excellent swimming. The best snorkeling is just a very short walk over to the ajoining Ulua Beach. The grounds are a beautiful tropical jungle with a very attractive pool area which was recently expanded to include additional lounging areas and more jacuzzi pools.

A terrific value is the Entertainment Book (a coupon type book purchased around the country) which offers a 50% discount for mountain or oceanview rooms, for a maximum of 7 nights. Only available low season, beginning April 15 on space available. *Mountainside and oceanside rooms $185-255, Mokapu Beach Club $350, One and two bedroom suites $475-1,200*

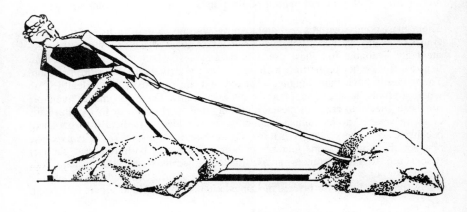

## MAUI INTER-CONTINENTAL WAILEA ★ (Hotel)

P.O. Box 779, Wailea, Maui, HI 96753. (808-879-1922) 1-800-367-2960, Honolulu 1-800-537-5589, Canada 1-800-268-3785.

The Maui Inter-Continental is situated on a low rocky promontory with a beautiful white sandy beach on either side. The restaurants in this lovely resort include the gourmet La Perouse, poolside Sun Spot, the Kiawe Terrace, the Kiawe Broiler, and Lanai Terrace. There is also a Sunday brunch and Aloha Mele Luncheon offered in the Makani Room. There are three pools, a seven-story tower and six low-rise buildings. The wonderful layout of this resort allows 80% of all guest rooms to have an ocean view. The pools are located in different areas of the property so that none are overly crowded. The units located nearest the beach afford wonderful private ocean views. The resort offers many complimentary daily activities as well as one of Maui's best luaus. A number of exciting annual events are sponsored by the resort. To name only a few, there is the Maui Art Expo each Jan. and Feb. as well as an annual hula and lei making competition each spring.
*Minimum $185, Moderate $250, Maximum $295, Dlx. $350, Suites run $350-950*
*Full American Plan or Modified American Plan available*
*Golf, tennis and honeymoon package plans also available*

## WAILEA SUITES RESORT

Anticipate a 1991 opening. This new facility will be reviewed in upcoming issues of THE MAUI UPDATE newsletter.

## THE FOUR SEASONS RESORT ★

3900 Wailea Alanui, Wailea, Maui, HI 96754. (808-874-8000) (National reservations 1-800-332-3442). 372 over-sized guest rooms (600 sq. ft) on eight floors encompassing 15 beachfront acres on Wailea Beach. A full service resort featuring two pools, two tennis courts, croquet lawn, health spa, three restaurants and two lounges. The public areas are spacious, open and ocean oriented. A very different atmosphere from other Maui resorts with the blue tiled roof and creamy colored building creating a very classical atmosphere. Even the grounds although a profusion of colors with many varied Hawaiian flora, are more structured in design with a vague resemblance what one might find at a Mediterranean villa. The focus of the resort is water. Throughout the resort's gardens and courtyards are an array of formal and natural pools, ponds, waterfalls and fountains.

Their guest policy features real aloha spirit, with no charge for use of the tennis courts or health spa and complimentary snorkel gear, smash or volleyball equipment are available for guest use. Guests services, which distinguish the Four Seasons from other properties, include their early arrival/late departure program. These guests have their luggage checked and are escorted to the Health Club where a private locker is supplied for personal items. The resort makes available for these guests an array of casual clothing from work-out gear to jogging suits or swim wear. Another unusual amentity is provided periodically to pool and beachside sunbathers who are offered a refreshing iced towel. For the convention planner, the Four Seasons features a 7,000 square foot ballroom, two banquet rooms and four conferences areas situated adjacent to a 3,000 square foot hospitality suite. The suite offers a large living room, two bedrooms, another living space designed for private meetings, kitchen, and a 1,000 square foot lanai.

Amenities for guests on the *club floor include a private lounge, 24 hr. concierge, complimentary breakfast, afternoon tea, evening cocktails and after dinner liqueurs. Numerous special package offers include a golf, romance or tennis package. Special family plan offers one room at the regular rate and a second room for children at $175 and a "Kids for All Seasons" program designed for hotel guests aged 5 - 12 years. Restaurants include the Cabana Cafe, Pacific Grill and Seasons. *Mt.v. $225, courtyard v. $250, o.v. $325, o.f. $350*
*\*Club Floor $385-425, Jr. suites $400-475, One bedroom suites $750-810*
*Children under 18 free, additional adults $35*

## THE PALMS AT WAILEA
150 luxury units are part of Phase I of construction on sixteen oceanview acres. Completion of these one and two bedroom units is anticipated for 1991 with an initially selling price projected at $300,000 - $600,000.

## WAILEA POINT VILLAGE
4000 Wailea Alanui, Wailea, Maui, HI 96753. 136 luxurious oceanview and oceanfront condominiums arranged in four-plexes which are laid out in a residential plan on 26-oceanfront acres. Privacy is maintained by a gate guard at the entrance. Unfortunately, no rental properties available here!

## POLO BEACH CLUB ★
20 Makena Rd., Wailea, Maui, HI 96753. (808-879-8847) Agents: Destination Resorts 1-800-367-5246, Colony Resorts 1-800-367-6046, Condo Resorts 1-800-854-3823, Hawaiian Apt. Leasing 1-800-854-8843 (1-800-472-8449 CA), Island Dreamscape 1-800-367-8047 ext. 216 U.S.

Additional persons (over 4) $20 each. Three night minimum. 71 apartments in a single 8-story building located on Polo Beach. The units are luxurious and spacious. Underground parking, pool area jacuzzi. This once very secluded area is soon to be "discovered." Construction next door to Polo Beach will result in the new 450 room Wailea Suites Resort opening in 1991.
*2 BR 2 bath (max 6) o.f. $325/275, o.v. $275/250    wkly/mthly discounts*

'ILIMA

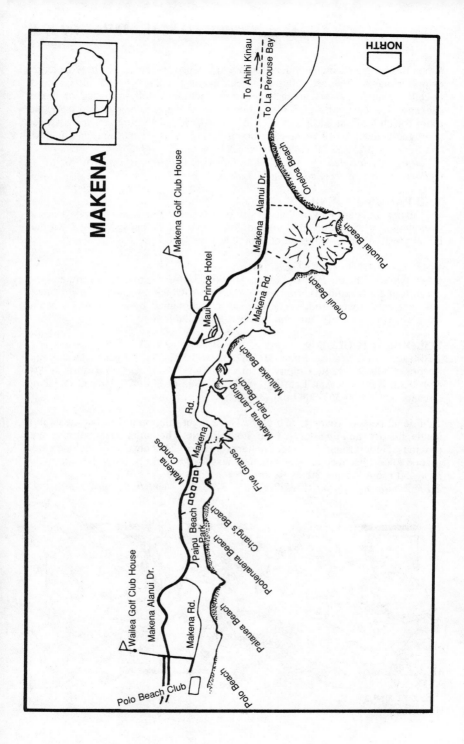

MAKENA

NORTH

To Ahihi Kinau
To La Perouse Bay

Oneloa Beach

Makena Golf Club House

Makena Alanui Dr.

Puuolai Beach

Maui Prince Hotel

Makena Rd.

Oneuli Beach

Mailuaka Beach

Makena Landing

Paipi Beach

Makena Rd.

Makena Condos

Five Graves

Poolenalena Beach

Chang's Beach

Paipu Beach Park

Palaua Beach

Wailea Golf Club House

Makena Alanui Dr.

Makena Rd.

Polo Beach Club

Polo Beach

# MAKENA

## INTRODUCTION

The area just south of Wailea is Makena, and one of the newer resort developments on Maui. The project began with the completion of an 18-hole golf course
in 1981. The Makena Surf condominium project opened in 1984. The Japanese
conglomerate, Seibu, has a magnificent new resort, the Maui Prince Hotel, located
at Maluaka Beach. Also in this area are several beaches with public access. They
include Oneloa, Puuolai, Poolenalena and Palauea beaches. Since the area is not
fully developed the end results remain to be seen.

**BEST BETS: *The Maui Prince Hotel* and *Makena Surf*** - Both are first class,
luxury accommodations on beautiful beaches.

## WHAT TO DO AND SEE

Here are the last really gorgeous and undeveloped recreational beaches on Maui.
Consequently, development in this area has met with a great deal of ongoing
controversy. The new paved road (Makena Alanui) runs from Wailea past the
Makena Surf and Maui Prince Hotel, exiting onto the Old Makena Rd. near the
entrances to Oneuli (Black Sand) Beach and Oneloa (Big Makena)-Puuolai (Little
Makena) Beaches. Past Ahihi Kinau Natural Reserve on Old Makena Rd. you will
traverse the last major lava flow on Maui, which still looks pretty fresh after some
250 years. The road is paved 1 1/2 miles beyond the Maui Prince and then a
rougher, rocky road extends to La Perouse Bay. (See BEACHES).

Hiking beyond La Perouse affords some great ocean vistas. You'll see trails made
by local residents in their four wheel drive vehicles, and fishermen's trails leading
to volcanic promintories overlooking the ocean. You may even spot the fishing
pole holders which have been securely attached to the lava boulders. The Hoapili
Trail begins just past La Perouse Bay and is referred to as the King's Highway.
It is believed that at one time the early Hawaiians made use of a trail that circled
the entire island and this may be the remnants of that ancient route. The state
Forestry and Wildlife Division and volunteers worked together recently putting in
place stone barricades to keep the four wheel drive vehicles and motorcycles from
destroying any more of the trail.

## ACCOMMODATIONS - MAKENA

Makena Surf    Maui Prince Hotel

**MAKENA SURF ★**
96 Makena Alanui Rd., Makena 96753 (808-879-1331) (1-800-367-8047 ext.510
U.S. or Canada 1-800-423-8733 ext.510.), Destination Resorts 1-800-367-5146,
Hawaiian Apt. Leasing 1-800-854-8843.

Located 2 miles past Wailea. All units are oceanfront and more or less surround Paipu (Chang's) Beach. These very spacious and attractive condos feature central air conditioning, fully equipped kitchens, washers and dryers, wet bar, whirlpool spa in the master bath, telephones and daily maid service. Two pools, and four tennis courts are set in landscaped grounds. Three historic sites found on location have been preserved.

| | |
|---|---|
| 1 BR (2) $240-265 / $180-205 | *Prices listed are 2 or 3 nights* |
| 2 BR (2)   285-310 /   210-235 | *Discounts for 4 nights or longer* |
| 3 BR (4)   375-405 /   300-330 | *Extra person $15/night* |

### MAUI PRINCE ★

5400 Makena Alanui, Makena, Maui 96753. (808-874-1111) Reservations: 1-800-321-6284. In sharp contrast to the ostentateous atmosphere of some of the Kaanapali resorts, the Maui Prince radiates understated elegance. Its simplicity in color and design, with an Oriental theme, provides a tranquil setting and allows the beauty of Maui to be reflected. The central courtyard is the focal point of the resort with a lovely traditional water garden complete with a cool cascading waterfall and ponds filled with gleaming koi. The rooms are tastefully appointed in cool neutral hues and equipped with the comfort of the guests in mind. The units have two telephones and a small refrigerator which is stocked with complimentary fresh fruit and sparkling water. Fresh flowers are provided in each and a yukata (summer kimono) is available for use during the guest's stay. A morning paper and 24-hour full room service add to the conveniences.

There is plenty of room for lounging around two circular swimming pools or in a few steps you can be on Maluaka (Nau Paka) Beach with its luxuriously deep, fine white sand and good snorkeling, swimming and wave playing. The resort is comprised of 1,800 acres including the Makena Golf Course, a 72-par 18 hole championship course designed by Robert Trent Jones, Jr. Restaurants include Prince Court serving Island Regional dishes (and one of the top three Sunday brunches), al fresco dining in Cafe Kiowai and the Japanese restaurant and sushi bar at Hakone.
*Partial o.v. $190, o.v. $220, o.v. prime $250, o.f. $290, Suites $350-700*

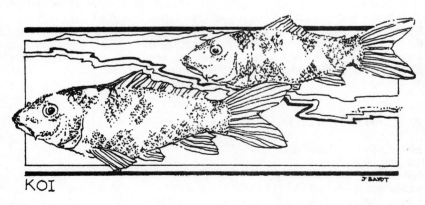

KOI

# WAILUKU AND KAHULUI

## INTRODUCTION

The twin towns of Wailuku and Kahului are located on the northern, windward side of the island. Wailuku is the county seat of Maui and Kahului houses not only the largest residential population on the island, but also the main airport terminal and deep-water harbor. There are three motel-type accommodations located around the Kahului Harbor and while the rates are economical, and the location is somewhat central to all parts of the island, we cannot recommend staying in this area for other than a quick stopover that requires easy airport access. This side of the island is generally more windy, overcast and cooler with few good beaches. Except for the avid windsurfer, we feel there is little reason to headquarter your stay in this area, however, there are good reasons to linger and explore.

## WHAT TO DO AND SEE

Kahului has a very colorful history, beginning with the arrival of King Kamehameha I in the 1790's from the big island of Hawaii. The meaning of Kahului is "winning" and may have had its origins in the battle which ensued between Kamehameha and the Maui chieftain. The shoreline of Kahului Bay began its development in 1863 with the construction of a warehouse by Thomas Hogan. By 1879 a landing at the bay was necessary to keep up with the growing sugar cane industry. Two years later, in 1881 the Kahului Railroad Company had begun. The city of Kahului grew rapidly until 1900 when it was purposely burned down to destroy the spreading of a bubonic plague outbreak. The reconstruction of Kahului created a full-scale commercial harbor, which was bombed along with Pearl Harbor on December 7, 1941. After World War II, a housing boom began, with the development of reasonably priced homes to house the increasing number of people moving to the island. The expansion has continued ever since.

Wailuku is the county seat of Maui and has been the center of government since 1930. It is now, slowly, experiencing a rebirth. It is often overlooked by visitors who miss out on some wonderful local restaurants and limited, but interesting shopping. A *"Walking Tour of Wailuku Town"* guidebook is available for $2 at the Wailuku Main Street office, 68 Market St., across the street from the old Iao Theater. Phone 244-3888.

*Market Street* in Old Wailuku Town is alive with the atmosphere of Old Hawaii. The area, rich in history, was built on the site of ancient Heiaus and witnessed decisive Hawaiian battles. Later the area hosted the likes of Mark Twain and Robert Redford. It is no wonder that such an area should re-emerge in the modern day with shops of a cultural nature. One-of-a-kind items can be found here, gathered from around the world and eras gone by. Such is the case with Old Wailuku Town and the cluster of interesting shops on the upper end of Market Street in an area known as Antique Row.

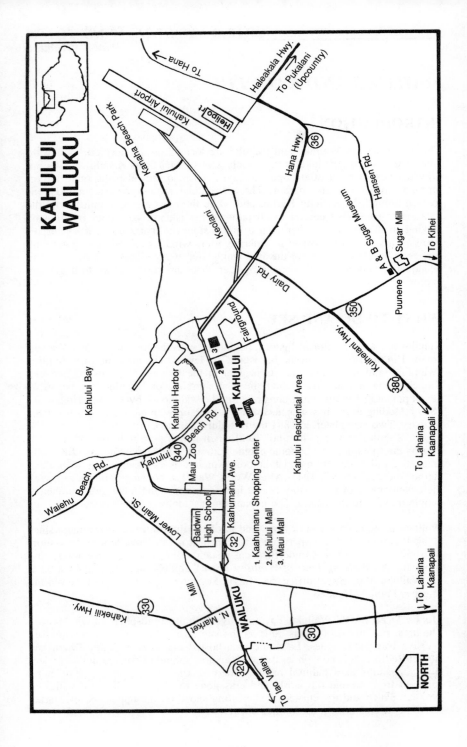

KAHULUI
WAILUKU

NORTH

1. Kaahumanu Shopping Center
2. Kahului Mall
3. Maui Mall

To Hana
Haleakala Hwy.
To Pukalani (Upcountry)
Kahului Airport
Heliport
Hana Hwy.
36
Hansen Rd.
Keolani
Kanaha Beach Park
A & B Sugar Museum
Sugar Mill
Puunene
To Kihei
Dairy Rd.
Fairground
350
Kahului Bay
Kahului Harbor
Kahului
KAHULUI
Kuihelani Hwy.
380
To Lahaina Kaanapali
Waiehu Beach Rd.
340
Maui Zoo Beach Rd.
Kaahumanu Ave.
Kahului Residential Area
Lower Main St.
Baldwin High School
32
Kahekili Hwy.
330
Mill
N. Market
WAILUKU
320
To Iao Valley
30
To Lahaina Kaanapali

118

Set against the lush backdrop of the Iao Valley and the West Maui Mountains, Antique Row is a small area offering a quaint alternative to the hustle-bustle vacation centers of Lahaina and Kihei. Surrounding this area is a multitude of wonderful and inexpensive ethnic restaurants. So don't limit your excursion to the few shops on the corner of Market and Main streets.

***Emura's*** at 49 Market Street has consistently proven to be the spot for the best buys of eel skin items. From wallets and purses, to shoes and attaches. Pay cash and get an extra discount!

The once Takata Market, a thriving butcher shop and grocery store in the 30's, is now home to Memory Lane and Traders of the Lost Art (158 Market St). Next door was the B. Hotta General Store and residence, which now houses an art gallery. Alii Antiques is across the street.

Operated by Tye Hartall, ***Traders of the Lost Art*** features a variety of native carvings and primitive ritual art, which he brings back regularly from the secluded Sepik River area of Papua in New Guinea.

Right next door is Joe Ransberger's shop, ***Memory Lane.*** Formerly a hair stylist in Beverly Hills and Lake Tahoe, Joe began his world travels in 1978 and landed on Maui four years later. A collector of unusual and one-of-a-kind items, Joe began selling a few of his pieces at the Maui Swap Meet. Business was brisk and Joe found he was able to purchase more and more of these collectible items which he stored away until he was able to open his shop on Market Street in August 1987. You'll find rare Japanese Netsukes and vintage Aloha shirts here!

***Alii Antiques*** opened its operation in 1988 next to Traders and recently moved to larger quarters just across the street. Favoring Asian art from the Ming and Ch'ing Dynasties, Alii Antique also offers items from the Mainland and Europe.

***Kaahumanu Church***, Maui's oldest remaining church was built in 1837 at High and Main Streets in Wailuku.

KAAHUMANU CHURCH

Hale Hoikeike in Wailuku houses the *Maui Historical Society* and is known as the *Bailey House* (circa 1834). To reach it, follow the signs to Iao Valley and you will see the historical landmark sign on the left side of the road. It's open daily from 9 am - 3:30 pm, and a small admission is charged. Here you will find the Bailey Gallery, (once a dining room for the female seminary that was located at this site), with paintings of Edward Bailey done during the 19th century. His work depicts many aspects of Hawaiian life during earlier days. Also on display are early Hawaiian artifacts and memorabilia from the missionary days. The staff is extremely knowledgeable and friendly. They also have for sale an array of Hawaiian history, art, craft and photographic books, all available at prices LESS than other Maui bookstores. Originally, the Royal Historical Society was established in 1841, but it was not reactivated as the Maui Historical Society until 1956. The museum was dedicated on July 6, 1957, then closed for restoration on December 31, 1973 and reopened on July 13, 1975. Of special interest are the impressive 20 inch thick walls that are made of plaster using a special missionary recipe which included goat hair as one ingredient. The thick walls provided the inhabitants with a natural means of air conditioning.

The *Maui Jinsha Mission* is located at 472 Lipo Street, Wailuku. One of the few remaining old Shinto Shrines in the state of Hawaii, this mission was placed on the National Register of Historic Places in 1978.

The *Halekii and Pihana State Monuments* are among Maui's most interesting early Hawaiian historical sites. Both are of considerable size and situated on the top of a sand dune. These temples were very important structures for the island's early Alii. Their exact age is unknown, although one resource reported that they were used from 1765 to 1895. The Halekii monument is in better condition as a result of some reconstruction done on it in 1958. Follow Waiehu Beach Road across a bridge, then turn left onto Kuhio Place and again on Hea Place. Look for and follow the Hawaii Visitors Bureau markers. Some say the Pihana Heiau (temple) was built by the menehunes (Hawaii's little people), others believe by the Maui chiefton, Kahekili.

*The Iao Valley* is a short drive beyond Hale Hoikeike. Within the valley is an awesome volcanic ridge that rises 2,250 feet and is known as the Iao Needle. A little known fact is that this interesting natural phenomena is not a monolithic formation, but rather what you are viewing is the end of a large, thin ridge. A helicopter view will give you an entirely different perspective! Parking facilities are available and there are a number of hiking trails. A recent addition is the *Tropical Gardens of Maui*. This botanical garden features the largest selection of exotic orchids in the Hawaiian islands. For a small fee you can stroll the grounds where they grow, and visit their gift shop filled with tropical flowers and Maui made products. Plants can be shipped home. Snack bar and picnic tables available. Phone 244-3085.

*The Heritage Garden/Kepaniwai Park* is an exhibit of pavilions and gardens which pay tribute to the culture of the Hawaiians, Portuguese, Filipinos, Koreans, Japanese and Chinese. Picnic tables and BBQ's available for public use. Located on Iao Valley Rd. Free admission, open daily. Public swimming pool for children is open daily from 9 am - noon and again from 1 until 4:30. Also a popular site for weddings and other functions, it is available for rent from the Maui Parks Dept. A $50 deposit is required and the Wailuku permit office (1580 Kaahumanu Ave., Wailuku, Maui (808) 243-7389) can provide the necessary forms.

*Pacific Brewing Company*, which produces Maui Lager beer, invites individuals or groups to tour their brewery. They are located on Imi Kala St. at the Millyard in Wailuku and open for tours 9 am - 2 pm. Call 244-0396 to schedule tours or for more information.

Just outside Wailuku on Hwy. 30, between Wailuku and Maalaea, is Waikapu, home of the *Tropical Plantation*. These fifty acres, which opened in 1984, have been planted with sugar cane, bananas, guava and other island produce. A ten acre visitor center includes exhibits, a market place, nursery and restaurant. There is no admission for entry into the marketplace or the restaurant, however, there is an $8 charge for admission to the grounds where there is a tram ride around the planted fields and various agricultural and Hawaiiana exhibits.

*Baldwin Beach* - See the section on BEACHES for Baldwin Beach and others in the area.

*The Maui Zoological and Botanical Gardens* are open 9-4 daily with FREE admission. Go up Kaahumanu to Kanaloa Street and turn by the Wailuku War Memorial Center. The zoo is on the right hand side. This is a zoo Maui style, with a few pygmy donkeys, sheep, goats, monkeys, Galapogos turtles, birds and picnic tables.

*The Kanaha Wildlife Sanctuary* is off Route 32, near the Kahului Airport, and was once a royal fish pond. Now a lookout is located here for those interested in viewing the stilt and other birds which inhabit the area.

A popular Saturday morning stop for local residents and visitors alike is *The Swap Meet* ★ held at the Kahului Fairgrounds (located off Pu'unene Hwy 35). You'll find us referring to this event for various reasons throughout this guide. For a fifty-cent admission (children free) you will find an assortment of vendors selling local fruits and vegetables, new and used clothing, household items and

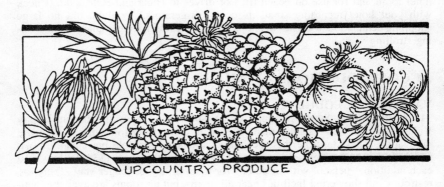

UPCOUNTRY PRODUCE

many of the same souvenir type items found at higher prices in resort gift shops. Here you can pick up some fantastic tropical flowers and for only a few dollars lavishly decorate your condo during your stay. Protea are seasonally available here too for a fraction of the cost elsewhere. This is also the only place to get true spoonmeat coconuts. These are fairly immature coconuts with deliciously mild and soft (to very soft) meat and filled with sweet coconut milk. We stock up on a weeks supply at a time. Another "must purchase" are some of the goodies from the Four Sisters Bakery! Free parking. Hours are 8 am - noon. NOTE: The fairgrounds may be changing locations soon, for information call 877-3100.

The *Alexander and Baldwin Sugar Museum* ★ is located at 3957 Hansen Rd., in Puuene. Puuene is on Highway 35 between Kahului and Kihei, the tall stacks of the working mill are easily spotted. The museum is housed in a 1902 plantation home that was once occupied by the sugar mills superintendent. Memorabilia include the strong-box of Samuel Thomas Alexander and an actual working scale model of a sugar mill. The displays are well done and are very informative. Monday thru Friday 9:30 - 4. Admission charge: $2 adult visitors, $1 adult Maui residents. Visiting students 6-17 years $1. Maui students 6-17, 50 cents. Children under 6 are free. Call 871-8058.

# WHERE TO SHOP

There are three large shopping centers in Kahului, all on Kaahumanu Street. The *Maui Mall* is only a two-minute drive from the airport. It offers two large grocery stores, Star and Safeway, and a large Long's Drugs which is great for picking up sundry items. They also have a variety of small shops and restaurants. The older, local style *Kahului Shopping Center* is lined with Monkey pod trees and is filled with local residents playing cards. Check out the Toda Drug Store that has a very reasonable luncheonette, or Ah Fook's grocery for their bentos. The largest shopping center is *Kaahumanu* with 47 shops and restaurants. Two major department stores, Sears and Liberty House, anchor this mall with the island's largest selection of clothing and gift shops in between. If you don't have accommodations with a kitchen, you might want to pick up a styrofoam type ice chest at one of these centers and stock it with juices, lunch meats and what not to enjoy in your hotel room and for use on beach trips or drives to Hana and Haleakala. (Check with your hotel regarding small in-room refrigerators.) Wailuku has no large shopping centers, but a cluster of shops down their Main Street makes for interesting strolling.

# ACCOMMODATIONS - WAILUKU

### VALLEY ISLE LODGE
310 North Market Street, Wailuku (808) 244-6880. Previously the old Happy Valley Inn, this new establishment opened in early 1990 following a complete renovation. They describe their accommodations as a simple but clean, casual, family style environment. Rooms start at $26 per night single ($34 double), $10 each additional person, with discounts for a seven day or longer stay. There are restroom and showering facilities near all rooms, but no rooms are available with

private baths. There is a TV lounge. Check in between 4 and 6 pm or by arrangement. They aim to be actively involved with the windsurfers and students who are looking for very basic, but more affordable accommodations.

## NORTHSHORE INN

2080 Vineyard St., Wailuku (808) 242-8999. New owners Chris and Katie Kunkel report that they have turned the old Wailuku Grand Hotel into a "clean, friendly, inexpensive and hospitable place to be enjoyed by visitors from all over the world." The Wailuku Grand was not previously included in our guides due to its deplorable condition. Chris from Keil, West Germany and Katie from Kent, Washington have been on Maui for three years.

Their Northshore Inn is located above one of our favorite local restaurants, Hazel's. They offer twenty-five remodeled rooms all with color TV and refrigerator. A lounge with T.V., VCR, and Japanese Shiatsu massage table is located upstairs. Walkways in the back lead to sitting areas and a barbecue nestled under an 80 year old Banyan tree. The garden also produces bananas, passion fruit, mangos and papayas for guests. We haven't seen it, but it might bear investigation for those on an extremely limited budget. Reservations accepted.
*Single $29, double $39, with weekly discounts.*

# ACCOMMODATIONS - KAHULUI

Maui Hukilau    Maui Beach
Maui Palms      Maui Seaside

One advantage to choosing this area for headquarters is its proximity to the Kahului Airport and its somewhat central location to all other parts of the island. The motels are clustered together on the Kahului Harbor.

## MAUI PALMS

(808-877-0051) Agent: Hawaiian Pacific 1-800-367-5004 They offer free airport pickup, in-room phones. *Room rates $61-74 / 51-64*

## MAUI BEACH

(808-877-0051) Agent: Hawaiian Pacific 1-800-367-5004. They feature air-conditioned rooms, free airport service. *Room rates $79-90 / 69-83*

## MAUI SEASIDE

1-800-367-7000 U.S., 1-800-654-7020 Canada, FAX (808) 922-0052. The older Maui Hukilau has been combined with the much newer Maui Seaside to form one property called the Maui Seaside. You might want to inquire when booking here about which of the buildings you will be in. The management, Sand and Seaside Hotels, tell us that all of their properties have been recently renovated and refurbished like new. *Room rates $55-85, with kitchenette $79-89*

# UPCOUNTRY and onward to HALEAKALA

## INTRODUCTION

The Western slopes of Haleakala are generally known as Upcountry and consist of several communities including Makawao and Pukalani. The higher altitude, cooler temperatures and increased rainfall make it an ideal location for produce farming. A few fireplace chimneys can be spotted in this region where the nights can get rather chilly. Accommodations are limited to two small lodges in Kula and a few cabins which are available with the park service for overnight use while hiking in the Haleakala Crater.

## WHAT TO DO AND SEE

Enroute to upcountry is Pukalani, meaning "opening to the sky," and it is the last stop for gas on the way to Haleakala. There are also several places to enjoy a hearty meal. (See RESTAURANTS).

*Haleakala* means "house of the sun" and is claimed to be the largest dormant volcano in the world. The crater is truly awesome and it is easy to see why the old Hawaiians considered it sacred and the center of the earth's spiritual power. There is a $3-per car charge for admission to the crater, U.S. residents age 62 and older enter free. The most direct route is to follow Hwy. 37 from Kahului then left onto Hwy. 377 above Pukalani and then left again onto Hwy. 378 for the last 10 miles. While only about 40 miles from Kahului, the last part of the trip is very slow. There are numerous switchbacks and bicycle tours doing the 38-mile down-hill coast. Two hours should be allowed to reach the summit. Sunrise at the crater is a popular and memorable experience, but plan your arrival accordingly. Many visitors have missed this spectacular event by only minutes. The Maui News, the local daily, prints sunrise and sunset times. The park offers a recording of weather and viewing conditions which can be reached by calling 871-5054. The ranger's number is 572-9306. A recording of hiking and camping information is available at 872-7749. Be sure you have packed a sweater as the summit temperature can be 30 degrees cooler than the coast and snow is a winter possibility. Mid-morning from May to October generally offers the clearest viewing. However, fog can cause very limited visibility and a call may save you a trip.

At the park headquarters, you can obtain hiking and camping information and permits. Day-hike permits are not required, however, they do request you complete a registration form at the trail head and deposit it in the box provided. The first stop is Kalahaku Overlook. Here you can see the rare silversword which takes up to 20 years to mature, then blooms once in July or August, only to wither and die in the fall. Keep your eye out for the many nene geese which inhabit the volcano.

Talks on Haleakala history and geology are presented several times daily in the summit building visitors center located at an elevation of 9,745 feet. It is open daily from 6 am - 3 pm (hours may vary). A short distance by road will bring you to the Sun Visitor Center located on the crater rim. This glassed-in vantage point (the Puu Ulaula outlook) is the best for sunrise and is the highest point on Maui. The rangers give morning talks here at 9:30, 10:30 and 11:30. The view, on a good day, is nothing short of awesome. The crater is seven miles long, two miles wide, and 3,000 feet deep. A closer look is available by foot or horseback (see RECREATION AND TOURS - Horseback riding.) A 2 1/2 hour hike down Sliding Sands Trail into the Haleakala Crater is offered by the park service every Saturday, Sunday and Tuesday. They depart at 10 am from the House of the Sun Visitor Center. A hike featuring native Hawaiian birds and plants is offered each Monday, Thursday and Friday. Again, check with the ranger information center (572-9306 or 572-7749) to verify trips, dates and times. A unique excursion begun by Cruiser Bob's and several other operations is a 38-mile downhill ride on a specially designed bike. (See RECREATION & TOURS - Biking).

The park service maintains 30 miles of well-marked trails, three cabins and two campgrounds. All are available only by trail. The closest cabin is about seven miles away from the observatory. Arrangements for these cabins need to be made 90 days in advance and selection is made 60 days prior to the dates requested by a lottery-type drawing. For more information, write: Superintendent, Haleakala National Park, P.O. Box 369, Makawao, Maui, HI 96768. Rates are $5 per night for each adult, and $2.50 for each child 12 and under. A $2.50 charge per person per night for firewood or a maximum of $15. A $15 cleaning fee is charged and will be refunded by mail after cabin inspection.

Short walks might include the three-fourth mile Halemauu Trail to the crater rim, one-tenth mile to Leleiwi Overlook, or two-tenth mile on the White Hill Trail to the top of White Hill. Caution, the thin air and steep inclines may be especially tiring. (See RECREATION AND TOURS -Hiking.)

*Science City* can be seen beyond the visitor center, but it is not open to the public. It houses a solar and lunar observatory, operated by the University of Hawaii, television relay stations, and a Department of Defense satellite station.

If time allows, there is more of Upcountry to be seen!

The Kula area offers rich volcanic soil and commercial farmers harvest a variety of fruits and vegetables. Grapes, apples, pineapples, lettuce, artichokes, tomatoes and, of course, Maui onions are only a few. It can be reached by retracking Hwy. 378 to the Upper Kula Road where you turn left. The protea, a recent floral immigrant from South Africa, has created a profitable business. *The Botanical Gardens* (878-1715) charges an admission of $2.50 for adults, and children 50 cents, open 9 am - 4 pm. The *Maui Enchanting Gardens*, on Hwy. 37 in Kula, charges $3.50 for adults, $1.50 for children for a self guided botanical tour. The *Sunrise Protea Farm* (878-2119) in Kula has a small, but diverse, variety of protea growing adjacent to their market and flower stand for shipment home. Dried assortments begin at about $25. Picnic tables available and no charge for just looking!

Be sure and stop in Keoke at *Grandma's Coffee House*. This wonderfully cozy, two year old restaurant is the place for some freshly made, Maui grown coffee, hot out of the oven cinnamon rolls or a light lunch. See Upcountry restaurants for more information.

***Poli Poli State Park*** is high on the slopes of Haleakala, above Kula at an elevation of 6,200 feet. Continue on Hwy. 377 past Kula and turn left on Waipoli Rd. If you end up on Hwy. 37, you've gone too far. The sign indicating Poli Poli is currently missing, so you could also look for a sign indicating someones home, it reads WALKER. It's another 10 miles to the park over a road which deteriorates to deep ruts and is often muddy. A 4-wheel drive is really a necessity for this road during wet weather. The park offers miles of trails, a picnic area, restrooms, running water, a small redwood forest and great views. A cabin, which sleeps up to 10, is available through the Division of Parks, P.O. Box 537, Makawao, Maui, HI 96768. For more information see the Hiking section in the recreation chapter.

Approximately 9 miles past the Kula Botanical Gardens on Hwy. 37, is the Ulupalakua Ranch. ***The Tedeschi Vineyards*** (879-6058), part of the 30,000 acre ranch, made its debut in 1974. The tasting room is located at the Ranch in the old jail and provides samples of their pineapple, champagne and red table wines. Free daily guided tours are offered between 10 am and 5 pm. The tour begins at the Tasting Room then continues to view the presses used to separate the juice from the grapes, the large fermenting tanks, and the corking and the labelling rooms.

If you continue on past the ranch on Hwy. 37 it's another very long 35 miles to Hana with nothing but beautiful scenery. Don't let the distance fool you. It is a good 3-hour trip, at least, over some fairly rough sections of road, which are not approved for standard rental cars. During recent years this road has been closed often to through traffic due to severe washouts. Check with the county to see if it is currently passable. If you're not continuing on, we suggest you turn around and head back to Pukalani and Makawao. Unfortunately, the Ulupalakua Road down to Wailea has been closed due to a dispute between the Ranch and the county. It is hoped that this or some other access between Upcountry and the Kihei/Wailea area will soon be developed. On the way down you can go by way of Makawao, the colorful "cowboy" town, and then on to Paia or Halemaile. Both have several good restaurants. (See RESTAURANTS - Upcountry).

TEDESCHI WINERY

## WHERE TO SHOP

The town of Makawao offers a western flavor with a scattering of shops down its main street, a few restaurants and numerous grocery stores. We recommend the **Komoda Store** for its popular bakery, but get there early! The Pukalani Shopping Center is a new mall with a grocery store, small shops and restauarants.

## WHAT TO SEE AND DO

The **Hui Noeau Visual Arts Center** may at first seem a little out of place, located at 2841 Baldwin Avenue, down the road from Makawao. However, there could not be a more beautiful and tranquil setting than at this estate, called Kaluanui, which was built in 1917 by famous Honolulu architect C.W. Dickey for Harry and Ethel Baldwin. The house was occupied until the mid-1950's and in 1976 Colin Cameron (grandson of Ethel Baldwin) granted Hui Noeau the use of Kaluanui as a Visual Arts Center. Near the entrance to the nine acre estate are the remains of one of Maui's earliest sugar mills. It utilized mule power and was the first Hawaiian sugar mill to use a centrifugal to separate sugar crystals. What were once stables and tack rooms are now ceramic studios. In addition to a gift shop, which is open year round, classes and workshops are also offered. Call for a current class schedule and, if you are on Maui during the first part of December, you will delight in the Country Christmas House held each year. Daily 9 am - 1 pm. 572-6560.

Fourth of July weekend is wild and wonderful in Makawao. Festivities include a morning parade through town and several days of rodeo events. Check the local paper for details.

## ACCOMMODATIONS - UPCOUNTRY

Accommodations are limited in Upcountry. Five and one-half miles past Pukalani is the Kula Lodge. Just beyond is the Silversword Inn which is currently closed. Both offer rustic cabins and restaurants. There are a small number of bed and breakfast facilities.

**KULA LODGE**
RR 1, Box 475, Kula, Maui, HI 96790 (808-878-1535). Offers five rustic chalet-like cabins with restaurant on the property. Prices are for bed and breakfast based on double occupancy. Additional guests $15. Full advance deposit with one week notice for refund.

*Chalet 1 ($125) and 2 ($115): Equipped with queen size bed, fireplace, lanai and stairs to loft with two twin beds.*
*Chalet 3 ($80) and 4($80): Queen size bed with ladder to loft with two futons.*
*Chalet 5 ($80): single story with double bed and studio couch.*

**SILVERSWORD INN**
Currently closed.

### PUALANI FITNESS RETREAT
PO Box 1135, Makawao, Maui, HI 96768. (808-572-6773). 1-800-782-5264. Pualani (which means heavenly flower) is located on the slopes of Haleakala. A maximum of eight guests experience a personalized program of swimming, aerobics, weight lifting, crater hikes and beach runs. Stress management, yoga, nutrition and cooking classes round out the schedule with gourmet vegetarian meals served. Owners Susan and William Linneman spent several years traveling throughout the U.S. and Europe researching this fitness retreat design. Rates are $1,800 per person per week for private, $1,600 per person shared, $1,500 per person for a couple sharing a room. Bed and Breakfast accommodations are also available $55 - $85 per night.

# *HANA*

## INTRODUCTION

If you do not plan an extended stay in Hana, you might consider at least an overnight stop at one of the facilities to break up the long drive to this isolated east coast of Maui. (Insiders secret! Hana can best be enjoyed before and after the throngs of visitors who daily make this drive, so plan a stay in Hana of several days or at least overnight!) Here is a different Maui from the sunny, dry resort areas on the leeward coast. The windward coast here is turbulent with magnificent coastal views, rain forests, and mountain waterfalls creating wonderful pools for swimming. However, DO NOT drink the water from these streams and falls. The water has a high bacteria count caused from the pigs which live in the jungle-like forests above. The beaches along the Hana Hwy. are unsafe for swimming.

The trip to Hana by car from Kahului will take at least three hours and plan on plenty of time to make some stops, enjoy these waterfalls up close and experience this unique coastline.

Accommodations vary from hotel/condo to campgrounds and homes at a variety of price ranges. The 7,000 acre Hotel Hana Ranch has achieved their goal of creating an "elegant ranch atmosphere." Several moderately priced condominiums and inexpensive cottages are also available.

Waianapanapa State Park, just outside Hana, has camping facilities and cabins. (See the Camping section for more information.) Ohe'o also has a tent camping area, bring your own drinking water.

Hana offers a quiet retreat and an atmosphere of peace (seemingly undisturbed by the constant flow of tourist cars and vans) that has lured many a prominent personality to these quiet shores. Restaurant choices are extremely limited and shopping is restricted to the Hasegawa General Store, the Hana Ranch Store or a few shops at the Hana Hotel.

## WHAT TO SEE AND DO

### PAIA - ALONG THE ROAD TO HANA

A little beyond Wailuku, and along the highway which leads to Hana, is the small town of Paia. The name Paia translated means "noisy," however, the origin of this name is unclear. This quaint town is reminiscent of the early sugar cane era when Henry Baldwin located his first sugar plantation in this area. The wooden buildings are now filled with antique, art and other gift shops to attract the passing tourist. (See RESTAURANTS for more information). The advent of windsurfing has caused a rebirth in this small charming town and a number of new restaurants have recently appeared over the last few years with more to come.

The *Maui Crafts Guild*, a group of local artisans own and operate this store. Pottery, koa furniture, weaving, wall sculptures, wood serving pieces, prints and basketry are featured. Very lovely, but expensive, hand-crafted items.

*Things from the Past* is housed in a former automotive garage and it is easy to spot on the Hana Hwy. with its friendly Hawaiian Santa out front. A conglomeration of items that will be of interest to just about anyone.

### KUAU RENTALS

794 Poho Place, Paia, Maui, HI 96779. (8088) 579-9400. Book a 2-bedroom unit on ocean next to Mama's Fish House and receive a 20% discount when you eat at Mama's. Color TV, BBQ, Washer/dryer, VCR and stereo. Maid service weekly. Seventh night is free! One bedroom units also available. *1 BR $60, 2 BR $125*

### HANA

Anyone who endures the three-hour (at least) drive to Hana deserves to sport the "I survived the Road to Hana" T-shirts which are sold locally. While it may be true that it is easy to fall in love with Hana, getting there is quite a different story. The drive to Hana is not for everyone, although many guide books claim otherwise. It is not for people who are prone to motion sickness, those who don't like a lot of scenery, those who are in a hurry to get somewhere or those who don't love long drives. However, it is a trip filled with waterfalls and lush tropical jungles (which flourish in the 340-inch average annual rainfall). Maps are deceiving. It appears you could make the 53-mile journey much faster than three hours but there are 617 (usually hairpin) curves and 56 miniature bridges along this narrow road. And believe it or not, each of these bridges has its own Hawaiian name! Even with recent repaving, most cars travel in the middle of the road, making each turn a possibly exciting experience, especially at night.

The Hana Hwy. was originally built in 1927 with pick and shovel, which may account for its narrowness, to provide a link between Hana and Kahului. There can also be delays on the road up to two hours if the road is being worked on. In days gone by when heavy rains caused washouts, it is said that people would literally climb the mud barricades and swap cars, then resume their journey. Despite all this, 300-500 people traverse this road daily, and it is the supply route for all deliveries to Hana and the small settlements along the way.

Now, if we haven't dissuaded you and you still want to see spectacular undeveloped scenery, plan to spend the whole day (or even better, stop overnight) in Hana. If you are driving, be sure to leave as early in the morning as possible. You don't want to be making a return trip on this road in the dark. Be sure to get gas, the last stations before Hana are in Kahului or Paia. With the exceptions of an occasional fruit stand, there is no place to eat and only limited stops for drinking water. Be sure you pack your own food and drink. Picnic's in Paia is a popular stop for a picnic lunch. For something a little more unusual try packing along some local style foods or a bento (box lunch). Takimaya's grocery story on Lower Market St. has an unbelievable assortment of cooked, pre-packaged food made fresh daily. Fried calamari, tako poki (raw octopus), kalbi ribs, baked yams, and much, much more. Packing some rain gear and a warm sweater or sweatshirt is a precaution against the sometimes cooler weather and rain showers. Don't forget your camera, but remember not to leave it in your car unattended.

We also might recommend that if you drive, select a car with an automatic transmission (or else be prepared for constant shifting). Another choice is to try one of the small van tours which go to Hana and leave the driving to them. Check to see whether the tours are operating their vans around the other side of the island. This will depend on the road conditions. It is also a road not recommended for rental cars. The scenery on this dry side is strikingly different from the northern coast rain forests. Good tour guides will also be able to point out the sights of interest along the way that are easy to miss! The only other alternative is to by-pass the Hana Highway by flying into Hana's small airport.

There are two good resources you might also consider taking along on your drive. Stop at the small booth on Dairy Road, across from the Shell service station. For about $20 you can rent a cassette player and narrated tape to follow along on your trip. *Maui's Hana Highway*, by Angela Kay Kepler runs about $5.95 at local bookstores and it's eighty information filled pages include plenty of full color photos of the area. Especially good for identifying the flora and fauna in this area.

*Ho'okipa*, about two miles past Paia, is thought by some to offer the world's best windsurfing. There won't be much activity here in the morning, but if you are heading back past here in mid to late afternoon when the winds pick up, you are sure to see numerous windsurfers challenging wind and wave. These waves are enough to challenge the most experienced surfers and are not for the novice except as a spectator sport. You'll note that on the left are the windsurfers while the waves on the right are enjoyed by the surfers! A number of covered pavilions offer shaded viewing and the beach, while not recommended for swimming, does have some tidepools (of varying size depending on the tidal conditions) for children to enjoy a refreshing splash. This beach is also a popular fishing area for local residents and you may see some folks along the banks casting in their lines.

The next area you will pass through is *Haiku*, which translated means abrupt break. It is not unusual to experience some overcast, rainy weather here. During 1989 this area had more rainy days than not! Enjoy the smooth wider highway here, the Hana Highway awaits you just ahead!

Two miles from the point where Hana Highway intersects Route 400 look for a small roadside trail marker by the Hoolawa Bridge. This area, known as ***Twin Falls***, offers a pleasant spot for swimming. The first pool has two waterfalls, but by hiking a little farther, two more pools of crystal clear water created by waterfalls can be easily reached. Remember, don't drink the stream water! Mosquitos can be prolific so pack bug spray.

There are no safe beaches along this route for swimming, so for a cool dip, take advantage of one of the fresh water swimming holes provided.

***Waikamoi Ridge*** - This picnic area and nature trail is about 1/2 mile past roadside marker #9 and has no restrooms or drinking water. This area is noted for its stands of majestic bamboo, and you are sure to see wild ginger and huge ferns.

***Puohkamoa Falls*** - Located near roadside Marker #11 is an area to pull off the road with parking only for a couple of cars. This small picnic area offers one covered (in the event of one of the frequent windward coast rain showers) table. A short tunnel trail through lush foliage leads to a swimming hole beneath the water fall.

***Kaumahina State Park*** - Just past roadside marker #12 you'll find this lovely park. This area overlooks the spectacular Honomanu Gulch, the rugged Maui coastline and in the distance a view of the Ke'anae Peninsula. Believe it or not this is about the half-way point to Hana. A good opportunity to make use of the toilet facilities.

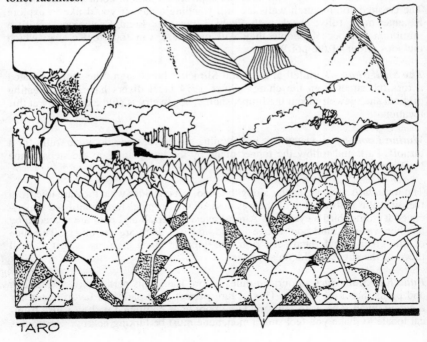

TARO

**The YMCA's Camp Ke'anae** - Offers overnight accommodations for men and women (housed separately). The rate is $5 a night. Arrival is requested between 4 pm and 6 pm. Bring your own food and sleeping bag. Phone (808-248-8355) for more information.

Just past Camp Ke'anae is the **Ke'anae Arboretum**. This free botanical garden is managed by the Department of Land and Natural Resources and is home to a myriad of tropical plants. A number of the plants have been labeled for your assistance in identification.

The **Ke'anae peninsula** was formed by a massive outpouring centuries ago from Haleakala. Today it is an agricultural area with taro the principal crop. The taro root is cooked and mashed and the result is a bland, pinkish brown paste called poi. Poi was a staple in the diets of early Hawaiians and is still a popular local food product which can be sampled at luaus or purchased at local groceries. Alone, the taste has been described as resembling wallpaper paste, (if you've ever tried wallpaper paste) although it is meant to be eaten with other foods, such as kalua pig. It is a taste that sometimes needs time to acquire. We have heard of island grandmothers sending fresh poi to the mainland for their young grandchildren. It is said to be extremely healthy, full of minerals and well tolerated by young stomachs. You'll see the fields filled with water and taro in varying stages of development. The Ke'anae Congregational Church, or **The Miracle Church,** in Wailua is a historical landmark with a fascinating history. In the mid-1800's the community was lacking in building material for their church. Quite suddenly a huge storm hit and, by some miracle, deposited a load of coral onto the beach. The crushed coral church walls are still standing today. If you'd like to explore Ke'anae more fully, check with Ekahi Tours (See Land Excursions). With a minimum number of people they will provide a van tour to Ke'anae which includes a tour of the poi factory.

**The Shell Shop** - Turn left at the Coral Miracle Church sign. The shop is located across the street from the church. Since 1974 local divers have been creating original shell jewelry from the limpet shell. The jewelry is sold exclusively at this location.

**Wailua Lookout** - Located just past Wailua on the roadside, look carefully for a turnoff. Park and follow the tunnel made by the hau plants, up the steps to the Lookout. The trek up is worth the excellent view.

The **waterfalls** are spectacular along the road, but consider what they are like from the air! We had no idea of the vastness of this tropical forest until we experienced it from a bird's eye view. Almost every waterfall and pool are preceded by another waterfall and pool above it, and above it there are yet others. The slice of this green wonderland seen from the winding Hana highway is just a small piece of the rugged wilderness above.

**Puaa Kaa State Park** - Fourteen miles before Hana, this park has two waterfalls and pools that are roadside. This picture perfect little park is a favorite stop for a picnic lunch. The waterfalls and large pool have combined with this lush tropical locale to make you feel sure a menehune must be lurking nearby. Restrooms and drinking water are available here too. Keep your eye out for mongoose. They have been "trained" by some of the van tour guides to make an appearance for a handout at some of these wayside stations. The best place to get a look at them is usually near the garbage cans. Toss a little snack and see if anyone is home.

With a little effort a sharp observer can spot the open ditches and dams along the roadside. These are the *Spreckles Ditches* built over 100 years ago to supply water for the young sugar cane industry. These ditches continue to provide the island with an important part of its supply of water.

*Waianapanapa* (pronounced WHY-A-NAHPA-NAHPA) *State Park* is four miles before Hana and covers an area of 120 acres. Translated it is said to mean glistening water. This area offers a number of historical sights, ancient heiaus (temples) and early cemeteries. You can spot one of the many stone walls used by the early Hawaiians for property boundaries, animal enclosures and also as home foundations. It is noted for its unusual black sand beach made of small smooth volcanic pebbles. The ocean here is not safe for swimming, but there is plenty of exploring! Don some mosquito repellent, tennis shoes are a good idea, and follow the well marked trails to the Waianapanapa Caves. The trail is lined with thick vines, a signal left by the early Hawaiians that this area was kapu (off limits). The huge lava tubes have created pools of cold, clear water. An ancient cave legend tells of a beautiful Hawaiian princess named Popoalaea who fled from Kakae, her cruel husband. She hid in the caves, but was discovered and killed. At certain times of the year the waters turn red. Some say it is a reminder of the beautiful slain princess, while others explain that it is the infestation of thousands of tiny red shrimp. Another three mile trail along the coast follows part of the ancient King's Hwy. which, in the days of the early Hawaiians, may have extended around the entire island. Several ancient heiaus can be found in this area. Another portion of the ancient highway can be found on the southern coastline past La Perouse Bay. Camping is allowed and there are rustic cabins available for rent (See ACCOMMODATIONS -Hana, which follows).

*The Helani Gardens* (248-8274) is a self-guided botanical tour by foot or car through some very dense vegetation. Created by Howard Cooper it opened in 1970 after thirty years of development. The lower area consists of five acres with

HELANI GARDENS

manicured grounds and a tropical pool filled with jewel colored koi. The upper sixty five acres are a maze of one-lane dirt roads through an abundant jungle of amazing and enormous flowering trees and shrubs. Wild plants from around the world are raised here and you'll see plenty of gigantic versions of your own house and garden plants from back home. Keep a sharp eye out for the large treehouse built by the Cooper grandchildren. This is really quite an enjoyable adventure, and one usually overlooked by those in a hurry to or from Hana and Ohe'o. Take your time and explore this unusual attraction. Admission $2 for adults, $1 children 6-16. Located about one mile before Hana. Picnic areas and restrooms available.

The last curve of the road will put you at **Hana's Gardenland**. There is no charge to browse through their flower displays and they thoughtfully provide picnic tables and a restroom. The plants sold here include the rare and beautiful and are available for shipping anywhere.

Now, back in the car for a drive into downtown Hana, but don't blink, or you might miss it. **Hana Cultural Center** (248-8622) opened in August of 1983. It contains a collection of relics of Hana's past in the old courthouse building and a small new museum. Open Monday through Saturday 11 am - 4 pm. Located near Hana Bay, watch for signs.

**Hana Bay** has been the site of many historical events. It was a retreat for Hawaiian royalty as well as an important military point from which Maui warriors attacked the island of Hawaii, and then were in turn attacked. This is also the birthplace of Ka'ahumanu (1768), Kamehameha's favorite wife. (See Beaches for more information.)

The climate on this end of Maui is cooler and wetter, creating an ideal environment for agricultural development. The Ka'eleku Sugar Company established itself in Hana in 1860. Cattle, also a prominent industry during the 20th century, continues today. You can still view the paniolos (Hawaiian cowboys) work at nearby Hana Ranch. There are 5,000 head of cattle which graze on 4,500 acres of land. Every three days the cattle are moved to fresh pastures. (Our family was thrilled when a paniolo flagged us to stop on the road outside Hana, while a herd of cattle surrounded our car enroute to fresher pastures.)

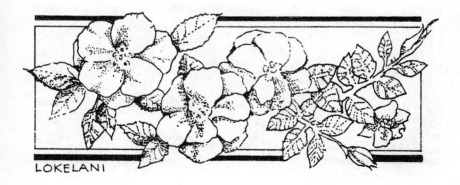

LOKELANI

Hana has little to offer in the way of shopping, however, the Hasegawa General Store offers a little bit of everything. It has been operated since 1910, meeting the needs of visitors and local residents alike. The cluttered, but well stocked shelves have even been immortalized in song! You may even run into one of the celebrities who come to the area for vacation. The Hana Ranch Store is open daily and the Hana Resort has a gift shop and boutique.

## SOME LOCAL HANA INFORMATION:

*St. Mary's Church:* (248-8030) Sat. Mass 6 pm, Sun. 8 am
*Wananalua Protestant Church:* (248-8040) Built in 1832. Services 10 am
*Hana Ranch Store:* (248-8261) 7:00 am-6:30 pm daily
*Hasegawa General Store:* (248-8231) 7:30-6 pm, Mon.-Sat., Sun. 9:30 am-3:30
*Hana Medical Center:* (248-8294) Emergencies 24 hours. Mon.-Fri. 8 am-noon
   and 2 pm - 5 pm. Sat. 8 am -noon. Closed Sunday.
*Bank of Hawaii:* (248-8015)  Mon.-Thur. 3-4:30 pm and Fri. 3-6 pm
*Library:* (248-7714) Tues.-Fri. 8-5 pm, Mon. 8-8 pm
*Post Office:* (248-8258) 8 am-4:30 pm, Mon.-Fri.

The oldest building in town, built in 1830, currently houses the laundry facility for the Hana Hotel.

On Lyon's Hill stands a large stone cross in memory of Paul Fagan. It was built by two Japanese brothers from Kahului in 1960. Although the access road is chained, the front desk of the Hotel Hana will provide a key. The short trip to the top will reward the visitor with a spectacular panoramic view of Hana Bay and open pasture of the Hana Ranch.

*Kaihalulu Beach* (Red Sand Beach) is located in a small cove on the other side of Kauiki Hill from Hana Bay and is accessible by a narrow, crumbly trail more suited to mountain goats than people. The trail descends into a lovely cove bordered by high cliffs and is almost enclosed by a natural lava barrier seaward. For more details see BEACHES.

*Hamoa Beach* - This gorgeous beach has been very attractively landscaped and developed by the Hotel Hana Maui in a way that adds to the surrounding lushness. The long sandy beach is in a very tropical setting and surrounded by a low sea cliff. As you leave Hana toward Ohe'o Gulch look for the sign 1 miles past the Hasegawa store that says "Koki Park - Hamoa Beach - Hamoa Village." Follow the road, you can't miss it.

You quickly pass fields of grazing world-famous Maui beef and reenter the tropical jungle once more. Numerous waterfalls cascade along the roadside.

After ten curvy, bumpy miles on a very narrow two-lane road, and a 45-60 minute drive, you arrive at one of the reasons for this trip, the Kipahulu Valley and *Ohe'o Gulch*. Popularly known in the past as the Seven Sacred Pools as a tool for attracting tourists, the original and proper name is trying to be reclaimed by the Park Service.

The 100 ft. *Wailua Falls* cascades beneath the narrow bridge (a great place for a photo) flowing over the blue-grey lava to create these lovely lower pools. The pools you see below the bridge are just a few of the more than 20 that have been formed as the water of this stream rushes to the ocean. The pools are safe for swimming so pack your suit, but no diving is allowed. Swimming off the black sand beach is very dangerous and many drownings and near drownings have occurred here.

The best time to enjoy these falls may be in late afternoon, when the day visitors have returned to their cars for the drive home. (Another good reason to make Hana an overnight trip.) The bluff above the beach offers a magnificent view of the ocean and cliffs, so have your camera ready. This area is of historical significance and signs warn visitors not to remove any rocks. A pleasant hike will take you to the upper falls. *Makahiku Falls* is 185 feet high and is a fairly easy half mile hike that passes through some cattle pastures. *Waimoku Falls* is another mile and a half. Three to four hours should be allowed for this hike which traverses across the stream and through a bamboo forest. Heavy rains far above in the mountains can result in flash floods. Avoid swimming in these upper streams or crossing the stream in high water. Check with the park rangers who keep advised as to possible flooding conditions. Also check with the park service (248-8251) to see if the free Saturday 9 am, ranger-guided hikes are available.

One interesting fact about Oheʻo is that many of the marine animals have evolved from saltwater origins. Others continue to make the transition between the ocean's salty environment and the fresh water of the Oheʻo stream. One of the most unusual is the rare oopu which breeds in the upper stream, migrates to the ocean for its youth and then returns to the stream to mature. After a glimpse of the many waterfalls, this appears to be a most remarkable feat. The ingenious oopu actually climbs the falls by using its lower front fins as a suction cup to hold onto the steep rock walls which form the falls. Using its tail to propel itself, the oopu travels slowly upstream.

The upper *Kipahulu Valley* is a sight visitors will never see. Under the jurisdiction of the park service, it is one of the last fragments of the native rain forests. The native plants in the islands have been destroyed by the more aggressive plants brought by the early Hawaiians and visitors in the centuries which followed. Some rare species, such as the green silversword, grow only in this restricted area.

Two miles further on is the *Charles Lindbergh grave*, located in the small cemetery of the 1850 *Kipahulu Hawaiian Church*. He chose this site only a year prior to his death in 1974, after living in the area for a number of years. However, he never envisioned the huge numbers of visitors that would come to Hana to enjoy the scenery and visit his gravesite. Please respect the sanctity of this area.

It is sometimes possible to travel the back road from Hana thru Upcountry and back to Kahului. This is Maui's desert region and it is a vivid contrast to the lush windward environs. While very parched, this route presents a hazard which can take visitors unaware. Flash floods in the mountains above, which are most likely November to March, can send walls of water down the mountain, quickly washing out a bridge or overflowing the road. The road is sometimes closed for months due to serious washouts. Check with the county to see the current status of this route. Car rental agencies post warnings that travel is not advised on this route for standard cars and that renters are responsible for all damage. If this route is passable (which at the time of printing was not) consider stopping at **The Kaupo General Store**. It has been operating for years and is open based upon the whim of the management. Take note of the many rock walls. This area supported a large native Hawaiian population and these walls served as boundaries as well as retaining walls for livestock, primarily pigs. The walls are centuries old and unfortunately have suffered from visitor vandalism and destruction by the range cattle. Cattle are now the principle area residents.

As you enter Upcountry and civilization once more, look for the **Tedeschi Winery**. Located at the Ulupalakua Ranch it offers tasting daily from 9 - 4. They began in 1974 and produced only a pineapple wine until 1983 when they harvested their first grapes. They also offer a champagne and a red table wine.

On the way back down you might stop at **The Botanical Gardens** which feature a close up look at the unusual protea flowers. Admission is charged. (See Upcountry for more information)

# ACCOMMODATIONS - HANA

Aloha Cottages
Hana Bay Vacation Rentals
Hotel Hana Maui
Waianapanapa State Park

YMCA Camp Ke'anae
Hana Kai Maui Resort Apartments
Heavenly Hana Inn

## ALOHA COTTAGES
PO Box 205 Hana, Maui, HI 96713 (808-248-8420). Six simple, clean two bedroom cottages with electric kitchens. Three units are oceanview. Rates are charged by number of bedrooms used and number of people. Rates from $55 to $80 for two which includes the 9% state tax. $10 each additional person. Owners Zenso and Fusae Nakamura want guests to enjoy themselves in Hana. Need a haircut? Then you might want to visit Zenso who also acts as local Hana barber.

## HANA BAY VACATION RENTALS ★
Stan Collins, PO Box 318 Hana, Maui, HI 96713 (808-248-7727) 1-800-651-7970. As an alternative to condominium and hotel living you might be interested in one of eight private homes available in and around Hana. They offer fully equipped one, two or three bedrooms homes with ocean or bay views. Prices range from $65 for the Kauiki Cabin to $150/night for a newly constructed Philippine Koa and Mahogany hand crafted duplex located 50 yards from Hana Bay. Payment in full in advance for 7 days or less, please include a SASE w/ 9.43% tax added. Stay 7 nights pay for 6! There is a 10% cancellation fee. Rates are $20 higher November 15th - March 30 and June 1 - August 30. We stayed in one of their oceanview homes and found it roomy, comfortable and clean. The surrounding yard was well maintained and a great opportunity for the kids to romp around while we enjoyed sitting on the porch with a peaceful view of Hana Bay.

# WHERE TO STAY - WHAT TO SEE
*Hana*

## HANA KAUHI
Box 296, Hana, Maui, HI 96713. (808-248-7029). Coleen Church and her family live on five acres near the Hana Airport. They invite guests to stay in their wooden duplex cottage which offers a downstairs one bedroom unit with double bed and a studio upstairs with two twin beds. Both feature full kitchens and baths. We've heard that this is a clean, well run and friendly operation. $50 for studio, $70 for one bedroom.

## HOTEL HANA MAUI ★
P.O. Box 8, Hana, Maui, Hi 96713 (808-248-8211) 1-800-321-HANA.

*Garden Accommodations $275, Sea Ranch Cottage Units $375*
*Garden One Bedroom Suites $475, Sea Ranch Cottage One Bedroom Suites $750*
*Compulsory meal plan of $85/day per person must be added to the above rates*
*Children 2 to 17 $45/day. No charge for children under two years of age*
*Hawaii state tax, food and room tax must also be added.*

This is the most secluded Maui resort, and a Hana landmark that has been called an island on an island. Five plantations were consolidated when Paul Fagan saw that the end of the sugar industry in Hana was close at hand. There had been 5,000 residents in Hana in 1941, and only 500 remained when he began the hotel and cattle ranch which rejuvenated Hana. Approximately 1/3 of Hana's population of over 1,000 are employed in some fashion by the hotel, ranch or flower nursery. Hotel Hana Ranch was opened for public use in 1947 and later renamed Hotel Hana Maui.

The 97-room hotel resembles a small neighborhood with the single story units scattered about the grounds. The rooms are simple, but elegant with hardwood floors, and tiled bathrooms with deep tubs. The resort prides itself on the fact that it has no televisions or room air conditioning. Newer additions are the 47 sea ranch cottages located oceanview at Kaihalulu Bay. These resemble the early plantation style houses. These cottages include oceanview and the majority offer spas on the lanais.

***The Historic Plantation House*** has been restored to its original elegance. Built in 1928, the 4,000 square foot building was the home of August Unna, Hana's first plantation owner. The surrounding four acres are filled with beautiful plants and trees that are more than 100 years old. The Plantation House is available as a guest home and offers two bedrooms and baths, a large living room with fireplace, dining room, library, bar and complete kitchen. To provide the latest technology for private business gatherings and meetings, it has been equipped with electronic data transmission equipment and audio-visual equipment that includes a large screen closed circuit television system. An adjacent pavilion and covered deck area add outdoor meeting areas. The site is also the location of the Hotel Hana Maui's weekly Manager's cocktail party.

Some exciting recreational opportunities await guests at the Hotel Hana-Maui. Under special licensing, the ranch is raising Chinese ringneck pheasant to be used for bird hunting, with dogs provided. Two hunters can be accommodated for either morning or afternoon excursions with a limit of four birds each. The slopes

of Haleakala are haven for wild boar, perhaps the most challeng[...]
game. Arrangements for hunting excursions can be made throug[...]
hunting is also available.

Overnight camping excursions begin with an experienced guide accompanying guests by jeep or horseback to the new foothill's campsite. Three large, high-walled tents with wooden floors provide sleeping and eating areas and flush toilets are also on location. An afternoon of hiking and exploring is followed by an evening barbecue under the stars. By then guests should all be ready for a good nights rest in sleeping bags on foam pads. A ranch breakfast is provided before the return to the hotel by noon. A minimum of four persons is required and children under the age of twelve must be with an adult. Non-campers can partake in a variety of hikes, or photo safaris.

Not to neglect the fishermen - there are plenty of opportunities available. With twenty four hours notice, a guided fishing excursion can be readied with all equipment included for pole fishing or shore casting. No license is required. A maximum of four fishermen per group.

Other activities include a weekly luau, many trails for hiking or horseback riding, or cookouts at Lehoula beach. A shuttle provides convenient transportation for the three mile trip to beautiful Hamoa Beach with private facilities for hotel guests. Tours are also available to Ohe'o Stream, or the plant nursery where the hotel raises mature indoor plants and flowers for use at the resort and to send to the mainland. Two swimming pools are located on the hotel grounds. They also offer a dining room as well as an informal family dining restaurant. (Restaurant dining and the weekly luau are available to non-hotel guests on a space available basis. Call for reservations.) Children's activities and overnight sitters are available.

Recent renovations include a bar with a large fireplace and an open deck with a quiet lounge adjoining for a peaceful atmosphere for conversation or reading. The new restaurant has a 35-foot ceiling with skylight, hardwood floors and a deck opening to a magnificent oceanview and excellent food. The library contains rare volumes of early Hawaiiana as well as popular novels. There is also a small boutique with resort fashions and jewelry in addition to a beauty salon. The "golf adventure" is three holes in the midst of the resort. The new Club Room has a television, and evening lectures are given here. A more peaceful and beautiful setting is difficult to imagine.

## PLANTATION HOUSES
Kumulani Rentals, PO Box 1190, Kihei, Maui, HI 96753. (808-879-9272) 1-800-367-2954. Swimming, riding, hiking and volleyball.
*1 BR o.v. (1-2, max 4) $125, Studio o.v. (1-2, max 2) $70*
*2 BR o.v. (1-4, max 4)    90, o.f. $150*          *No minimum stay*
*3 BR o.v. (1-5, max 6)   150*                      *$10 charge for extra persons*

## WAIANAPANAPA STATE PARK
P.O. Box 1049, Wailuku, Maui, HI 96753. (808-244-4354) The State Park Department offers cabins that sleep up to six people. The units have electric lights and hot water, showers and toilet facilities. The kitchen is equipped with a range,

ut no oven, and a refrigerator. Bedding is provided and clean linens and towels are refreshed every three days. No pets are allowed and bring your own soap! A five-day maximum stay is the rule and guests are required to clean their units before departure, leaving soiled linens. A 50% deposit is required for reservations and they are booked way ahead (6 months to one year). Children are considered those ages 11 and under, adults are counted as being 12 years and above. A pro-rated list of rates will be sent to you by the Parks Department on request. Included here are a few sample prices. The beach is unsafe for swimming, however, there are some interesting trails, pools, and lava tubes. The beach is not sand, but actually very small, smooth black pebbles. Mosquito repellent is recommended, even for a short walk through the pool area.

*1 adult-$10/day, 2 adults-$14, 4 adults-$24, 6 adults-$30*
*1 adult 1 child-$10.50, 2 adults 2 children-$18, 2 adults 4 children-$20*

### YMCA CAMP KE'ANAE
In Ke'anae. (808-248-8355) Bring your own sleeping bag and food. Separate facilities for men and women. Accommodations are dormitory style. *$8 a night*

### HANA KAI MAUI RESORT APARTMENTS
P.O. Box 38, Hana, Maui, HI 96713 (808-248-8426) or (808-248-7346) 1-800-346-2772. 13 oceanfront units on Hana Bay, fully furnished including kitchens. Maid service is daily and there are laundry facilities available. Attractive and well kept grounds with ornamental lava rock pool, no swimming pool.
*Studio (2, max 3) $80, 1 BR (2, max 5) $95       Extra persons $10/night*

### HEAVENLY HANA INN
P.O. Box 146, Hana, Maui, HI 96713. (808-248-8442) 4 modern units in a single Japanese-style inn. Each unit is 2-bedroom, with bath, dinette and lanai that sleeps 2 - 6 people. Payment in full is requested in advance to hold reservations. NO CREDIT CARDS. *Units begin at $60 for two, up to $100 for four*

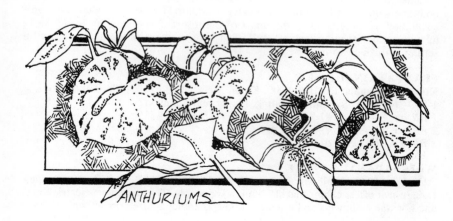
ANTHURIUMS

# BOOKING AGENTS

## ALL ABOUT HAWAII
PO Box 1843
Beaverton, OR 97075-1843
1-800-336-8687 or 1-800-274-8687
(503) 644-800

| | |
|---|---|
| Kaanapali Villas | Kamaole Sands |
| Paki Maui | Sands of Kahana |
| Luana Kai | |

## AMFAC HOTELS
Amfac Resorts Hawaii
PO Box 8520
Honolulu, Hi 96830-0520
1-808-945-6121   1-800-227-4700

Royal Lahaina Resort

## ASTON HOTELS & RESORTS
2255 Kuhio Avenue
Honolulu, Hi 96815
1-808-922-3368   1-800-367-5124
From Canada 1-800-445-6633

| | |
|---|---|
| Kaanapali Villas | Napili Point |
| Paki Maui | Sands of Kahana |
| The Mahana | Kaanapali Shores |
| Maui Hill | Maui Lu Resort |
| Kamaole Sands | Maui Park |

## CLASSIC RESORTS
50 Nohea Kai Drive
Lahaina, Maui, HI  96761
1-800-642-6284 (808-667-1400)
From Canada call collect

Kaanapali Alii
Lahaina Shores
Puunoa

## COLONY RESORTS
32 Merchant St.
Honolulu, HI 96813
1-808-523-0411   1-800-367-6046

| | |
|---|---|
| Kahana Villa | Sands of Kahana |
| Napili Shores | |

## CONDOMINIUM RENTAL HAWAII
2439 S. Kihei Rd., Suite 205A
Kihei, Maui, Hi 96753
1-808-879-2778   1-800-367-5242

| | |
|---|---|
| Hale Kamaole | Maui Parkshore |
| Hale Pau Hana | Sugar Beach |
| Kihei Akahi | |

## GENTLE ISLAND HOLIDAYS ★
PO Box 1441, Kihei, Maui, HI 96753
1-808-877-3945 1-800-544-6050

Bookings for the majority of condos
with a broad selection of units. We
found them very congenial and
helpful, with some units below
standard front desk prices.

## CONDO RESORTS INTERNATIONAL
3303 Harbor Blvd. Suite K-8
Costa Mesa, CA 92626
1-800-854-3823

| | |
|---|---|
| Coconut Inn | Napili Point |
| Kahana Reef | Napili Shores |
| Kaanapali Alii | Papakea |
| Kaanapali Shores | |
| Kahana Villas | Paki Maui |
| Kapalua Villas | Polo Beach Club |
| Laule'a | Puunoa Estates |
| Mahana | Sands of Kahana |
| Napili Kai Club | The Whaler |

(Bookings are for package plans
which include rental car)

## DESTINATION RESORTS
3750 Wailea Alanui
Wailea, Maui, HI 96753
1-800-367-5246   1-808-879-1595

Makena Surf
Polo Beach Club
Wailea Condominiums

## HAWAIIAN APT. LEASING
479 Ocean Avenue, Suite B
Laguna Beach, CA 92651
1-800-472-8449 CA
1-800-854-8843 U.S. except CA
1-800-824-8968 Canada

| | |
|---|---|
| Kaanapali Alii | Laule'a |
| Kaanapali Royal | Milowai |
| Kaanapali Shores | Mahana |
| Kahana Vilas | Makena Surf |
| Kahana Sunset | Maui Eldorado |
| Kamaole Sands | Maui Sunset |
| Kapalua Villas | Napili Bay |
| Kihei Bay Surf | Papakea |
| Kihei Bay Vista | Polo Beach Club |
| Kihei Holiday | Sands of Kahana |
| Kihei Kai Nani | Sugar Beach |
| Kihei Resort | The Whaler |
| Maui Kaanapali Villas | |
| Maalaea Yacht Marina | |

## HANA BAY
## VACATION RENTALS ★
PO Box 318
Hana, Maui, HI 96713
1-808-248-7727
Stan Collins has a great alternative
to condo vacationing. Choose one of
his Hana cottages or homes.

## HAWAIIAN PACIFIC RESORTS
1150 South King St.
Honolulu, HI 96882
1-800-367-5004
FAX 1-800-477-2329

| | |
|---|---|
| Maui Beach | Surf and Sands |
| Maui Palms | Resort Isana |

## HAWAIIANA RESORTS INC.
1270 Ala Moana Blvd.
Honolulu, HI 96814
1-800-367-7040
1-800-423-8733, ext. 159 Canada

Kaanapali Royal
Maui Kaanapali Villas
Laule'a (Maui Beach Club)
Kihei Bay Vista

## KAANAPALI VACATION
## RENTALS  PO Box 998
Lahaina, Maui, HI 96761
1-800-367-8008
1-800-423-8733 ext. 515 Canada

| | |
|---|---|
| Hale Mahina | Lahaina Shores |
| Kaanapali Alii | The Mahana |
| Kaanapali Royal | Sands of Kahana |
| Kapalua Villas | The Whaler |
| Maui Eldorado | |
| Maui Kaanapali Villas | |

## KIHEI MAUI VACATIONS ★
P.O. Box 1055
Kihei, Maui, HI 96753
1-808-879-7581 1-800-542-6284 U.S.
1-800-423-8733 Ext. 4000 Canada

| | |
|---|---|
| Maalaea Yacht Marina | |
| Kauhale Makai | Maalaea Kai |
| Kihei Akahi | Makena Surf |
| Kihei Alii Kai | Menehune Shores |
| Kihei Bay Surf | Milowai |
| Kihei Garden Estates | |
| Kihei Holiday | Maui Vista |
| Kihei Kai Nani | Wailea Condos |
| Luana Kai | Homes-Cottages |
| (a good range of S. Maui properties) | |

## KUMULANI
PO Box 1190, Kihei, Maui, HI 96753
1-800-367-2954

| | |
|---|---|
| Hale Kamaole | Mana Kai Maui |
| Island Surf | Maui Kamaole |
| Kamaole Sands | Maui Sunset |
| Kauhale Makai | Wailea Ekahi |
| Kihei Akahi | Wailea Ekolu |
| Laule'a | |

## MAALAEA BAY RENTALS ★
RR 1 Box 389
Wailuku, Maui, HI 96793
1-808-244-5627  1-800-367-6084

| | |
|---|---|
| Hono Kai | Maalaea Kai |
| Kanai A Nalu | Makani A Kai |
| Maalaea Banyan | Milowai |
| Maalaea Yacht Marina | |

## MAUI 800
Tom Morrow & Assoc. Inc.
P.O. Box 1506
Kahului, Maui, HI 96732
1-808-244-5627 (244-7012)
1-800-367-5224

Coconut Inn        Maui Eldorado
Hale Napili        Maui Hill
Hotel Hana Maui    Maui Islander
Kahana Sunset      Maui Kai
Kahana Village     Napili Bay
Lahaina Shores     Napili Kai
Laule'a            Noelani
(Plus 60 other hotels and
condos on Maui. Bookings
for the other islands also)

## MAUI CONDOMINIUMS
PO Box 1089
ALDERGROVE, BC,
CANADA V0X 1A0
1-800-663-6962 Canada

Aston Maui Park  Maui Islander
Hale Kamaole     Kamaole Sands
Hono Koa         Napili Bay
Island Sands     Paki Maui
Kihei Akahi      Papakea
Kihei Bay Surf   Royal Kahana
Kihei Garden Estates
Kihei Holiday    Sugar Beach
Kihei Kai Nani   Whaler
Mahana
Maui Beach Hotel
Maui Palms Hotel
Maui Kaanapali Villas
Valley Isle Resort
Village By the Sea

## MAUI NETWORK LTD.
PO Box 1077
Makawao, Maui, HI 96768
1-800-367-5221
1-800-423-8733 ext. 260 Canada

Kahana Outrigger
Kihei Beach Resort
Kuleana
Whaler

## MELE ADVENTURES
Ron and Judy Brown
888 Wainee St. #100
Lahaina, Maui, HI 96761
1-800-657-7714
Vacation rentals in addition to
arranging activities and tours.

## OIHANA PROPERTY MANAGEMENT
840 Alua
Wailuku, Maui, HI 96793
1-808-244-7684 or 1-808-244-7491
1-800-367-5234

Kana'I A Nalu    Kamoa Views
Kauhale Makai    Kealia
Leinaala         Maalaea Banyans
Island Sands     Maui Vista

## OUTRIGGER HOTELS HAWAII
2335 Kalakaua Ave.
Honolulu, HI 96715-2941
1-800-367-5170
FAX 1-800-4564329

Kaanapali Beach Hotel

## OVER THE RAINBOW, INC.
186 Mehani Circle
Kihei, Maui, HI 96753
(808) 879-5521
Specializes in assisting the disabled
traveler with any physical limitations.
Booking accommodations, tours,
personal care.

## PLEASANT HAWAIIAN HOLIDAYS
P.O. Box 5020
Westlake Village, Ca. 91359-5020
1-800-242-9244

Kahana Beach     Maui Marriott
Kamaole Sands    Royal Lahaina
Inter-Continental

Bookings are for package plans of 7
to 14 nights on one or more islands
and includes airfare and rental car.

## RSVP
1575 W. Georgia St., 3rd Floor
Vancouver, BC Canada V6G 2V3
1-800-663-1118 U.S. and Canada

| | |
|---|---|
| Hale Kamaole | Kamaole Sands |
| Hono Koa | Maui Park |
| Kaanapali Royal | Maui Sunset |
| Kaanapali Shores | Maui Vista |
| Kahana Sunset | Menehune Shores |
| Kahana Villa | Napili Bay |
| Kahana Village | Napili Shores |
| Kamaole Sands | Paki Maui |
| Kauhale Makai | Papakea |
| Kihei Akahi | Puamana |
| Kihei Alii Kai | Sands of Kahana |
| Kihei Holiday | Sugar Beach |
| Kihei Resort | The Whaler |
| Kuleana | |
| Mahana | (And a number of |
| Maui Hill | other properties) |
| Maui Kaanapali Villas | |

## RAINBOW RENTALS
PO Box 1893, Kihei, Maui, HI 96753
1-800-451-5366

| | |
|---|---|
| Island Sands | Koa Lagoon |
| Kamaole Sands | Luana Kai |
| Kauhale Makai | Maalaea Surf |
| Kealia | Maui Vista |
| Kihei Alii Kai | Sugar Beach |
| Kihei Resort | |

## RAINBOW RESERVATIONS INC.
PO Box 11453
Lahaina, Maui, HI 96761-6453
1-808-667-7858  1-800-367-6092

| | |
|---|---|
| Hololani Resort | Napili Shores |
| Kahana Outrigger | |
| Luana Kai | Papakea |
| Maui Sands | Pohailani |
| Valley Isle Resort | |

## REAL HAWAII
1341 Pacific Ave.
Santa Cruz, CA  95060
1-800-367-5108 (408-423-9923)
Hale Ono Loa

## RESORTS PACIFICA
505 Front St. #206
Lahaina, Maui, HI  96761
1-800-367-5637 (808-661-7133)
Papakea

## RIDGE REALTY RENTALS ★
888 Wainee St. #207
Lahaina, Maui, HI 96761
1-808-667-2851 1-800-367-8047 ext.
133, 1-800-423-8733 ext. 133 Canada
The Ridge (Kapalua)

## KONDO KING
## RESORT VACATIONS
1321 Whytecliff Rd.
Palatine, IL  60067 (312-358-1315)
Condo rentals and vacation with car
packages.

## VILLAGE RENTALS
Azeka's Place, P.O. Box 1471
Kihei, Maui, HI 96753
1-800-367-5634  1-808-879-5504

| | |
|---|---|
| Kalama Terrace | Sugar Beach |
| Kauhale Makai | Wailana Sands |
| Kihei Garden Estates | |
| Kihei Resort | Wailea Condos |
| Nani Kai Hale | |

## WHALER'S REALTY ★
Whaler's Village, Suite A-3
2435 Kaanapali Parkway
Lahaina, Maui, HI 96761
1-808-661-8777  1-800-367-5632

| | |
|---|---|
| Golf Villas | Kaanapali Villas |
| Kaanapali Alii | Mahana |
| Kaanapali Royal | Papakea |
| Kaanapali Plantation | |
| Kaanapali Shores | The Whaler |

(They offer high quality
condos at fair prices)

## WINDRIGGER MAUI
## VACATIONS
1-800-345-6284 (808-871-7753)
Vacation rentals $35 and up on the
north, south and west shores.

# Restaurants

## INTRODUCTION

Whether it's a teriburger at a local cafe or a romantic evening spent dining next to a swan lagoon, Maui offers something for everyone. We're confident that you will enjoy exploring Maui's diverse dining options as much as we have!

The majority of restaurants in the Maalaea to Makena, and Lahaina to Kapalua areas have been included, and for the adventurer or budget conscious traveler, take special note of the wonderful local dining opportunities in Kahului and Wailuku.

Needless to say, we haven't been able to eat every meal served at every restaurant on Maui, but we do discuss with a great many people their experiences in order to get varied opinions. We also welcome your comments. (See READER RESPONSE.)

Following this introduction, the restaurants are first indexed alphabetically and then also by food type. The restaurants are then divided by geographical area, separated by price range, and listed alphabetically in those price ranges. These are: "INEXPENSIVE" under $10, "MODERATE" $10 to $18, and "EXPENSIVE" $18 and above. As a means of comparison, we have taken an average meal (usually dinner), excluding tax, alcoholic beverages and desserts, for one person at that restaurant. The prices listed were accurate at time of publication, but we cannot be responsible for any price increases. For quick reference, the type of food served at the restaurant described is indicated in Italic type next to the restaurant name. Sample menu offerings are also included as a helpful guide. An important post-script here is the rapidity with which some island restaurants open and close, change names, and raise prices. Our quarterly newsletter, THE MAUI UPDATE, will keep you abreast of these changes. (See ORDER INFORMATION).

Several new resorts are scheduled to open in Wailea in 1990 and 1991 and we have included at least a brief sketch of some. All will be updated in our quarterly newsletter as they open.

Our favorite restaurants are generally either a real bargain for the price, or serve a very quality meal, and are indicated by a ★.

# BEST BETS

## TOP RESTAURANTS

Our criteria for a top restaurant is excellence of food preparation and presentation, a pleasing atmosphere, and service that anticipates or responds promptly to one's needs. While the following exemplify these criteria they are also all "deep pocket" restaurants, so expect to spend $70 - $100 or more for your meal, wine and gratuity for two. Generally, anything you have will be excellent. Remember, even the best restaurants may have an "off" night, but these are seldom. Also, chefs and management do change, rendering what you may have found to be excellent on one occasion quite different the next. However, the following have proven to be consistent through the years. Enjoy your meal, enjoy being a little bit spoiled, and remember those muu-muus are great for covering up all those calories!!

Bay Club, Kapalua Resort Hotel
David Paul's Lahaina Grill
Hakone, Maui Prince Hotel
La Perouse, Hotel Inter-Continental
Plantation Veranda, Kapalua Resort Hotel
Prince Court, Maui Prince Hotel
Raffles', Stouffer Wailea Beach Resort
Sound of the Falls, Westin Maui Resort
Spats II, Hyatt Regency at Kaanapali
Swan Court, Hyatt Regency at Kaanapali

## TOP RESTAURANTS IN A MORE CASUAL ATMOSPHERE

While these restaurants are less expensive, it is still easy to spend $60 or more for dinner for two. They serve a superior meal in a less formal atmosphere.

Avalon, Lahaina
El Crab Catcher, Whaler's Village at Kaanapali
Haliimaile General Store, Haliimaile
Island Fish House, Kihei and (Kahului, now Mickeys)
Kapalua's Garden Restaurant, Kapalua Resort Hotel
Kapalua Grill and Bar, Kapalua
Longhi's, Lahaina
Mama's Fish House, Paia
The Villa, Westin Maui
Tasca, Lahaina

## LA CUISINE FRANCAIS

Maui's French restaurants all fall in the "champagne" price range. Personally, we find that each has its own special merits that make for an enjoyable evening and dinner and that one cannot be singled out as "the best."

Chez Paul, Olowalu     Gerard's, Lahaina     La Bretagne, Lahaina

## BEST DINNER CRUISE

Stardancer

## BEST SEAFOOD
*All of the top restaurants plus:*
Fresh Island Fish, Maalaea (casual eatery and fresh seafood market)
Mama's Fish House, Paia
The Villa, Westin Maui
Island Fish House, Kihei
Gerard's, Lahaina

## BEST BUFFETS
Buffets are a good way to enjoy a great meal with a wide selection of food at a moderate price. And you may not have to eat for the next two days! The best Sunday champagne brunches are Prince Court in South Maui, and Sound of the Falls in West Maui. Swan Court features the best daily breakfast buffet. The others are all excellent, just not quite as extravagant. The best priced Sunday brunch is at the Kaanapali Beach Hotel. Moana Terrace has the best deal on a dinner buffet offering an early bird special. The following are listed in alphabetical order.

The Garden Restaurant (Kapalua Bay Resort) features an artistic presentation and unusual variety of gourmet specialties in an open-air setting for their Sunday brunch. The Mayfair Buffet features hot entrees that change weekly. $16.50 adults, no champagne.

Makani Room (Maui Inter-Continental Hotel, Wailea) has a fine Sunday brunch that is in a little more casual setting. 9 am-1 pm $26.

Moana Terrace (Marriott Hotel) has evening buffets $17.50 or arrive between 5 and 6 pm and enjoy a $4 discount.

Palm Court (Stouffer Wailea Beach Resort) serves a dinner buffet nightly priced at $22 and a daily champagne breakfast buffet for $13.50.

Prince Court ★ (Maui Prince Hotel, Makena) features a spectacular display of over 160 food choices at the champagne Sunday brunch, each arranged as a work of art. Each taste as good as they look! 9:30 am-1:30 pm $25.

Raffles' (Stouffer Wailea Beach Resort, Wailea) has a true taste extravaganza for their champagne Sunday brunch with Tony Van Steen at the piano. You'll be sure to see some dishes here that you've never sampled before. 9 am-2 pm $23.50.

Royal Ocean Terrace (Royal Lahaina Resort, Kaanapali) entices guests with their lavish salad bar, hot pasta bar, and sushi bar. Don't miss out on the desserts too! 9 am-2 pm, $19.50.

Sounds of the Falls ★ (Westin Maui, Kaanapali) serves a marvelous Sunday champagne brunch, especially for the price! Some people come just for the sushi bar! 9 am-2 pm, $19.75.

Swan Court ★ (Hyatt Regency, Kaanapali) features a lovely breakfast buffet daily in an elegant atmosphere. 6:30-11:30 am, until 1:30 pm on Sun. $13.50.

## BEST SALADS
Chinese chicken salad and Gado Gado salad both available at Avalon restaurant in Lahaina.

## BEST SALAD BAR
Royal Ocean Terrace Restaurant at the Royal Lahaina Resort.

## RESTAURANTS
*Best Bets*

### TOP "LOCAL" RESTAURANTS (Kahului/Wailuku)
We have delighted in exploring the many small, family-owned "local" restaurants in Kahului and especially in Wailuku. The food in these establishments is not only plentiful and well prepared, but also very inexpensive. The service is often better and friendlier than at many of the resort establishments.

Aki's Hawaiian Food and Bar (Hawaiian)
Fujiya's (Japanese)
Hazel's (Home style)
Mama Ding's (Puerto Rican)
Saeng's Tai Cuisine (Thai)
Sam Sato's (Japanese/noodles)
Siam Thai (Thai)
Tasty Crust (Home style)
Tokyo Tei (Japanese)
The Vietnam Restaurant (Chinese and Vietnamese)

### BEST GOURMET LUNCHEON
The Class Act restaurant at the Maui Community College in Kahului serves a five course gourmet lunch twice weekly offered by the Food Service students. Available only during the school year.

### BEST PIZZA
Pizza Circus, phone 871-1133, in Kahului, has inside seating and pizza to go. The crust is the secret ingredient in making this pizza an outstanding choice.

### BEST SANDWICHES
Paradise Fruits in Kihei with it's "health conscious" menu. Also, the island's best shakes and fruit smoothies.

### BEST HAMBURGER
Best hamburger with a view goes to Cheeseburger in Paradise. For a fabulous view-less burger check out Fat Boy's in Honokowai.

### GOOD AND CHEAP
One of the best meals for the price may be an Early Bird Special at one of Maui's better restaurants. The restaurants sometimes offer these only during low season, so check to see current availability and price. Hours vary with the restaurant, but usually begin between 5 and 5:30 and end between 6 and 6:30. The following are our personal favorites and offer quality meals. Generally, the meals are the same ones that you would pay more for an hour later, but you are limited in your selections.

Island Fish House, Kihei
Moana Terrace, Marriott Hotel, Kaanapali (buffet)
Kihei Prime Rib and Seafood House

### BEST HAWAIIAN
Aki's, located on Market Street in Wailuku is small, quaint, and very inexpensive. In West Maui check out the Lahaina Cafe at 505 Front St.

## BEST FAST FOODS
Azeka's Market and Paradise Fruits in Kihei offer good food. In West Maui, the hamburgers and plate lunches at Fat Boy's in Honokowai are "ono."

## BEST SHAVE ICE
Is shave ice becoming a lost art on Maui? It may seem so! Finding it can be difficult. Check Rainbow Snacks and Seeds at the Lahaina Shopping Center near Nagasako market in West Maui, and Azeka's snack shop in Kihei.

## BEST RIBS
Chris's Smokehouse features great ribs at their Lahaina restaurant. In Kihei, stop at Azeka's market for their specially marinated ribs that you can cook.

## FAVORITE BREAKFASTS
Pioneer Inn has it hand's down as our favorite breakfast stop. The French toast, made of Portuguese Sweet Bread, is fabulous. The rustic atmosphere is reminiscent of by-gone days, and the prices reasonable. All of the waitresses are terrific, but if you're lucky enough to have Ma serve you, you'll get an especially warm aloha along with your meal. In an atmosphere of tropical delight a breakfast buffet is served daily at Swan Court. The macadamia nut pancakes are light and fluffy, and cooked to order. On the other side of the island in Wailuku is Tasty Crust, an inexpensive local restaurant where unusual crusty pancakes are their specialty.

## MOST OUTRAGEOUS DESSERT
The Lahaina Provision Company's Chocoholic Bar at the Hyatt Regency Kaanapali is a chocolate lover's dream come true. This dessert buffet features soft, self-serve ice cream and an array of toppings. Hot caramel sauce, hot fudge, hot coffee-chocolate sauce, strawberries, shaved chocolate, chipped chocolate, chocolate kisses, grated chocolate, chocolate mousse, and the list continues. You can make the trip through as many times as you or your waistline can tolerate.

## BEST BAKERY
Our favorite Lahaina bakery is cleverly named, The Bakery, and an early stop will ensure you the best selection of their wonderful French pastries. Cheese and luncheon meats are also available. In central Maui don't miss a stop at the Homemade Bakery at 1005 E. Lower Main Street, their bread pudding is fantastic. And be sure to stop at the Four Sisters Bakery in Wailuku at Vineyard Street at Hinano. Melen, Mila, Beth and Bobbie arrived from the Philippines nine years ago. Their father had operated a Spanish Bakery in Manila for 15 years before moving the family to Maui. Not a large selection, but the items are delicious and different. One sweet bread is filled with cinnamon pudding, a sponge cake with a thin layer of butter in the middle of two moist pieces. The butter rolls are very good and the Spanish sweet and cinnamon rolls delicious. They sell their items only at this location and at the Swap Meet each Saturday morning. Hours are Mon.-Thursday and Saturday and Sunday 5 am until 8:30 pm, Friday until 10 pm.

**BEST VEGETARIAN** The only strictly vegetarian fare is available at The Vegan restaurant in Paia.

**BEST LUAU**   (See luau section which follows)

# *ALPHABETICAL INDEX*

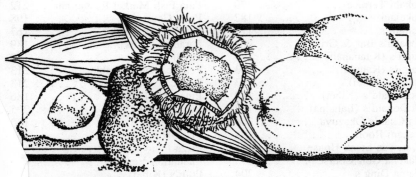

AVOCADO, MANGO, COCONUT, PAPAYA

# FOOD TYPE INDEX

# RESTAURANTS
*Food Type Index*

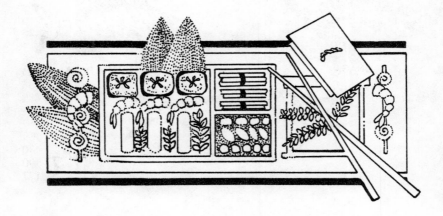

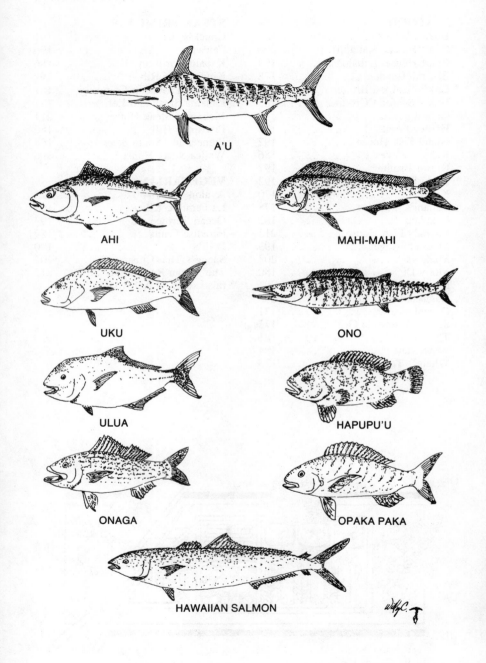

A'U

AHI

MAHI-MAHI

UKU

ONO

ULUA

HAPUPU'U

ONAGA

OPAKA PAKA

HAWAIIAN SALMON

# A FEW WORDS ABOUT FISH

Whether cooking fish at your condominium or eating out, the names of the island fish can be confusing. While local shore fishermen catch shallow water fish such as Goatfish or Papio for their dinner table, commercial fishermen angle for two types. The steakfish are caught by trolling in deep waters and include Ahi, Ono, and Mahi. They sometimes provide a healthy struggle before being landed. The more delicate bottom fish include Opakapaka and Onaga which are caught with lines dropped as deep as 1,500 feet to ledges or shelves off Maui's west shoreline. Here is a little background on the fish you might find on your dinner plate.

A'U - The broadbill swordfish averages 250 lbs. in Hawaiian waters. Hard to locate, difficult to hook, and a challenge to land. Considered a steakfish.

AHI - The yellow fin tuna (Allison tuna) is caught in deep waters off the Kaua'i coast. The pinkish red meat is firm yet flaky. This fish is popular for sashimi. They weigh between 60 and 280 pounds.

ALBACORE - This smaller version of the Ahi averages 40 - 50 pounds and is lighter in both texture and color.

AKU - This is the blue fin tuna.

EHU - Orange snapper

HAPU - Hawaiian sea bass

KAMAKAMAKA - Island catfish, very tasty, but a little difficult to find.

LEHI - The Silver Mouth is a member of the snapper family with a stronger flavor than Onaga or Opakapaka and a texture resembling Mahi.

MAHI - Although called the dolphin fish, this is no relation to Flipper or his friends. Caught while trolling and weighing 10-65 lbs. this is a seasonal fish which causes it to command a high price when fresh. *Beware*, while excellent fresh, it is often served in restaurants having arrived from the Philippines frozen and is far less pleasing. A clue as to whether fresh or frozen may be the price tag. If it runs less than $10 it is probably the frozen variety. Fresh Mahi will run $16 - $20 a dinner. This fish has excellent white meat that is moist and light. It is very good sauteed.

MU'U - We tried this mild white fish at the Makawao Steak House and were told there is no common name for this fish. We've never seen it served elsewhere in restaurants.

ONAGA (ULA) - Caught in holes that are 1,000 feet or deeper, this red snapper has an attractive hot pink exterior with tender, juicy, white meat inside.

ONO - Also known as Wahoo. ONO means "very good." A member of the Barracuda family, its white meat is flaky and moist. It is caught at depths of 25-100 fathoms while trolling and weighs 15 to 65 pounds.

'OPAE - Shrimp

OPAKAPAKA - Otherwise known as pink snapper and one of our favorites. The meat is very light and flaky with a delicate flavor.

PAPIO - A baby Ulua which is caught in shallow waters and weighs 5 -25lbs.

UKU - The meat of this grey snapper is light, firm and white with a texture that varies with size. It is very popular with local residents. This fish is caught off Kaua'i, usually in the deep Paka Holes.

ULUA - Also known as Pompano, this fish is firm and flaky with steaklike, textured white meat. It is caught by trolling, bottom fishing, or speared by divers and weighs between 15 and 110 pounds.

# LUNCHEON SHOWS

Aloha Mele Luncheon at Maui Inter-Continental Hotel features Jesse Naokee and Auntie Emma Sharpe. The schedule for this special afternoon show varies, so call the hotel. Cost is $16.50. Decked out in their finest aloha wear, you'll find a real local following for the popular Emma and Jesse. This show demonstrates the spirit of Hawaii as many of the audience join the stage for some wonderful spontaneity. In the lobby beginning about 10:30 there are displays of Hawaiian made crafts from Niihau shell leis, to leis and carved koa wood bowls.

# LUAUS AND DINNER SHOWS

For a local luau, check the Maui News. You may be fortunate to find one of the area churches or schools sponsoring a fund raising luau. The public is welcome and the prices are usually half that of the commercial ventures. You'll see spontaneous, local entertainment.

Most of the luaus are large, with an average of 400 - 600 guests, with one of the smallest being the Old Lahaina Luau with only 200. Most serve traditional Hawaiian foods. The entertainment ranges from splashy broadway-style productions to a country barbecue, or a more authentic Hawaiian dance and song. In general there are a few standard things to be expected at most luaus. These are shell leis, photos, an imu ceremony, playing the Hawaiian wedding song, kalua pig, poi, and coconut pudding. Upon arrival there may or may not be some waiting in line before it's your turn to be greeted with a shell lei and a snapshot of your group (available for purchase after the show).

It is very difficult to judge these luaus due to their diversity. While one reader raves about a particular show, another reader will announce their disappointment with the same event. Read carefully the information provided keeping in mind that the performers do come and go and also keep up on changes in our newsletter THE MAUI UPDATE.

### OVERALL BEST BETS:
*Best Imu Ceremony* - Rodney Arias at Stouffer's
*Best Dinner Show* - Hyatt Regency
*Best Haupia* (coconut pudding) - Kaanapali Beach
*Best Atmosphere and Luau - South Maui* - Maui Intercontinental
*Best Atmosphere and Luau - West Maui* - Old Lahaina Luau
*Best Luau Food* - Maui Intercontinental
*Best Luau Show* - Royal Lahaina

### HANA RANCH
"Hana Ranch Luau" is held at Lehoula Beach on Fridays at 6 pm. Open to non-hotel guests based on availability for $50. Guests are transported via hay wagon or van to the beachfront luau location. A very local and family oriented production. Many of those involved in the entertainment are folks you might see working in another capacity around the hotel. Phone 248-8211.

## HYATT REGENCY
In Kaanapali. The "Drums of the Pacific" is more a dinner show than a luau. Of all the shows we saw on Maui, this was the most like a Las Vegas review. A 5:30 arrival was suggested, which meant standing in line until 6 pm. Pictures were taken while waiting prior to being admitted to the grounds, where you were greeted with a lei and taken to a table by your server. The grounds are outside on the perimeter of the Hyatt, but there is no ocean view. A welcome mai tai drink (regular or non-alcoholic) was served (additional drinks were charged). The dinner selections that night included ono in lemon/ginger, 1/2 teriyaki chicken, 8-10 oz prime rib and a combination plate of char sui, ono, Polynesian chicken and teri beef. All except the combo were served with potato, carrots and broccoli. The combo was served with rice. A very brief imu ceremony was presented on the edge of the grounds with people viewing from their seats as best they could. The pig was then paraded through the center aisle. Salads arrived in attractive wooden pineapple bowls filled with mixed greens, shrimp, tomato and topped with papaya seed dressing and a spear of fresh papaya and pineapple. A sample dish of poi and pig was passed around. While dinner was served we were entertained with Hawaiian melodies. A fashion show displayed the variety of ways the parieu can be worn (and yes they were available for sale). Cocktail only seating began at 7 pm and then began a very fast paced one hour show. The costumes were excellent, and the dances performed were from various polynesian isles. They offered not one, but an impressive three fire dancers. This show was very professionally done, with excellent performers and a host that resembled a young Wayne Newton. The food was a good option for those preferring non-luau fare, and all in all, it was a show worth seeing. Prices $40 for adults, juniors $32.

## JESSE'S LUAU GARDENS
1945 S. Kihei Rd. Jesse Nakooka performs three nights a week at this Polynesian buffet in Kihei. Jesse has been involved with luaus for many years. The food for this luau is provided by Island Fish House. Located next to Island Fish House in Kihei, it is offered Wednesday, Friday and Sundays. If you're in the area about 9:30 in the morning you can stop by to see the imu being buried (no charge). They recommend you arrive at the outdoor luau gardens about 5:30. They offer valet parking and a shell lei greeting is followed by a photo session. We'd recommend arriving a little earlier to get a seat nearer the front. This is a big luau and the seats are packed tightly together along tables covered with tapa looking cloths. They offer an open bar, a mai tai punch or a tropical non-alcoholic punch. Musical entertainment happens until about 6:30 when it is time for the imu ceremony. As often happens the imu ceremony takes place in an area that isn't big enough for everyone to see, especially the little folk. Dinner is served on paper trays and includes the kalua pig, kalbi ribs, chicken and fish. The theme of the show is dances of old and new Hawaii. Jessie sings a few songs, and the five female and two male dancers perform. Two fire dancers are also featured. It is a long show, lasting until almost nine, of which at least half involves people coming up on stage from the audience for various and assorted reasons. Birthdays, anniversary people, honeymooners are all greeted and Jessie asks where they are from and where they are staying. Another large group comes up to learn a lengthy hula. Enough already! We came to see a show, not to see the audience perform. This gets a fair rating for food and poor rating for the show. Adults $39, Children 3 - 12 are $20, under age 3 is free.

# RESTAURANTS
*Luaus*

## KAANAPALI BEACH HOTEL
At Kaanapali. This is the newest luau to join the bunch and the most unusual in many ways. No arrival photographs are taken, no leis given and no wedding chant performed. The first stop is an outdoor area near the beach where fresh pineapple tidbits were served for appetizers along with several tables selling leis and jewelry. The young dancers mingle about talking to the guests, chatting about their different costumes. At 6 pm the imu ceremony held followed by a mini-parade which escorts the pig back to the indoor eating area. A rum punch is offered. It should be first explained that the Kaanapali Beach Hotel is actively involved in educating their staff with the only Hawaiiana program for employees. It therefore follows that this luau is designed to provide the guests with the feeling of ohana (family) and is handled much the same way a large family might enjoy an evening. Many of the performers are young, many are employees or relatives of hotel employees, and all are very good. The show is a Hawaiian show, no fire dancers, Tahitian or other South Sea performers. This is a show that fully expresses the beauty of Hawaii and its culture and people. The food is expertly prepared in the hotel's kitchen. A menu of chicken long rice, kalbi ribs, breaded mahi, salads and, of course, poi invites guests to try some of Hawaii's favorite foods. Quite frankly the food was a step up from many of the larger and longer running shows. The haupia (coconut pudding) was creamy and rich, unlike the usually white gelatinous versions, and the kalbi ribs merited a trip back for thirds. The show was choreographed with dances depicting the various changes in the history of the Hawaiian hula. A nice touch is the invitation they extend to various young performers for a "one night stand." A brother and sister combination, coincidentally siblings of one of the performers, did a lovely hula for us. And guess what? Not one person from the audience got on stage! They do hope to eventually move the show to an outdoor luau grounds and we agree that would be an improvement to a good, truly Hawaiian show. Friday only. Adults $37.50, children half price. Maximum occupancy 250 people. Phone 661-0011.

## MAUI INTERCONTINENTAL WAILEA
The Intercontinental has an outdoor setting that is both beautiful and spacious with a sublime ocean view. The stage is set up to offer the ocean and beautiful sunset as backdrop. An open bar is available with drinks other than fruit punch at extra charge. Dinner moves swiftly through several buffet lines serving mahi, chicken hekka, kalua pig and perfectly cooked teriyaki sirloin steaks hot off the luau gardens barbecue, and overall the best food at any of the island's luaus. The best dessert table was also found here with a selection of macadamia cream pie, coconut cream pie, banana cream pie, coconut macaroons, and pineapple cake squares. The show begins just as the sun is setting with Ka Poe o Hawaii and Paradyse and Brother Kauluwehe as host. Not an elaborate show, but one featuring a good range and quality of entertainment. Firedancer Prince Tevaga is one of the best. For overall food, entertainment and setting, this is the best of the luaus in South Maui. Tuesday and Thursday, Adults $39, Children 5 - 10 $19.50, under 5 free. Phone 879-1922.

## MAUI LU
In Kihei. "Remember the good times of old Hawaii?" queries the Maui Lu. Well if you don't, they suggest you try their traditional luau on Saturday night. $32 for adults and $19 for children. Phone 879-5881

## MAUI PRINCE HOTEL

While Makena does not offer luau entertainment, you might want to stop in for the Hawaiian Hula show each Friday evening from 6:30-8:30 pm in the central courtyard. Call ahead to verify. And for fine Hawaiian melodies visit George Paoa in the Molokini Lounge.

## OLD LAHAINA LUAU

The Old Lahaina Luau at 505 Front Street in Lahaina is situated right on the beach and offers the most beautiful luau setting in West Maui. A celebration of aloha in the traditional Hawaiian style is emphasized with guests greeted by shell leis and offered a choice between table setting or mats on the ground. You will note the many young Hawaiians that form their helpful and friendly staff dressed in colorful Hawaiian garb. One beautiful young lady demonstrates and sells leis by the water's edge. This luau is one of Maui's smallest, with a maximum of 225 people. Following the imu ceremony it's time to visit the buffet where selections include a pleasing array of half a dozen salads and entrees such as chicken, fish and of course kalua pig! There are four buffet lines which speed the guests through smoothly. New in 1990 is the addition of an open bar. Get your cameras ready as the show begins with the dancers arriving on stage in a torch-lit procession. This symbolizing the Tahitian's arrival in Hawaii. A strikingly beautiful maiden appears strolling along the beach, appearing forlorn as her people have migrated away. The show is strictly Hawaiian, no fire dancers here. Auntie Eileen and Piilani Jones perform the duties of show hostesses. For food, atmosphere and good Hawaiian entertainment this gets the overall rating for the best luau in West Maui. Held Tues.-Sat. at 5:30. Phone 667-1998. Adults $42, Children $21.

## ROYAL LAHAINA RESORT

In Kaanapali. We arrived at 5:15 for a 45 minute wait until gate opening. Apparently, arrivals of people were staggered. It's worth the wait for a good seat. Ready-made mai tai's, fruit punch, open bar and lei beads were the first order of business. The luau grounds are much like those at the Hyatt. Near the ocean, but without an oceanview and seating at padded picnic table benches. While people were settling in, Alakai'i, the hostess, invited folks up for a hula lesson. They also had a little humorous presentation about the luau foods to be served. Just the right amount of audience interaction. The imu ceremony was the shortest. (Some people

hadn't even reached the pit before it was over.) Dinner began at 6:45 with four tables and 8 lines allowing people to flow quickly through. Real glass glasses for drinks were a pleasant surprise, although the coffee cups were plastic. Large wooden trays offered plenty of room to pile on the teri beef, kalua pig, pineapple chicken, lomi lomi salmon, poi and salad bar. Desserts were unmemorable as evidenced by the many tables littered with uneaten cake. The show begins with the legend of the naupaka. A beautiful story of a lost love performed not on the main stage, but on a side stage with the sunset sky for an impressive backdrop. Costumes were excellent and the history of the hula included the influences that the movies of the 1950's had on this unique Hawaiian dance. Alakai'i is the core of the show and she was warm, funny and truly appeared to enjoy herself, and the fire dancer Tevesie was very good. This was the best luau show, but expect mediocre fare for your meal. Seven days a week. Phone 661-3611. Adults $39.58, children 12 and under $19.79 (includes tax). Editors note: As we go to press we hear that this show has been completely revised with a more modern, Las Vegas style. If you go let us know about the new show!

## SHERATON MAUI HOTEL

At Kaanapali. In the summer of 1989, just a day or two before they were to celebrate their tenth anniversary (the longest running Maui luau), the Sheraton luau, to the surprise of many, shut down. Again to the surprise of many it reopened with some management changes. After nearly a dozen luaus they all start looking a lot alike. What is unusual here at the Sheraton is an opportunity to learn a little more personally about some Hawaiian culture and activities. Following the lei greeting there is a chance to mingle with the staff to test your ability at twirling the poi balls or throwing the spear between two sticks imbedded into the ground and learn how to tie a ti leaf skirt. An open bar runs until 7 pm after which rum punch is available. The dinner buffet includes pineapple chicken, island snapper, teri beef steak and of course, kalua pig. The entertainment stars the very talented Barry Kim and includes the hula of the past (Kahiko) and the present (Auana), along with dances of the South Pacific. The outdoor setting here is a plus, set on the Sheraton's lush grounds. A rapid weather change during our visit here offered an opportunity for us to discover what happens when it rains at a luau! The show goes on! Brightly colored rain ponchos are passed to the guests while the dancers get wet! Adults $39.50, Children $17. Phone 661-LUAU.

Also at the Sheraton is their Discovery Room dinner show. Dinner menu offers a limited and pretty standard fare with Chicken Kaanapali at $28.50, teriyaki steak $31.50, prime rib $29.75. First dinner seating begins at 6:30 pm and seating continues until about 8 pm. Dinners include appetizers, soup or salad and tea or coffee. A $5 charge for the polynesian show, Maui On My Mind is added onto the entree price. The show begins at 8:30 and lasts approximately one hour. Non-diners can pay $15 which includes two drinks.

## STOUFFER WAILEA BEACH RESORT

Offered Monday and Thursday in their outdoor luau gardens. Seating at roomy round tables following the prescribed ritual of beads and photo taking. Arrive early for the best seats. Rodney Arias, the host, IS the show. He is a warm, gregarious guy with a beautiful tenor voice. His imu ceremony is excellent, both humorous and informative on cooking techniques and the differences between the old and new methods of preparing kalua pig. Even during dinner he mingles and visits with the guests. The buffet line moved quickly with two tables set up with two lines each. A nice selection of salads were followed by entrees that included mahi, teri beef, kalua pig, and chicken hekka. Desserts included banana cream pie, haupia, and pineapple upside down squares. An open bar was available and waitresses served fruit, mai tai or pina colada punch from large pitchers. The setting was lovely with the full ocean view. Our recommendation is that they move the stage so that the ocean and sunset form a backdrop for the stage. As it is now, you have the view to your back while the show is performed. The performers were good, although not the best. Audience entertainment was kept to a minimum with a couple of women and men volunteering to try their hand (or rather hips) at the hula. $38 adults, $21 for children. Phone 879-4900.

## TROPICAL PLANTATION BAR-B-Q
## SUNSET HAYRIDE & COUNTRY PARTY
Each Monday and Wednesday the charcoals are fired up for an all-you-can-eat BBQ feast. Grilled steaks, BBQ chicken, chili, corn on the cob, corn bread, fresh fruits, and a dessert table which includes a variety of pies. Tropical drinks, fruit punch, or iced tea. Entertainment includes plantation cowboys and cowgirls who sing and dance to some Hawaiian and country favorites, as well as a tour of the plantation. Adults $38, children $28, small fee for roundtrip transportation from Lahaina, Kaanapali, Kihei, Wailea and Kahului. Phone 244-7643.

## PRIVATE GROUP LUAUS
Toll free 1-800-367-8047 ext. 540 U.S. Maui Activity Guide, PO Box 1209, Lahaina, Maui, HI 96761. (808) 244-1586. Family gatherings, weddings or whatever, Private Group Luaus can handle catering for twenty or more and with or without a luau show. Done on your property or they can also arrange space.

***DID YOU KNOW?*** Luaus are definitely not low-calorie dining options. So eat and enjoy, but just in case you are interested, here is the breakdown! Kalua Pig 1/2 cup 150 calories, Lomi Lomi salmon 1/2 cup 87 calories, Poi 1 cup 161 calories (but who could eat that much!), fried rice 1 cup 200 calories, fish (depending on type served) 150-250 calories, chicken long rice 283 calories, haupia 128 calories, coconut cake 200-350 calories, Mai Tai 302 calories, Pina Colada 252 calories, fruit punch 140 calories, Blue Hawaii 260 calories, Chi Chi 190 calories.

GRATED RIPE COCONUT

# LAHAINA

## INEXPENSIVE

### AMILIO'S DELICATESSEN *American*
Lahaina Square (above Lahaina Shopping Center - north end of town) (661-8551). HOURS: Breakfast/lunch and dinner. COMMENTS: Take out or eat in. Deli sandwiches, plate lunches and pizza.

### THE BAKERY ★ *French/American*
911 Limahana (turn off Honoapiilani Hwy. by Pizza Hut) (667-9062). HOURS: Mon.-Sat. 7 am-5 pm, Sun. 7 am-noon. SAMPLING: Whole wheat cream cheese croissants also ham, or turkey stuffed croissants, or small sandwiches such as turkey dijon. Huge fresh fruit tortes, fudge, and fresh breads are made here daily. COMMENTS: There is no seating area. Arrive early in the day to insure getting the best selections. It's well worth the stop if you are a lover of pastries. The selection is magnificent and tasty too!

### BELLA PIZZA *Italian*
930 Wainee Street, Lahaina (661-8128). HOURS: 8 am-11 pm

### BLACKIE'S BAR *Mexican*
On Hwy. 30 between Lahaina and Kaanapali (667-7979). Look for the tree house structure with the orange roof. HOURS: Daily, 10 am-10 pm SAMPLING: Burritos, enchiladas, tacos, nachos, smoked hot dogs, good burgers and Louisiana hot links. COMMENTS: Incredible photos of boat wrecks and memorablilia line the walls. Jazz on Fri. and Sun., 5-8 pm. Not a family restaurant.

### BLUE LAGOON *American*
Located on the lower level of the Wharf Shopping Center (661-4181). HOURS: 11 am-9:30 pm. SAMPLING: Sandwiches include prime rib, BLT, Turkey or club $4.95 - $5.95, soup runs $1.95 - $2,95, entrees offer fish and chips, BBQ ribs, shrimp platter, cajun mahi $6.95 - $14.95 and include rice or potato or fries and Texas toast. COMMENTS: This is a "self-serve" style restaurant. Place your order at the counter and pick it up on a tray and find a table in the courtyard.

### BURGER KING *American*
South end of Lahaina by the Banyan Tree. HOURS: 6:30 am-1 pm serving breakfast, lunch, and dinner. SAMPLING: Breakfast sandwiches, salad bar, and the usual burgers at prices slightly higher than the mainland.

### CHEESEBURGER IN PARADISE ★ *American*
811 Front St., Lahaina (661-4855) HOURS: Breakfast 7:30-10:30 am. Lunch/dinner served 11 am-11 pm SAMPLING: A new addition is their breakfast menu which offers an array of omelettes, macadamia nutcakes, French toast and Superbrowns (hashbrowns topped with bacon and cheese). For lunch select from the Ahi Grille, Akamai Sandwich (cheddar and cream cheese, grilled onions and mushrooms on a whole wheat roll), chicken or green salads, BLT and of course cheeseburgers $5-7, fries $3. COMMENTS: A casual dining atmosphere located

in the upstairs loft-like setting with open-air dining and wonderful views of the Lahaina Harbor and Front Street. Very good hamburgers, just figure you are paying extra for the view. Future plans for outlets in Kihei, Kona, Honolulu and other mainland paradise spots.

**CHRIS'S SMOKEHOUSE** ★ *American*
New location on Honoapiilani Hwy. near Pizza hut. (667-2111). HOURS: Lunch 11 am-4 pm; dinner 5-10 pm. SAMPLING: BBQ ribs, steak, fish, and chicken. Also take-out. Same menu as in their former Lahaina Square Shopping Center location. Ribs, chicken, N.Y. steak charbroiled fish $9.95 - $13.50. May be opening for breakfast in the near future.

**CHRISTINE'S FAMILY RESTAURANT** *American/Local*
At Lahaina Square (661-4156). HOURS: Tues.-Sat. 7 am-9:30 pm, Sun. and Mon. 7 am-2 pm. Breakfast served anytime, lunch and dinner menus are the same. SAMPLING: Plate lunches $5.50 - $7.25, burgers from $4.50, sandwiches from $2.60, dinners $9 - $10. COMMENTS: Very popular spot for local residents.

**COLONEL SANDERS' CHICKEN** *American*
Lahaina Shopping Center (661-3422).

**COUNTRY KITCHEN** ★ *European*
Lahaina Square Shopping Center (near Foodland) (661-3330). HOURS: Lunch 11-2, dinner 5-9 pm Mon.-Sat. SAMPING: Complete dinners $13.95 include soup and salad (green bean-egg), tea or coffee and dessert with a choice of two entrees nightly that might include osso bucco, roast duck, roasted baby pheasant, prime rib, fish, bouillabaise. On the lighter side or for lunch there is a kula or country salad, ($3.50 - $4.95), fresh mussels ($6.95) pasta ($5), seafood platter ($7.50) or kona top sirloin ($11.95) COMMENTS: This very unpretentious establishment, tucked away in the shopping center above the Lahaina Shopping Center, is being discovered like wild fire. Not only are the dinner prices incredible, but the food is superb. In addition to the complete dinners you can stop in for a baked Alaska flambe cognac for four ($18)! Because of its limited seating and rapidly growing popularity, reservations are a good idea.

**DENNY'S** *American*
Lahaina Square Shopping Center (667-9878). HOURS: 24 hours a day. SAMPLING: Breakfast served anytime. Burgers and local style dishes. Dinners include steak, seafood and chicken $6.79 - $10.79.

**GOLDEN PALACE** *Chinese*
Lahaina Shopping Center (661-3126). HOURS: Lunch 11-2, dinner 5-9. Also take out. SAMPLING: A large variety of selections with Chinese and some Szechuan dishes. Affordably priced from $4 - $11. COMMENTS: Beer, cocktails, and Chinese wine are also available.

**GREAT JANE'S EATERY** *American*
888 Wainee St., Lahaina (661-5733). HOURS: Mon.-Fri. 7-4, Sat. 7-2, closed on Sun. SAMPLING: Breakfast offers only a bagel or wheat toast. Deli style sandwiches and salads. COMMENTS: 4 tables outside, 2 inside. Food to go.

### HAPPY DAYS ★ *American*
Lahainaluna Street a block up from Front St. (667-6994). HOURS: Breakfast 7-11, lunch/dinner 11-9 pm. SAMPLING: 1/4 or 1/3 pound burgers, hot dogs, soup and sandwiches from $3.95. Homemade soup. COMMENTS: As its name implies, this place takes one back to the 1950's. Their fountain serves up favorites such as "real flavored" cokes, egg creams, hand-dipped milkshakes and malts. Yum!

### HARD ROCK CAFE *American*
Lahaina Center. Scheduled opening summer 1990.

### HARPOONER'S LANAI ★ *American/Local*
On Front St., wharfside at the Pioneer Inn, (661-3636). HOURS: Daily, breakfast 7-11 am, lunch 11:30 am-2 pm. SAMPLING: Eggs with ham, links, Portuguese sausage, or bacon for $4.25. Coconut, banana, macadamia nut, or blueberry pancakes at $3.75. French toast runs $3.25. Lunches might include Portuguese bean soup at $2.50, or a hamburger at $4.25. COMMENTS: Our usual order is their French toast which is four half-slices of Portuguese Sweet Bread that are thick and custardy. The atmosphere is rustic, casual, and makes one feel almost as if the clock had been turned back.

### LAHAINA BROILER *Seafood/American*
887 Front Street (661-3111) HOURS: Breakfast 7-10 am, Lunch 11-2, dinner 5-8:30 pm SAMPLING: Lunch sandwiches, salads or entrees $3.95 - $9.25. Dinner entrees served with seafood chowder or bean soup or tossed salad. Chicken curry or charbroiled sirloin, cajun shrimp, salmon steak, NY steak $12.75 - $17.95. COMMENTS: New renovations have opened up this wonderful oceanfront setting. Done in hues of grey, green and mauve, several readers report that new isn't always better and they miss not only the old atmosphere, but the old food and service. Cocktails.

### LAHAINA TREEHOUSE AND SEAFOOD BAR RESTAURANT
Lahaina Marketplace on Front St. (667-9224). HOURS: Lunch 11 am-5 pm SAMPLING: Salads, sandwiches, entrees $4.95 - $7.95. COMMENTS: Pupus such as hot wings and spring rolls also available $1.95 - $5.95.

### LANI'S PANCAKE COTTAGE ★ *American*
Wharf Shopping Center, across from the Banyan Tree (661-0955). HOURS: Breakfast and lunch 6:30 am-3:30 pm. SAMPLING: In addition to the popular breakfast fare they offer entrees such as fish and chips or chili burgers $4.50 - $6.50. COMMENTS: Very busy during breakfast hours, offering seating indoors or on the patio.

### MARIE CALLENDAR *American*
Located at the Cannery Shopping Center, Kaanapali side of Lahaina (667-7437). HOURS: 7 am-10 pm Mon.-Thurs. and Sun. Until 11 pm on Fri. and Sat. SAMPLING: Breakfast omelettes served all day from $6 - $7.50 or choose a fresh fruit waffles. Lunches include pasta, hamburgers, salads, and sandwiches. Dinner entrees begins daily at 4 pm, priced $10.95 - $15.95 include teri steak, ribs, chicken or meatloaf. Daily specials. COMMENTS: Great pies, but the service is often slow, and the food nothing special.

## MCDONALD'S   *American*
Located at Lahaina Shopping Center. HOURS: Open for breakfast, lunch, and dinner. SAMPLING: The usual for McDonald's with a few added items such as Saimin. COMMENTS: Indoor eating and also a drive-through.

## MOOSE McGILLYCUDDY'S   *American*
844 Front St., upper level of Mariner's Alley, a small shopping alley at the north end of town (667-7758). HOURS: Breakfast from 7:30 am-11 am., also lunch and dinner. SAMPLING: The really big eater may want to tackle a moose omelette consisting of 12 eggs with two types of cheese, bacon, sausage, mushrooms, sprouts, spinach, and onion for $16.95. Regular omelettes $5.95. Half dozen types of burgers $4.25 - $7.50. Pizza's, salads, sandwiches, Mexican selections and a kid's menu. Ice cream and non-alcoholic drinks. COMMENTS: A very affordable, not overly exotic atmosphere. Moose operates an alcohol awareness service and offers a designated driver program. The selected individual wears a special badge and gets served free non-alcoholic drinks. Very popular place for their breakfasts, $2.69 for their early bird special of eggs, bacon, potatoes, rice or toast and orange juice served 7:30 - 9 am. Their early bird dinners are just as popular. Evening music on the wild side. Large screen sports TV.

## PANCHO & LEFTY'S ★   *Mexican*
658 Front Street at the Wharf Shopping Center on the courtyard level in the rear (661-3956). HOURS: Daily 11 am-10 pm. COMMENTS: This is one of those unexpected surprises that you'd like to have happen more often. Located in the back corner, lower level of the Wharf Shopping Center, this site has been a different restaurant almost every year. The margaritas were the first surprise - flavorful and definitely not the usual watered down version. The tostadas and guacamole were both great. The Mexican hamburger was not only good, but might be up for a nomination as the island's best burger. The shopping center quiets down in the evening, so it is pretty much the local folk that find their way here. If you're hungry for something from south of the border, give this spot a try. Their house specialty is frajitas served with chicken, beef or seafood. Prices run $6.95 - $12.95.

## PIZZA HUT   *Italian*
127 Hinau, Lahaina, (661-3696).

## PIZZA PATIO   *Italian*
658 Front Street at the Wharf Shopping Center on the courtyard level in the rear (667-2523) HOURS: 11 am-10 pm. SAMPLING: Pizza available with standard, whole wheat or sourdough crusts beginning at $6.50 for a small chesse up to $18.95 for a 16" with everything. One of the best selections of pizza toppings to be found include everything from fresh maui pineapple to chopped clams, jalapeno rings and pine nuts or sunflower seeds. Calzones, pasta and specialty sandwiches such as Italian roast beef with bell peppers or homemade meatball. COMMENTS: Outdoor patio or indoor dining. Wine and beer available.

## SEASIDE INN ★   *Hawaiian/ethnic/local style*
Front St. across from the Cannery Shopping Center (661-7195). HOURS: Bentos available from 6:30 am, hot lunches and okazuya service from 10 am-2 pm.

Dinner 6-9:30 followed by Karoake entertainment. SAMPLING: Hot lunches from $4 with items changing daily. (Okazuya is a buffet style where you pay for each serving you choose). Bentos (plate lunches) run $4 - $5. COMMENTS: Jane Nagasako of Nagasako Supermarket, operates this oceanview location. One of the few places in Lahaina for a good local plate lunch.

### SIR WILFRED'S ESPRESSO CAFE    *French/American*
The Cannery Shopping Center, Kaanapali side of Lahaina (667-1941). HOURS: Breakfast until 11 with a lunch/dinner menu served until 9:30 pm. SAMPLING: Breakfasts include Eggs Sir Muffin, Eggs Florentine or Quiche priced $2.95 - $3.95. Daily lunch specials include sandwiches and salads from $3.55 - $4.95. COMMENTS: A very small, pleasant eatery. Great coffee and espresso. Gourmet coffees available for purchase.

### SONG'S    *Chinese/Hawaiian*
658 Front Street at the Wharf Shopping Center (667-1990). HOURS: Daily for lunch and dinner. SAMPLING: This is okazayu style buffet with chicken adobo, kim chee, stir fry vegetables and more. After viewing the selections in the display case, your plate is dished up for you. Only a single table near the door offers seating, or get yours to go. Saimin, lomi lomi, stuffed cabbage, prices $1.50 - $5.

### SOUTH SEAS/SNUG HARBOR    *American/seafood*
At Pioneer Inn, harborside (661-3636). HOURS: Open only for dinner. SAMPLING: Broil-your-own ground beef, 1/2 chicken, beef stir fry, or pork chops $9.95 - $18.95 served with salad bar, baked beans, rolls. COMMENTS: Seating at South Seas is poolside at the very rustic Pioneer Inn Hotel. The Snug Harbor has indoor seating at the same location, same menu. Very casual atmosphere and they do offer a children's menu.

### SUNRISE CAFE    *French*
693A Front St.,(661-3326). HOURS: 5:30 am-midnight. SAMPLING: Breakfasts around $5, lunch and dinner $6.95. Two selections daily might include chicken, fish or beef. Quiche, cheesecake, espresso. COMMENTS: This is a very small, quaint eatery with food available to go.

### TAKE HOME MAUI    *American*
143 Luakini (turn by the Baldwin House off Front Street) (661-7056). SAMPLING: Fresh fruit smoothies, toffuti, sandwiches ($4.25ish), ice cream and sodas in the freezer. Fruits and Hawaiian coffee are among the items to be shipped or taken home. COMMENTS: This is a fast food type stop, limited seating. Fruit smoothies are delicious!

### THAI CHEF    *Thai*
Lahaina Shopping Center (667-2814). HOURS: Lunch 11-2:30 Mon.-Fri., dinner 5-10 nightly. SAMPLING: Entrees such as Thai crisp noodles, green papaya salad and garlic squid are $5 - $10. COMMENTS: We were disappointed with the quality and size of the portions. The chicken pieces were mostly dark meat and not trimmed of gristle. None of the dishes we ordered were memorable and it was an expensive meal for the amount of food served.

**VILLAGE PIZZERIA**   *Italian*
At 505 Front St. (661-8112). HOURS: Lunch and dinner daily SAMPLING: Pizza is available in Neapolitan style (thin crust) or Sicilian (thick crust). A 14" plain pizza starts at $10, a combo of 4 items $14.25.

# MODERATE

**ALEX'S HOLE IN THE WALL**   *Italian*
834 Front St. (661-3197). HOURS: Lunch 11 am-2:30 pm Dinner 6-10 pm, Mon.-Sat. SAMPLING: For lunch it's salada, pasta or sandwiches in the $6 range. A variety of seafood, veal or chicken dishes and pasta entrees $16.00. Veal and scampi priced $23. Spaghetti, linguine and fettucini ala carte $7.50 - $13. Complete dinners include soup or salad and bread and butter. COMMENTS: They have made some menu changes and are now open for lunch. Our dinner was well prepared, although after a day on the beach the portions were a little small for hearty appetites. A popular place for local residents, but we tend to prefer our Maui dining to include some of those magnificent ocean vistas or warm sunsets instead of dark inside atmosphere. Located up a flight of stairs, difficult for anyone handicapped.

**BENIHANA**   *Japanese*
658 Front Street at the Wharf Shopping Center on the upper level. (667-2244). HOURS: Lunch 11:30-2 pm, dinner 5-10 pm. SAMPLING: Lunch selections include chicken, filet mignon, daily specials $7.25 - $9.50 or a bento lunch at $7.50. Complete dinners with chicken, steak or seafood $13.25 - $26.75. COMMENTS: This well known mainland chain has joined the ever increasing number of teppanyaki style restaurants on Maui. This method of cooking involves a great deal of flourish and preparation by a chef who cooks your meal at the table in front of you. Part of the price of the dinner is for the show!

**BETTINO'S**   *American/local*
At 505 Front St. (661-8810). HOURS: Breakfast begins at 7 am. Also open for lunch and dinner. Bar opens at 8 am. SAMPLING: Lunches, including sandwiches and burgers are priced $4.50 - $6. Dinners of Mahi Mahi, chicken Mediterranean or short ribs run $13 - $20. Childrens menu $6.95 - $8.95. COMMENTS: This is a little off the tourist track, and a no frills type of place, but who needs them with a window table right next to the ocean. One of their specialties, Shrimp Bettino offered an abundance of huge shrimp and plenty of pasta for $16.95. The ribs and chicken parmesan were equally good dinner values. Our readers report the best Eggs Benedict ever here too!

**CHART HOUSE** ★   *American*
1450 Front St. (661-0937). HOURS: Dinner daily 5:00-9:30 pm SAMPLING: Fresh fish, chicken, top sirloin, scallops, teriyaki beef kabobs, $13.95 - $32.95. COMMENTS: Entrees are served with an unusual, all you can eat salad bar. A large bowl filled with mixed greens, tomatoes, onions, eggs and cucumbers is served at your table along with smaller bowls filled with croutons, fresh pineapple and several kinds of homemade dressing. If it's not enough, ask for more. The bread is fresh and homemade with a choice of white sourdough or molasses

"squaw" bread. There is a comfortable open air atmosphere with lots of wood and lava rock. The limited number of oceanview tables are a hot commodity and require that you arrive when they open. They also have a waitress who stops by to take orders from their seafood appetizer bar. Potatoes and rice are available at an extra charge. They provide one of the best keiki menus we've seen. The menu serves as a coloring book and a small pack of crayons are thoughtfully provided. The children's selections include four very reasonable dinners, $2.95 for ground sirloin, chicken teriyaki or beef kabobs at $4.95, and a small sirloin steak at $6.45. The entree portions were large. Included with the meal was salad, rice or potato and a soft drink. A paradise for parents! The adult entree prices may be slightly higher than other restaurants, but the cut of prime rib was enormous and the fresh fish portion very ample. They don't take reservations so there can be a long wait. They also offer another restaurant in Kahului.

**HARBOR FRONT** *International/seafood*
Wharf Shopping Center, top level (667-7822). HOURS: Lunch 11:30-2 pm weekdays, dinner 5-10 pm daily. SAMPLING: Fish and chips for lunch at $6.95 or seafood combo at $10.95. Dinner entrees offer fish and chips, chicken piccatta or bombay, peppersteak flambe $10.95 - $21.95. COMMENTS: No harbor view here, and the eating is inside. We have heard good things, but on the occasions we've visited, we were never too impressed. For a fun dessert they prepare bananas flambe and cherries jubilee at your table.

**KIMO'S** ★ *American/seafood*
845 Front St. (661-4811). HOURS: Lunch 11:30 am-2:30 pm. Dinner daily 5-10:30 pm. SAMPLING: Lunch offers grilled roast beef with cheese, burgers and salads $5.95 - $7.95. Their fresh fish varies daily as does its price. Baked scallops, koloa pork ribs, shrimp Tahitian or top sirloin $11.95- $15.95. Keiki dinners, or for less hungry adults, include hamburgers and chicken sandwiches priced $5.50 - $5.95. COMMENTS: No reservations, so plan on a wait. Put in your name, have the hostess give you a time frame, and just enjoy prowling the Lahaina shops. They have a waterfront location and, if you're really lucky, you'll get a table with a view. Opinions vary greatly about Kimo's. Some really like it and others really don't. Our experience has been very good service and well prepared fresh fish. They also have a bar on the lower level that serves a young crowd and an ocean view for sunset.

**KOBE'S STEAK HOUSE** ★ *Japanese*
136 Dickenson (667-5555). HOURS: Dinner from 5:30. SAMPLING: Teriyaki chicken $10.90, Teppan shrimp $16.90, lobster $23.00, sushi bar. Dinners include soup, shrimp appetizer, vegetables, rice, and green tea ice cream. COMMENTS: A sister of the Palm Springs and Honolulu restaurants, they offer teppan cooking (food is prepared on the grill in front of you) and the show is as good as the meal. They have a small parking lot adjacent to the building. Keiki menu. Reservations are recommended.

**LAHAINA COOLERS** *American*
180 Dickenson St. (661-7082). HOURS: Breakfast 7-12 am lunch/dinner noon-12 pm. SAMPLING: Pizza (single size), fettucini $8 - $10. COMMENTS: They have a great slogan, "Because Life is Too Short to Eat Boring Food." While it is

true enough that their menu isn't boring we have a problem with the price per serving. (A single serving pizza for $9 - $10? Not even on Maui!) The breakfast includes omelettes or "exotic" pancakes $3 - $3.50. The lunch and dinner menu is the same, however, a few luncheon specials and a pizza/salad combination make lunch a better option than dinner. Full bar service and tables inside or out.

**LONGHI'S ★** *French/Continental*
888 Front St. (667-2288). HOURS: 7:30 am-10 pm; dessert served until 11 pm. SAMPLING: The menu is given orally by the waiter. A possible selection might include zucchini frittata, filet mignon with bernaise or prawns amaretto and save room for the desserts. A'la carte dinners run $16 - $22 for seafood, meat and poultry, slightly less for pasta dishes. Salads and fresh vegetables $4 - $12. COMMENTS: This is another of those Maui restaurants that people either love or they hate. A consumer report was done on Maui dining by a local magazine a couple of years back and Longhi's came in third in the category of "best service" and first in the category "worst service." In any case, they have become a near legend in Lahaina. The setting is casual, with lots of windows open to the bustling Lahaina streets. Personally, we have enjoyed our meals here, although the oral menu can sometimes mean you spend a little more than planned! An early breakfast before the visitors arrive in Lahaina is a great way to start the day. They offer espresso and a good wine selection. Valet parking nightly.

**MOLLY B'S** *Homestyle*
1312 Front St. (667-5558) HOURS: Breakfast, lunch and dinner 7 am - midnight daily. SAMPLING: Homestyle cuisine includes pot roast or meatloaf entrees served with real mashed potatoes. Dinners include soup or salad, starch and vegetable. Priced $10 - $18. COMMENTS: Molly Wilson has merged her successful catering business, An Absolute Affair, with this newly remodeled restaurant. While many new Lahaina restaurants are jumping into the Asian/Pacific cuisine, Molly's is taking a unique step back to good old fashioned homestyle dining in a comfortable atmosphere. Private parties. Catering services. Entertainment. Check for early bird dinner specials! Opened too late to be reviewed.

**MUSASHI** *Japanese*
Lahaina Shopping Center (667-6207). HOURS: Lunch 11 am-2 pm, dinner 5-10 pm. SAMPLING: Beef or chicken teriyaki, sukiyaki, nabe. Lunches run $5.25 - $10. Dinners include soup, side dish, pickles and rice with chicken or salmon teriyaki, catch of the day, or Unaju (barbecued eel over rice) $11.95 - $16.55.

**OCEANHOUSE and SEAFOOD CAFE** *American/Seafood*
831 Front St. (661-3359). HOURS: Seafood cafe serves lunch 11 am-3 pm and dinner 4 pm-9 pm. Oceanhouse serves dinner only 6 pm-10 pm. SAMPLING: Dinner features a large array of seafoods, meats and poultry. Samples include BBQ ribs, veal piccata, baked stuffed prawns, and cioppino $17.95 - $32.95 served with chef's "daily selections." Upstairs is the more casual Seafood Cafe which offers a limited selection of sushi, salads, appetizers, pasta and pizzas $5.75 - $9.95 and entrees, served with rice or potatoes, vegetable and bread include fresh catch of the day, fish and chips, several chicken and steak selections. COMMENTS: Located in the middle of Lahaina with a tremendous ocean view. The Seafood Cafe is especially nice for an evening cocktail, sunset and pupus.

## OLD LAHAINA CAFE    *Hawaiian*

505 Front Street. (661-3303) HOURS: Lunch 11-3. Dinner 5:30-10 pm. SAMPLING: Lunch offers sandwiches such as Portugese sausage, fresh fish or seafood melt along with daily specials that might include fried calamari sandwich or teriyaki filet $5.95 - $7.95. In addition to the limited dinner menu of stir fried chicken, filet mignon and fresh fish they offer daily dinner specials which might include shrimp scampi, baked stuffed chicken or scallops marinara $9.95 - $15.95. COMMENTS: Nothing "old" about this place. Located on the beachfront with a breezy atmosphere. Their luau plate is very popular. Operated by the same folks that run the beachfront luau.

## SAM'S PUB/SAM'S UPSTAIRS    *American/Seafood*

505 Front Street (667-4341) HOURS: Dinner 5:30-10 pm. SAMPLING: Sam's is really two restaurants in one. Sam's Pub is downstairs and offers light dining of hamburgers, Greek salad, fish and chips, 11 am-midnight. Sam's Upstairs offers more complete meals such as chicken in herb butter, veal chop or scampi $19.95. COMMENTS: From Sam's Upstairs you can get a great view of the imu ceremony below at the Old Lahaina Luau! They currently offer a special two for one dinner on Tuesdays.

## TASCA ★    *Mediterranean*

608 Front St. (661-8001), HOURS:11:30 am-midnight. SAMPLING: A variety of dishes in mini, medium or maximum size portions. COMMENTS: The three operators bring with them a wealth of experience from some of Maui's best. Michel Roy spent eight years at the El Crab Catcher and brother Richard Gagliardi (both originally from Montreal) was at Chez Paul for five years. The third partner is Gilles Sandras from Tahiti who was a part of the short-lived Champagne restaurant at this same location.

As they explain it, "Somewhere between the hors d'oeuvres and the full course meal - lies the culinary kingdom of *Tapas*. Those bits and bites served in Spanish bars that have become increasingly popular in cities across the United States. This idea began long ago in a Spanish cafe, the community gathering place. A place to share tales and ales. To protect your drink from dirt and dust, the bartender would put a top on it, or tapa. On that Tapa he would then put some of the bars favorite snacks for you to nibble on. Herein lies the concept for Tasca; certainly not a new idea, but one worth recreating."

The front of this restaurant is deceiving since it stretches in a long and narrow path to the back courtyard. Choose a quiet seat in the garden or one in the front to watch the passers-by and ponder the selections. Oysters on the half shell, salade Nicoise, Escargot, saffron chicken, baby back ribs, paella, ratatouille, fettucini Alfredo or broiled ahi. All are offered in medium and large portions, and some, such as prawns Mediterranean or breaded calamari are in small portions. Mix and match to create a balance for you and your group. We can recommend the opakapaka, the ribs, the chicken and a few others!   Each dish is delicately seasoned with an interesting blend of spices and herbs. Large portions run $10.95 - $19.95 for a single item. Not the place to come with a small pocket book and a huge appetite, but a delightful atmosphere and some delicious foods.

## THE TIGER  *Korean*

730 Front St., (661-0112). Dinner served 6 pm-12 am daily. SAMPLING: Marinated beef, seafood or chicken is brought to your table for a cook-it-yourself meal. Prices run from $16.75 to a dinner for two at $39.75. Side orders for exotic tastes include oxtail and rice soup or grilled tongue. COMMENTS: Downstairs in the Tigers Den you'll find lunch plates and sandwiches priced in the $5 range from 11 am-2 pm and live entertainment from 10 pm.

## WHALE'S TALE  *American/Seafood*

Across from the Pioneer Inn at 666 Front St., upstairs (661-3676). HOURS: Lunch served from 11:30-2:30, dinner from 5:30 pm. SAMPLING: Lunches run $4.95 - 9.95 for an assortment of sandwiches, salads and fresh fish. Dinner selections include fish, ribs, chicken and steak priced $10.95 - $15.95. COMMENTS: Entertainment daily from 4 pm until closing. Reservations accepted.

# EXPENSIVE

## AVALON ★  *Contemporary Asian*

844 Front St., (667-5559). HOURS: Daily 12-12. SAMPLING: A number of the dishes may be prepared to your liking, mild, medium or spicy. In addition to daily chef's specials there is fresh clams, steamed with garlic black bean sauce $17.95, Indonesian chicken stir-fry $19.95, and giant prawns, Eastern scallops or fresh fish prepared five different ways ($19.95 - $22.95, fish at market price). Side orders of fresh cut shoestring fries or brown rice is available. One of many good reasons to come to Avalon is for salad. Not just any ordinary salad mind you, but something special. The Gado Gado salad ($9.95) comes from the island of Bali and is a tasteful blend of romaine lettuce, cucumbers, tofu, steamed vegetables on a bed of brown rice and topped with peanut sauce. Equally delicious is the Chinese tofu salad at $8.95. Both are huge portions. And, if you, dare some Caramel Miranda for dessert. COMMENTS: The name Avalon, according to Celtic or Gaelic legend, is the West Pacific island paradise where King Arthur and other heroes went following death. The look here is 40's Hawaiian with antique aloha shirts adorning the walls, ceiling fans whirring and wonderful multi-colored oversized dishes. This one is a favorite of ours and we don't visit Maui without stopping by at least once!

## CHEZ PAUL ★  *French*

Five miles south of Lahaina at Olowalu (661-3843). HOURS: Dinner 5:30-10:30 pm. SAMPLING: Dinners run $18.95 for a vegetarian fare to $24.95 for most selections. The menu includes veal, fish, duck, and beef in wonderful French sauces. Dinners include French bread, soup or salad and two vegetables. Pate, escargot, and shrimp are available as appetizers. Save room for some very special desserts. COMMENTS: This small restaurant has maintained a high popularity with excellent food and service. It's not surprising they have won numerous dining awards. The wine list is excellent, although expensive. Wines are also available by the glass. Two seatings for dinner are offered, 6:30 or 8:30. Reservations are a must. Visa and Mastercard are accepted.

## DAVID PAUL'S LAHAINA GRILL ★ *New American*

127 Lahainaluna at the Lahaina Hotel. (667-5117) SAMPLING: The menu is described as New American Grill Cuisine with a Southwestern flair and offers "yakiniko" style dining or cooking it at your table. The menu changes, but you might enjoy a salad of Peking roasted squab ($12), appetizers such as fire sticks ($14.95) these are also known as "Southwestern Yakiniku" served with chicken, beef, shrimp or fish, marinated in spicy oils and then grilled over Kiawe wood and served with lobster relish. Soups ($7- $9) such as carrot with saffron. And a la carte entrees that range from grilled free range chicken breast to tequila shrimp with fire cracker rice or Norwegian salmon baked in corn husks ($19.95 - $35). COMMENTS: David Paul was formerly with the Black Orchid on O'ahu and chef at Chez Michelle and the Hyatt Regency on O'ahu. Opening on the verge of our going to press we sent our roving island reporter, Joy, to check out this newest arrival to Lahaina. She was delighted! Sampling the Ahi Carpaccio she reports that the cilantro was the perfect touch in this delicate dish which was served with corn bread sticks and herb butter. The spicy crab cakes were an adventure in taste and texture and she and her escort emitted frequent "Ahhh's", and "Ummm's". The fire sticks served on a miniature grill were also excellent and the fish was the best she had ever had. In addition, no details have been overlooked in decor and the staff was fresh, energetic and smiling. Their menu states that "it is our goal to provide the finest food and service found anywhere in the Islands. We strive for perfection. Some items may take a little longer than usual, please be patient, sit back and relax. It is worth the wait." And on Maui, why rush anything! For a very special evening, don't overlook this new restaurant! It is top top on our list to try!

## GERARD'S ★ *French*

In the lobby of the Plantation Inn at 174 Lahainaluna Rd. (661-8939). HOURS: Dinner only. SAMPLING: Dinner selections in the $15 - $25 range. COMMENTS: Although partial to Maui's sunset ocean views while dining, sitting beneath a mango tree on the veranda of the Plantation Inn while dining at Gerard's is hard to beat. Just a short walk up Lahainaluna Road off Front Street in Lahaina, you are swiftly moved worlds away to a picturesque atmosphere reminiscent of a Gone with the Wind era. Crisply attired in green and peach the restaurant offers indoor or outdoor seating. A wine list features a range of

moderate to expensive selections from California, France and the Pacific Northwest. On a recent evening we began with a wonderful rich crab bisque followed by Ulua, the fresh fish and the evening special, a calamari with pasta dish. Our guidelines for a restaurant review center around the fresh fish. In other words a restaurant is as good as its fish! And Gerard's did not disappoint. The entrees arrived, complimented with miniature vegetables. The calamari was a healthy portion in a wonderfully seasoned sauce. The fish was prepared with a spinach and bernaise sauce and not only was it excellent, but it was the best fish meal in recent memory and a good size filet too! The desserts, wonderful of course, vary nightly. Try the chocolate decadence complimented by a cafe au lait or a cappucino. The perfect evening was rounded out with background guitar music by talented Corey Carson who plays three nights a week. (By the by, his father just happens to be Johnny Carson.) Gerard's was a winner of the 1989 Paradise Publications Best Bet award for fine dining.

**LA BRETAGNE RESTAURANT FRANCAIS ★** *Contemporary French*
3 Mokuhinia Place, off Front St. between Banyan Tree and 505 Front St. (661-8966). HOURS: Dinner only 6-10 pm. SAMPLING: The menu is off to a great start with a tantalizing selection of appetizers $7 - $11, and soups or salads $4 - $10. The house specialties include rack of lamb with shallots, fresh seafood in puff pastry (delicious!), wild mushrooms Provencale. Entrees run $17 - $27 and are served with fresh vegetables and fresh French bread. A wine list offers a limited number of popular French and American selections by the bottle or glass. COMMENTS: The little country house was built in 1920 by Sheriff Kaluakini who served as King Kamehameha's right hand man. The home opened for dining in 1978 and intimately seats only 50. It's a little tricky to find off Front Street, but your search will be rewarded. The flowered upholstery on the chairs and the brocade wall coverings all add to the wonderful atmosphere here. Combined with personable service at a relaxed pace, it feels like you are dining in the home of a friend. In the kitchen you'll find owner/chef Claude Gaty who prides himself on his self-taught cooking ability. Born and raised in France, resettled in the U.S. and after a 1983 vacation to Maui returned to stay. Reservations are highly recommended, and while dress is casual, shorts aren't advised.

**STARDANCER** *American*
The 150-foot yacht floating in the water off West Maui. (871-1144) HOURS: Dinner only, two seatings six nights a week. COMMENTS: Accommodating up to 149 guests, they offer a complete buffet each evening. Style of preparations will change weekly but will include fresh local seafood and steaks or prime rib, along with an open bar. Two cruise seatings are offered at 5 pm and 7:30 pm ($50 per person). Then plenty of time for the guests to dance the night away to the sounds of Raw Silk with music of the 1940's to 1980's. If dancing isn't your style, there are plenty of panoramic ocean views, starlit nights and fresh air on their top deck. They aim to reach the slightly older crowd and elegance has been achieved with a striking art deco style. Shuttles depart from Lahaina Harbor, Pier 4. At 10 pm, Tues.-Sat., this elegant floating dinner club transforms itself into the High Spirits nightclub. Dinner guests are invited to stay on in the lounge, and non-diners can take the shuttle service which runs every 20 minutes (10 pm-2 am) and a $5 round trip fare is charged. COMMENTS: The boat was nearly new when it was purchased on the east coast from a dinner cruise operation and some major

money was spent in renovations. The interior, we were told, took six months to hand paint to form the mauve marble-like quality. We chose the first sitting which proved to be not only less crowded, but had the advantage of a Hawaiian winter sunset. While the food wasn't five star, it was well prepared and a far cry from any other dinner cruise we've experienced on Maui. (The carrot cake for dessert was GREAT.) Plus no wind in your face, no plastic tray to balance on your lap, no engine noise! The service was pleasant and prompt. The entertainment was a trio that played during the first hour of the cruise and then moved upstairs and there was plenty of room on that beautiful black marble dance floor! It's a pleasant and romantic evening and we can finally offer our readers a dining cruise that we can recommend.

# KAANAPALI

## INEXPENSIVE

### BEACH BAR    *American*
Westin Maui, located near the beach (661-8939). HOURS: 9-6 pm. SAMPLING: Hot dogs, sandwiches, or tropical fruit salad. Ice cream too!

### GARDEN BAR    *American*
Westin Maui. (667-2525) Located atop the center island of the pool complex. HOURS: 9 am-11 pm.

### KAANAPALI BEACH COFFEE SHOP    *American/Hawaiian*
Kaanapali Beach Hotel (661-0011). HOURS: Breakfast from 6 am; lunch from 11 am, and dinner from 5 pm. COMMENTS: Cafeteria style with $5 - $7 plate lunches. A good value, but lacking in atmosphere.

### KAU KAU BAR    *American*
Poolside at the Maui Marriott. HOURS: Breakfast 7-10 am, lunch and snacks 10-5 pm. SAMPLING: Coffee from 6 am, full lunch menu includes cheeseburgers and sandwiches.

### OHANA BAR AND GRILL    *American/Italian*
Embassy Suites Resort (661-2000). HOURS: 11 am-8 pm daily. SAMPLING: If you are a guest of the hotel, this is where you'll find your morning breakfast cooked to order. They also offer poolside lunch and dinner selections. Specialties from the grill include tiger shrimp in marsala wine $16.95, pizza and calzone from $8.95, Filet steak with pesto butter $17.95, chef's special pasta $10.95, fettucini Alfredo $8.95. Also a selection of wines.

### RICCO'S    *Italian*
Whaler's Village Shopping Center (661-4433). HOURS: 10-10 pm daily. SAMPLING: Deli dog $2.95, pasta dishes such as lasagne, fettucini $6.95 - 8.25, stromboli $7.25, calzone $6.95, small cheese pizza starts at $8 - 12 for a large size.

**ROYAL OCEAN TERRACE** ★ *American*
Royal Lahaina Hotel (661-3611). HOURS: Daily for breakfast lunch and dinner 6 am-10 pm. Sunday brunch served 9-2 pm. SAMPLING: A daily breakfast buffet is $10.50 or order off the menu. Champagne brunch on Sunday $19.50 features omelettes cooked to order, sushi, fresh pasta, tropical fruits, breakfast pastries, exotic breads and more. Lunch choices include sandwiches and salads $5.95 - $8.50. The Royal Ocean Terrace Food Fare is a smorgasbord of at least 25 salads and is $9.50 a la carte or $4 with any dinner. Winner of our Best Bet for Maui salad bars! Dinners include steak, chicken, lamb and seafood from $10.50 - $16.95. COMMENTS: This is a very attractive, airy restaurant. The night we dined their salad bar had 26 items and included lettuce or spinach salad, pasta salad, an excellent "new potato" salad, shrimp salad, fresh vegetable salad and more. To accompany your green salad, they offer the standard dressings as well as carafes of unusual oils and vinegars to combine on your own, i.e. grape seed oil, walnut oil, raspberry vinegar. The dessert bar tempts you on your arrival and you might want to save room for one of their special ice cream drinks! For the lighter eater they offer several sandwich selections as well.

**YAMI YOGURT** *American*
Whaler's Village (661-8843). SAMPLING: Sandwiches such as cheese and egg salad and garden or fruit salads in the $3 - $4 range. COMMENTS: No seating in the restaurant, but a few tables outside. Call ahead and order a picnic lunch.

# MODERATE

**BLACK ROCK TERRACE** *American*
Poolside at the Sheraton Hotel (661-0031). HOURS: Continental breakfast, lunch 11-3:30 pm, and dinner 5:30-9:30 pm. SAMPLING: Early bird dinners 5:30-6:30 pm nightly. Sandwiches begin about $6. Dinners include salad bar and run $14.50 - $21.50. Entrees include stuffed prawns, roasted pork loin island style, or coconut shrimp. Dinner size salads are available $8.95 - $9.95 or a selection of sandwiches and omelettes from $7.50. Children's portions available.

**CAFE KAANAPALI** *American*
Whalers Village (661-4944). HOURS: Breakfast 7 am-2:30 pm, lunch 11-4 pm, and dinner 4-10 pm. Light menu available late in the lounge. SAMPLING: Lunch menu available at dinner as well as a selection of complete dinners $7.95 - $12.95 such as NY steak, lime chicken or shrimp basket include french fries, rice or mashed potatos. Fresh fish specials nightly. Pasta, Mexican selections, sandwiches, salads and cafe burgers are complimented by a complete soda fountain. Sandwiches, salads and burgers are in the $6 - $9 range. COMMENTS: A moderately priced restaurant with a variety of menu selections which is a welcomed addition to Whaler's Village.

**CHICO'S** *Mexican*
Whaler's Village Shopping Center (667-2777). HOURS: Daily for lunch 11:30-2:30, taco bar 11:30-midnight and dinner 5-10:30. SAMPLING: Dinner combination from $9.50, fajitas from $9.95, fresh fish $16.95. Children's menu items $4.95. COMMENTS: A good variety of Mexican fare with their freshly made flour tortilla shells enticing as you enter.

### CHOPSTICKS    *Oriental*
Royal Lahaina Resort (661-3611). HOURS: Dinner nightly 6-9:30 pm. SAMPLING: Features Chinese, Japanese, and Thai foods. "The Place to Graze." Nutty chicken $5.95, sweet and sour pork $3.95, coconut shrimp, ginger beef, hot and spicy chicken wings $7.25. COMMENTS: So what is grazing? Chopsticks is the first restaurant on Maui to introduce the concept of grazing and it really is a wonderful idea for the adventurous eater. Items from Mainland China, Micronesia, Hong Kong and Thailand are served in appetizer size portions. The menu is a la carte and in the style of a passport. Each visit is validated and five validation stamps equals one complimentary selection. The light eater can be satisfied easily, while the more hearty appetite can pick and choose a banquet. This is a good opportunity to try something new and unusual, without being committed to an entire meal of it.

### COOK'S AT THE BEACH ★    *American*
Westin Maui, north side of the swimming pool (667-2525). HOURS: Breakfast 6:30-11, lunch and dinner 11-11 pm. SAMPLING: Egg dishes from $8.75, cereals, pancakes, waffles and bagels from $3.50. Lunch selections include two soups, four salads and sandwiches or burgers $7 - $10. Dinner offers pizza and homemade pasta, selections from the kiawe grill such as hamburgers, rib eye steak chops or sandwiches from the traditional club to a steak sandwich with Maui onions and Kula tomatoes $7.95 - $10.25. An all you can eat dinner buffet features a selection of salads, cold meats and hot entrees along with prime rib, and dessert for $19.75 except holidays. Vegetarian choices, wok cooked dishes and fresh fish range $8.75 - $16.25. A keiki (children's) menu available too! COMMENTS: An unusual opportunity to enjoy gourmet foods while dining in swimwear! A great family restaurant with a varied assortment certain to please everyone and fairly reasonable prices for a resort!

### EL CRAB CATCHER ★    *American/Seafood*
Whaler's Village Shopping Center (661-4423). It's a little hidden on the ocean side. HOURS: Lunch 11:30-3 pm, poolside pupus 3-10 pm and dinner 5:30-10:30 pm. SAMPLING: Lunch offers include French dip, burgers, salads and sandwiches. Dinner entrees are served with soup or salad, vegetable and bread basket. Fresh island fish and its preparation varies daily. Dinner selections include rack of lamb, scallops champagne, shrimp tempura, teri fish $17.95 - $19.95. Light diners can select the cajun mahi, grilled lamb salad, or prime rib sandwich $12.95 - $13.95. COMMENTS: This was once an athletic club and the pool remains beachside around which you can dine or enjoy a sunset and cool drink. Some have been known to even take a dip! They do a wonderful job with their fish preparation. Save room for one of their scrumptious desserts. This is a very popular place and reservations are a good idea unless you want to wait an hour or more for restaurant dining.

### LEILANI'S ★    *American/Seafood*
Whaler's Village Shopping Center, on the beach (661-4495). HOURS: Lunch 11:30-4, dinner 5-10:30 pupu bar 4 pm-midnight. SAMPLING: Ginger chicken $10.95, quarter pound cheeseburger $5.50, and Malaysian shrimp or double cut lamb chops $16.95. Fresh fish priced daily. COMMENTS: This restaurant is a branch of the Kimo's and the Kapalua Bar and Grill operations. They also offer

a limited children's menu, and some luscious desserts. They don't take reservations or personal checks, but do honor major credit cards. A reliable stop for a good meal. Their Seafood Bar is located on the lower level with lighter fare and is open to the beach.

**LOKAHI TERRACE**   *American/Buffet*
Embassy Suites Resort (661-2000). HOURS: Dinner only Tues.-Sat. 6 pm-9 pm. SAMPLING: A fixed price of $14.50 (plus tax and tip) at this ample dinner buffet featuring prime rib, chicken, and fresh fish. Children 6-12 $12, under age 5 free.

**LUIGI'S PASTA PIZZERIA**   *Italian*
Kaanapali Resort, by the golf course (661-3160). HOURS: Daily 11:30 am-midnight, bar until 2 am. SAMPLING: Lunch selections include pasta or pizza as well as sandwiches such as submarines, French dip or club. Dinner selections $8 - $17 include pizza, calzone, spaghetti, fettucini, scampi and steak.

**MANGO JONES**   *Hawaiian/American*
Kekaa Drive. 667-6847 HOURS: Dinner only, pupus from 4 pm. SAMPLING: From the "lava rock broiler" sample pepper steak, breast of chicken or veal chop. "Island house specials" include BBQ chicken, Azeka style ribs, Mango Jones curried chicken (cooked in a creamy coconut curry sauce), or seafood selections such as scampi, hamma jang seafood stew (shrimps, scallops, fish, mussels and vegetables simmered in Mango's own brew). Prices for full dinners $15.99 - $21.99. For desserts try the house special, a torch mango, which is a mango flambe (in season) or banana flambe, or Haleakala crepes featuring bananas, topped with macadamia nuts, ice cream, hot chocolate sauce and shredded coconut. COMMENTS: Formerly The Peacock restaurant, this hillside location overlooking the golf course and Kaanapali resorts was transformed into an Italian villa called La Scala. Shortly thereafter it was taken over by Jon Applegate, Maui restauranteur with three Luigi's restaurants. It opened too late for review, but if their food is prepared as creatively as their menu describes, this will be a welcome new addition to Kaanapali.

**MAUI ROSE**   *American/Seafood*
Embassy Suites Resort (661-2000).HOURS: Dinner 5:30-10:30 pm Sunday brunch 10 am-2 pm SAMPLING: Sunday brunch, $17.95 adults and $12 children price tag includes prime rib, mahi mahi, sashimi and sushi and pastry galore. The dinner side salads are priced $5.95 - $7.95, soups about $3, single serving pizza at $6, and other selections such as fresh fish, tiger shrimp, veal, rabbit, NY steak, prime rib and lobster are $15.95 - $26.95. COMMENTS: A gracious floral setting adjacent to Kaanapali Beach provides a relaxing eating environ. Our service was inept, but we noticed other tables receiving better service from other waithelp. We sampled the Ono, which can be difficult to prepare, and it was moist and tender. It was accompanied by miniature vegetables. The special evening salad was prettier than it was tasty. The island fish appetizer was very attractive and a pleasant surprise. The tiger prawns included five large ones and were very good. The soup of the day was an uninspired tomato. Sunday brunch is reportedly quite a nice affair and priced at a "suite" price of $17.95.

## MOANA TERRACE ★ *American/Buffet*

Maui Marriott Hotel (667-1200). HOURS: 6:30 am-11 pm.SAMPLING: Daily 6:30-11 am their breakfast buffet runs $11.95, $5 children 5-12, under five are free. Omelettes and eggs prepared to your liking and Belgian waffles. Items off the menu also. Following breakfast their all day dining menu is available. Lighter dining options include gourmet burgers ($9.75), cobb salad ($8.75) or pasta dishes about $9. A la carte entrees range from chicken piccata ($13.95) to hot rock shrimp and swordfish ($18.95).Lunch is 11:30-5 pm Mon.-Sat. with sandwiches and salads. They feature a nightly theme buffet served 5-9 pm with a discount "Sundowner Buffet Special" for early diners (5-6 pm). The evening buffets run $17.50 adults, $13.50 for early bird diners, and $6.50 for children 12 and under. Hawaiian fare is featured Fridays, seafood on Saturday, Sunday, Tuesday and Wednesday you'll find prime rib, Monday journey to China and Thursday to Italy. Complete dinners from the menu include soup or salad with your entree, which may include pasta primavera $14.20, chicken teriyaki $15.90, or prime rib $17.90. COMMENTS: Dinner and Sunday brunch reservations recommended. Their early bird buffet special is a great value, call to confirm the hours it is offered.

## MOBY DICK'S *American/Seafood*

At the entrance to the Royal Lahaina Resort (661-3611). HOURS: Dinner nightly 6-9:30 pm. SAMPLING: Seafood buffet Friday nights for $18.95 entices with shrimp, scallops, fresh fish, lobster, pasta and more. Dinners are served with salad or homemade clam chowder and the "Chef's accompaniments." Entrees include chicken tarragon $17, island prawns $21, fresh fish $20.95 - 21.95, and New York steak $18.50.

## NANATOMI *American/Japanese*

Kaanapali Golf Course Club House (667-7902). Lunch served 11-3, dinner and sushi bar 5-10 pm, karaoke 10 pm-1 am. SAMPLING: Lunches include tuna, turkey, ham, or beef sandwiches from $6 - $7 or kamaaina selections such as Yakisoba, chicken katsu, or beef curry. Dinners are served with miso soup, rice and tsukemono and include "All American" selections such as lobster tail, scampi or NY steak and Japanese favorites such as shabu shabu (beef and vegetables in a broth with dipping sauces), tempura selections and sushi dinners $12.95 - $25.95. Really hungry? Try their seven course dinner for two $77.13, includes dessert! COMMENTS: Keiki menu offers two choices at $6.95. Weekend entertainment and evening karaoke bar.

## PAVILLION *American*

Hyatt Regency Hotel, lower level (661-1234). HOURS: Daily for breakfast, lunch and dinner. SAMPLING: Breakfasts begin at $5.75 for griddle cakes, plain omelettes $6.50. Their all you can eat breakfast buffet served 6:30-11:30 am is $10.95 for adults, $5.95 children and features juices, fresh fruits, pastries, cereals, breakfast meats, egg dishes and tropical crepes. Luncheon selections offer hot or cold sandwiches $6.95 - $9.50, specialties such as shrimp tempura or an array of salads that range from fruit to Thai beef or Chinese chicken $6.95 - $14.50. Dinners offer several salad or sandwich choices, daily "wok" specials and Asian Pacific entrees such as kalbi ribs, Thai curried chicken or Indonesian nasi goring $9.95 - $15.75. COMMENTS: Located near the pool. Keiki dinner menu also available.

## THE RUSTY HARPOON ★ *American*

Whaler's Village (661-3123). HOURS: Daily breakfast 8-11 am, lunch 11:30-3, and dinner 5-10 pm, tavern menu served noon to midnight. SAMPLING: Dinner entrees include fresh fish, linguine, stir fry or teri chicken, Korean ribs, prime rib, steaks $11.95 - $17.50 and are served with rice or pasta and vegetable. Pizza $7.95, sushi $6.95 - $9.95, salads $2.95 - $8.95, burgers $7.50. Good selection of appetizers, try the bucket of clams! COMMENTS: Currently offering an early bird special 5-6 pm and a late afternoon special from 3-5 pm. This is a great spot for family dining with a diverse and affordable menu. Keiki menu available as well!

## TIKI TERRACE *American/Hawaiian*

Kaanapali Beach Hotel (661-0011). HOURS: Breakfast 7-11 am, Mon.-Sat. Dinner 6-9:30 pm, champagne brunch served Sundays 9 am-2 pm. SAMPLING: Breast of chicken waina, fresh catch from the Kai (preparations change daily), Scampi a la Lahaina, price range $15 - $21. Soup or salad included with all entrees. COMMENTS: Complimentary hula show nightly with Hawaiian entertainment throughout the evening while you dine.

## VILLA RESTAURANT ★ *American/Seafood*

Westin Maui (667-2525). HOURS: Dinner only 6-10 pm. SAMPLING: Seafood is the specialty! Dinners run $20 - $26, steak and lobster $38. Their daily fresh catch is prepared grilled, sauteed or poached and served with clam chowder or Hawaiian field greens, fresh vegetables and potato or double fried rice. Type of fish is subject to availability, priced $22 - $23. Other entrees are a la carte and include filet mignon, NY steak, sauteed prawns, chicken breast. COMMENTS: One of our best bets for fresh island fish and a beautiful setting. All tables look out onto the lagoon where swans, and exotic ducks float along peacefully.

# EXPENSIVE

## DISCOVERY ROOM *Continental/American*

Sheraton Hotel, atop picturesque Black Rock (661-0031). HOURS: Breakfast 6:30-10:30. Dinner daily. SAMPLING: Breakfast buffet is served until 10:30 a la carte service until 11 am. Dinner menu offers are limited and pretty standard with Chicken Kaanapali at $28.50, teriyaki steak $31.50, prime rib $29.95. On Tuesday evenings they feature an international buffet dinner. First dinner seating begins at 6:30 pm and seating continues until about 8 pm. Dinners include appetizers, soup or salad and tea or coffee. A $5 charge for the Polynesian show, Maui On My Mind is added onto the entree price. The show begins at 8:30 and lasts approximately one hour. Non-diners can pay $15 which includes two drinks.

## LAHAINA PROVISION ★ *American*

Hyatt Regency Hotel (661-1234). HOURS: Lunch 11:30-3 pm and dinner 6-10:30 pm. The lounge is open from 11:30 am-11:30 pm. SAMPLING: Luncheon items run $7 - $12 for sandwiches or hot entrees. Dinner selections include fresh fish, T-bone steak or lamb chops $12 - $28. COMMENTS: This restaurant is cleverly perched above the pool and on the edge of one of the Hyatt's waterfalls. This place may be a favorite if you're a chocolate lover. They have a CHOCOHOLIC BAR (served 6-11 pm) that features rich ice cream with an incredible choice of terrific temptations to top it. If you have dinner, it's an additional $3.95, but you can come later for dessert only and indulge for $5.95. Reservations are recommended for dinner or dessert. Major credit cards are accepted.

### LOKELANI ★ *American/Seafood*
Maui Marriott Hotel (667-1200). HOURS: Dinner served 5:30-10 pm. SAMPLING: All of the fish is market priced daily and your selection may be sauteed or charbroiled with lemon butter or fine herb sauce or the Chef's sauce of the day. ($22.95 - $25.95) Other choices include sauteed scallops curacao ($22.95), or shrimp and scallop fettucini ($20.95) All selections are served with fresh bread, cup of seafood chowder or a garden salad. Scampi, sauteed fresh fish, scallops, or crab legs share the menu here. Dinners include salad, cup of soup, and freshly baked bread. Prices run $15.95 - $19.95. COMMENTS: The food was well prepared. An after dinner delight was the coffee, which was accompanied by a tray of wonderful additions - raw sugar, cinnamon sticks, anise covered M & M's, chocolate chips, lemon peel, and fresh whipped cream. They are currently featuring an early bird dinner special, 6-7 pm, $14.95 for entrees which include fresh fish or prime rib. Reservations recommended. Major credit cards accepted.

### NIKKO'S JAPANESE STEAK HOUSE *Japanese*
At the Maui Marriott Hotel (667-1200). HOURS: Dinner only. SAMPLING: Samari Sunset Menu is served only from 6 pm-6:30 pm daily and prices are considerably less, for example chicken ($12.95), scallops ($16), sukiyaki steak ($14.95) are served with steamed rice, teppan-yaki vegetables, Japanese green tea and green tea ice cream. Children's menu for ages 10 and under $8.95. After that the menu selections increase in price with a sesame chicken dinner ($18.95), filet mignon ($24.95), shrimp ($23) or scallops ($23). Children's menu goes up to $10.95. COMMENTS: Part of the price is the "show." The chef works at your table and is adept at knife throwing and other dazzling cooking techniques. The menu also adds that a 15% service charge will be added to your check. Reservations recommended. Major credit cards accepted.

### SOUND OF THE FALLS ★ *French/Oriental Influences*
Westin Maui (667-2525). HOURS: Champagne brunch served Sunday brunch 10-2 pm. Dinner 6-10 pm. SAMPLING: Sunday brunch features an elaborate buffet with sushi, egg, fish, poultry and meat dishes priced at $21.75. The Sound of the Falls merits our Best Bet for brunches in West Maui. For the sushi lover it is worth the price alone for all you can eat! The dinner menu has some varied and unusual combinations such as whole boned baby hen on fried noodle cakes with poha berry chutney ($26), kiawe wood grilled scallops on spiced eggplant sauce ($28), sesame-roasted rack of lamb on a guava and merlot sauce ($32) or sauteed veal loin with shiitake mushrooms and lemongrass cream. Appetizers, soups and salads priced $5.50 - $20.50. If the pocketbook allows, select the beluga malossol caviar poi blinis ($75)! We didn't sample it, but the party of four next to us each ordered one! Desserts are fit for royalty and beautifully presented. Decadent coupe cafe chocolate, a combination of chocolate and coffee ice cream with kahlua chocolate sauce served with whipped cream, shaved chocolate and a wedge of chocolate brownie, or on the lighter side, a pineapple sorbet with sliced strawberies and a dash of Grand Marnier. Add to all this a very lengthy wine selection. COMMENTS: A beautiful setting and impeccable service, highly recommended.

### SPATS II ★   *Italian*
Hyatt Regency Hotel (661-1234). HOURS: Dinner only. SAMPLING: Appetizers, soups (minestrone or cioppino), and salads (warm lobster salad, buffalo mozzarella) $5 - $12.50. Pastas are available in two sizes, appetizer or entree. Fettucini $9.50/$17.50, canneloni $11.50/$20. Entree offerings vary from duo pescatore (medallions of two fresh Hawaiian fishes with pesto sacue and tomato concass) to veal chop pizzaoiola (veal sauteed with sun-dried tomatoes, pine nuts, white wine and basil, topped with melted scamorza cheese) $19.75 - $30. COMMENTS: The interior resembles more an elegant British pub than a restaurante Italiano. The food and service were both spectacular. The a la carte pasta appetizers proved sufficient for a child's portion or would be adequate for a light eater. Excellent fare. Reservations recommended. "Top 40" music is played in their lounge from 10 pm-2 am. Sun. - Thurs., and until 4 am on Fri. and Sat. Dress code requires slacks, shirts with collars and covered shoes.

### SWAN COURT ★   *Continental*
Hyatt Regency Hotel (661-1234). HOURS: Open for breakfast 6:30-11:30 a.m and for dinner. SAMPLING: Their breakfast buffet is $13.50 and includes fresh-squeezed orange juice, macadamia pancakes with a variety of toppings, French toast, cereals, yogurt, fresh fruits, and a good choice of hot breakfast foods. The menu items run $7 - $15. Dinner salads and soups $5 - $8. Entrees are a la carte $23 - $32 and include veal chop saute armagnac, champagne chicken, or fresh island fish Eichenholz (baked on a plate of oakwood, with capers and mushroom garnish). Save room for dessert! Chocolate afficionados must try the Swan Court mousse! It is a chocolate swan filled with chocolate mousse accompanied by little clouds of whipped cream topped with macadamia nuts-rich enough for 2! ($7) COMMENTS: The atmosphere and the view of the Swan Court are worth it for a special treat! Our "best bet" for a daily breakfast buffet. The dinners are excellent and many unusual preparations are offered. The service here makes one feel very pampered. Our only criticism is the proximity of other tables and noise seems to carry. Reservations, especially for dinner, are advised. Major credit cards accepted.

# KAHANA/NAPILI/KAPALUA

## BAR ONLY

### KAPALUA BAY RESORT - BAY LOUNGE ★
Enjoy a fabulous sunset in this elegant setting. Pupus served are complimentary. The tropical drinks run $4 - $6 with some the size of small fishbowls. A pupu menu is available which offers sushi, sashimi or shrimp cocktails. Live soft background music is provided.

## INEXPENSIVE

### DOLLIE'S    *American/Italian*
4310 Honoapiilani Hwy. at the Kahana Manor (669-0266) HOURS: 10 am-midnight. SAMPLING: Pizzas, sandwiches, and pasta $5-$10 COMMENTS: You'll find one of the best selections on Maui with more than 40 domestic and imported beers. Espresso too! Food to go. Very popular spot.

### FAT BOY'S ★    *Local Style*
On lower Honoapiilani Hwy near the Kaanapali Shores resort. (669-6655) HOURS: Mon.-Fri. 10 am-9 pm, Sat.-Sun. until 7 pm. SAMPLING: Plate lunches $4.95 - $5.95 offer chicken curry, kalua pig, chicken or beef teriyaki, roast beef and gravy, fish plate, sweet and sour chicken and come with steamed rice and macaroni salad. Besides the famous "fat boy" burger, there are teriyaki steaks or chicken sandwiches, Philly steak sandwich, fish sandwich and teri burger. Side orders of teriyaki chicken salad, green salad and french fries. Daily specials.

COMMENTS: Don't blink or you might pass right by this special restaurant and you definitely don't want to do that! This is the best local style restaurant in West Maui and a one-of-a-kind on the island. It's been open only a year and business is booming. Nothing fancy, just good food and plenty of it. Stroll on in, place your order at the counter and watch it being whipped up before you in their very orderly kitchen. Then saunter on out to the picnic table in front. There are other good burger spots and some have a view better than Honoapiilani Road, but it just might be time to discover what the locals know! The fat boy burger wins hands down as the best Maui burger for the money ($3.99). One of those that takes two fists and a handful of napkins to enjoy. All the menu items are dynamite, but that burger makes me want to hop a plane and get one to go! A burger, fries and a large soda will set you back only $6.19 and fill up even the hardiest of eaters.

### GAZEBO    *American*
Poolside at the Napili Shores. HOURS: Breakfast 7:30-noon, lunch noon to 3 pm. SAMPLING: Breakfast offers omelettes and an assortment of tropical pancakes or "eggs 'n things" from $3. Lunch selections include burgers, sandwiches and salads $3.95 - $5.50. COMMENTS: Popular with Maui residents for the friendly atmosphere with a wonderful ocean view.

**HONOLUA GENERAL STORE**   *American*
Above Kapalua as you drive through the golf course. HOURS: 6:30 am-8 pm.
Daily for breakfast and lunch. COMMENTS: The Kapalua Hotel refurbished this
once funky and local spot. The front portion displays an assortment of Kapalua
clothing and some locally made food products. Breakfasts include pancakes, eggs
and such. Lunches include four local plate lunches daily which might include stew
or teri chicken $4.75 - $5.50 or a smaller portion called a hobo which is just a
main dish and rice for $2.95. Sandwiches and burgers from their deli too.

**MARKET CAFE**   *American/Italian*
Kapalua Bay Hotel Shops (669-4888). HOURS: Daily for breakfast, lunch and
dinner. Open 8 am-9 pm. SAMPLING: Breakfast served daily until 11 am,
Sundays until 1 pm. Omelettes, tropical pancakes, hot oatmeal with blueberries
$4.95 - $6.95. Lunch items include house specials such as three cheese lasagne
or a beef tenderloin sandwich, grilled mahi or chicken sandwiches, soups and
salads $6.95 - $9.96. Dinner is served from 5 pm with selections leaning to an
Italian flare with pasta, fish, chicken and veal dishes $11.95 - $17.95. Offerings
include veal piccata, eggplant parmigiana, shrimp parmigiano and linguini
marinara. Beer & wines. COMMENTS: This small restaurant has good, affordable
fare, and is part of a market that carries some unusual imported foods.

**RICCO'S**   *Italian*
On Lower Honoapiilani Hwy. next to AAAAA storage. PHONE: 661-4433.
HOURS: 11 am-10 pm daily. SAMPLING: Pizza, sandwiches, burgers, salads and
house specialties such as beef prime rib bones, half a chicken or St. Louis style
full or half rack spare ribs in Ricco's smokey BBQ sauce.

# MODERATE

**BEACH CLUB**   *American*
Kaanapali Shores Resort (667-2211). HOURS: Breakfast 7-11 am, lunch 11:30-
3:00 and dinner from 5:30-9 pm. SAMPLING: Dinners range from seafood and
chicken to steak. Bar service. COMMENTS: Currently their early bird special
features prime rib, chicken or mahi for $10.95.

**CHINA BOAT**   *Cantonese/Szechuan/Mandarin*
4474 L. Honoapiilani (669-5089). HOURS: Lunch daily 11-3 dining nightly 5-10
pm. SAMPLING: Shrimp with vegetables, beef with broccoli, and some hot and
spicy dishes as well for $8.95 - $24.95. COMMENTS: Newly remodeled with a
new menu as well. They offer karaoke entertainment in the evenings.

**ERIK'S SEAFOOD GROTTO**   *American/Seafood*
4242 Lower Honoapiilani Hwy., on the second floor of the Kahana Villa Condo
(669-4806). HOURS: Dinner daily 5-10 pm. SAMPLING: Dinners include
chowder or salad, potatoes or rice, and bread. Halibut steak, baked stuffed prawns,
Louisiana catfish, rack of lamb, or lobster thermidor run $12.95 - $18.95.
COMMENTS: Check for early bird dinner specials. This is one of those restaurants that we don't hear much good about, but then we don't hear anything bad
either! Kind of a middle of the road dining experience.

## KAHANA KEYES

Valley Isle Resort in Honokowai. (669-8071). HOURS: Dinner only. SAMPLING: Early bird specials from 5-7 pm for $8.95 - $11.95 include lobster tempura, prime rib, steak and lobster or steak and crab with salad bar, rice or potatoes, rolls and butter. Regular dinner offerings of seafood, beef or chicken run $9.95 - $15.95.

## KAHANA TERRACE  *American*

Sands of Kahana Resort (669-0400). HOURS: Breakfast 7:30-11, lunch 11:30-2:30, and dinner 5:30-9. SAMPLING: Pupu menu at bar. Lunch menu offers sandwiches, salads, hamburgers $5 - $8. Complete dinners with entrees of N.Y. steak, Maui ribs, chicken $8.95 - $16.95. A limited keiki menu. COMMENTS: Very quiet dining, this restaurant seems to be primarily used by the resort guests.

## KAPALUA GRILL AND BAR ★  *American/Continental*

200 Kapalua Drive, just across the road from the Kapalua Hotel and a short drive up Kapalua Drive (669-5653). HOURS: Lunch 11:30-3 pm and dinner 5-10 pm. SAMPLING: Lunches include burgers and hot sandwiches, for $5.95 - $9.95 or lunch specials which change daily. The dinner selections are available on a daily basis. Offerings might include prawns and chicken sauteed with lemon, capers and white wine, fresh Norwegian salmon, roack of lamb, roast duckling in a black peppercorn and brandy sauce or osso bucco $13.95 - $20.95. An admirable wine list begins at about $16 with more extravagant selections such as a '45 Chateau Lafit Rothschile at $990. COMMENTS: Tank tops are okay daytime attire, but not appropriate for evening. This is a sister facility to Leilani's and Kimo's, but its menu is a little more gourmet. A golf course and ocean view add to the pluses of this restaurant. It's a popular restaurant, so you might want to call ahead.

## ORIENT EXPRESS  *Thai/Chinese*

Napili Shores Resort, one mile before Kapalua (669-8077). HOURS: Dinner 5:30-10 pm. SAMPLING: Ginger beef, garlic shrimp, seafood in clay pot, and spinach pork are priced $7.95 - $14.95. COMMENTS: This restaurant is run by the same folks who operate Chez Paul. The Thai food is the best in West Maui.

## PINEAPPLE HILL  *Continental*

Up past Napili on the "freeway to nowhere," turn left for Kapalua and you will see the entrance (661-0964 or 669-6129). HOURS: Dinner 5:30-10 pm with cocktails beginning at 4:30. SAMPLING: Dinners include a house salad, vegetables and rolls. Entrees begin at $8.95 for linguine with herb sauce up to $27 for steak and lobster. In between are selections such as shrimp Tahitian, New York pepper steak, rack of lamb and fresh island fish. A limited, but affordable and varied wine list. COMMENTS: This was once the home of plantation manager David Fleming. He was one of Maui's early agricultural pioneers who helped establish mango, liche, pineapple and other exotic plants and trees. Just past Kapalua is the beach park bearing his name. He completed the plantation house in 1915 and planted those beautiful Norfolk Pines which line the drive. It opened for dining in the early 1960's. Pineapple Hill has one of the loftiest settings for sunset viewing. We recommend enjoying cocktails out on the front lawn while watching the sun descend. Recent renovations have freshened up the interior. Several recent reports indicate that both serve and food is very good, although in the past we have had mixed reviews.

**POOL TERRACE RESTAURANT AND BAR**  *International*
Kapalua Hotel, poolside (669-5656). HOURS: Lunch and dinner 11 am-9 pm.
SAMPLING: Soups include daily special, baked onion or gazpacho, $5 - $6 and
salads vary from pasta or caesar at $6.50 to the Kapalua seafood salad with
chilled crab claws and shrimp for $24.50. Sandwiches of tuna, turkey, or roast
beef along with burgers $5.50 - $7. Pizzas and fajitas $8.50 - $12. House
specialities include cashew chicken, shrimp and scallops, or beef broccoli $10 -
$15. Espresso. Premium wines availabe by the glass from their Cruvinet machine.
COMMENTS: The beautiful location of this casual poolside setting features an
ocean view from every seat.

**SEA HOUSE**  *American/Local style*
Beachfront at Napili Kai Beach Club (669-6271). HOURS: Breakfast 8-11, lunch
12-3, and dinner 6-9. SAMPLING: Breakfast offers omelettes, corned beef hash
or local style with Portuguese sausages. Lunch includes sandwiches & salads that
run $3 - $8. Light suppers include cheeseburgers, little neck clams, tempura or
chow mein $7.50 - $13.95. Dinners $14.95 - $19.75, include scampi, lobster, roast
duck, and rack of lamb. Rice or baked potatoes, and vegetable included. Hawaiian
music performed Mon., Wed., Thurs., & Sat., 8-10 pm

# EXPENSIVE

**THE BAY CLUB ★**  *French/Seafood*
At Kapalua near the entrance to the resort (669-8008 after 5 pm, 669-5656 before
5 pm). HOURS: Lunch 11:30-2 and dinner 6-9:30 pm. SAMPLING: Lunch
selections include bay scallops, fruits of the sea in fresh artichoke, omelette of the
day, fettucini pescatore, or sandwiches such as a club, monte cristo, or shrimp and
cheddar $5.75 - $10.50. Coffee $2. A la carte dinners include fresh fish prepared
six exotic fashions $24 - $26, shellfish, veal, filet mignon, lamb chops $21 - $26.
Soup and salads $3.50 - $5. Coffee $3. Espresso available. COMMENTS: The
restaurant is situated on a promontory overlooking the ocean, a perfect spot from
which to enjoy the scenic panorama along with pupus and cocktails. You might
want to indulge in one of their ice cream libations, such as a Bay Lounger (dark
rum, fresh pineapple and ice cream) or the Bay Club Delight (kahlua, Grand
Marnier, Amaretto and ice cream) $5 ish. Extensive wine list. The dress code
requires swimsuit coverups for lunch and, in the evening, long sleeve dress shirts
or jackets for men and no denim. A pianist serenades you through dinner adding
a romantic touch. With the changes in ownership over the last few years, service
has suffered. We hope that the current management can restore the impeccable
service and food, because for us, this has always been a very special restaurant.

**THE GARDEN RESTAURANT ★**  *Continental*
Kapalua Bay Resort Hotel (669-5656). HOURS: Breakfast 7-10 am, lunch 12-2:30
Mon.-Sat., 10 am-2:30 Sun., and dinner nightly 6-10:30. SAMPLING: Breakfast
fare $5 - $7, Dinners are a la carte, priced $21 - $25, and include grilled capon,
Hawaiian lobster, rack of lamb, prime rib, or roast duckling. The Mayfair buffet
served Sunday, without champagne, is $16.50. COMMENTS: Enjoy continental
dining in this semi-open tropical setting. Excellent food and service make this a
pleasant dining experience. Reservations recommended, a must for Sunday brunch.

## PLANTATION VERANDA ★   *Continental*
Kapalua Bay Hotel (669-5656). HOURS: Dinner only, 6:30-9:30. SAMPLING: Appetizers include Hawaiian spiny lobster with squid ink pasta and roasted red bell peppers or seafood crepe with king crabmeat, spinach, and saffron. $10 - $14. Soups and salads $4 - $7. Entrees include island fish poached with saffron, tomatoes and Hawaiian red shrimp or sauteed with artichokes and morels with a white butter sauce. Other selections feature scallopini of Lanai Axis venison sauteed with Peter Herring liqueur and sun-dried cherries, served with petit dumplings or boneless Bob White quail filled with apples and raisins, deglazed with port wine $23 - $27. COMMENTS: This formal setting offers elegant dining. Fine wine by the glass is available from their cruvinet. Enjoy your dinner serenaded by harp music. Jackets are required and reservations are advised.

# KIHEI

## INEXPENSIVE

### AZEKA'S SNACK SHOP   *Local Style*
Azeka's Place Shopping Center on South Kihei Rd. HOURS: 9:30-4 pm Mon.-Sat. SAMPLING: Plate lunches $3.10 - $4.15, tuna sandwiches $1.05, teriburger $1.80, hamburger $1.10. COMMENTS: The teriburger contains marinated beef and is quite tasty. Among the plate lunch selections are teriyaki beef or meat loaf, served with rice and macaroni salad and packed in a styrofoam carton for convenience "to go." Several picnic tables offer limited outdoor seating.

### BONANZA   *American*
Kukui Shopping Center (879-2525) HOURS: 6-11 am for breakfast, 11 am-4 pm for lunch, dinner 4 pm-9 pm. SAMPLING: This is a combination fast food/all-you-can-eat restaurant. You place your entree order for chicken, steaks etc. ($6.79 - $9.99) and then help yourself to beverages ($1.29 with free refills) and their "freshtastic" food bar. Four types of soup (including one low-cal), rolls, and side dishes such as pastas, salads (fruit, vegetable and green), potato skins, and nachos are followed up by five desserts which might include brownies, apple crisp, pudding, jello and soft serve ice cream. Not for the light eater, they also offer daily breakfast specials. COMMENTS: Traveling with a pack of ravenous teenages? Then this is the place to come!

### DENNY'S   *American*
Kamaole Shopping Center. HOURS: 24 hours daily. SAMPLING: Burgers, local style dishes and dinners which include steak, seafood and chicken $6.79 - $10.79. Breakfast served anytime.

### FRESH ISLAND FISH   *American/Seafood*
Maalaea Harbor.(244-9633 or 242-6532) HOURS: 10 am-9 pm Mon.-Sat. SAMPLING: Check their menu for fish and chips ($6.95), Opakapaka ($9.95), Lobster medalion saute ($12.95). COMMENTS: This restaurant/seafood market is one of the best places to pick up really fresh island fish to prepare in your own kitchen.

**GREAT FETTUCINI, ETC.**   *American/Italian*
Aston's Kamaole Sands Resort at 2695 S. Kihei Rd. (874-8700) HOURS: 6 am-9:30 pm. SAMPLING: Lunches range $3 - $7, dinner specialties include stir fried shrimp or chicken, shrimp marinara or scampi and fettucini made fresh daily on the premises. Prices for complete dinners run $5.95 - $13.95. COMMENTS: Opened in early 1990, they feature casual poolside breakfasts and lunches and candlelit dinners. Poolside and suite service is also available. An innovative twist is that the waiters and waitresses perform magic tricks to the delight of the kids.

**HENRY'S BAR AND GRILL**   *American*
Lipoa Shopping Center (879-2849) HOURS: 10 am-2 am SAMPLING: Grill your own steak or chicken served with salad bar and roll $7.75. COMMENTS: Four televisions, one a big screen, for sporting events or stop by to see your favorite soap opera!

**INTERNATIONAL HOUSE OF PANCAKES**   *American*
Azeka's Place Shopping Center on South Kihei Rd. HOURS: Sun.-Thurs. 6 am-midnight, Fri. and Sat. 6 am-2 am. SAMPLING: Breakfast, lunch, and dinner choices served anytime. The usual breakfast fare and sandwiches. Dinners run $5.95 - $10.95 and include soup or salad, roll and butter. COMMENTS: A children's menu is available. Very crowded on weekends.

**ISLAND THAI**   *Thai*
Azeka's Shopping Center (874-0813). HOURS: Mon.-Fri. 11-3 for lunch, daily 5-9 for dinner. SAMPLING: Appetizers, soups, salads and entrees such as chicken cashew basil, evil prince chicken or garlic cabbage $4.75 - $6.95. COMMENTS: The prices are pretty decent, but the portions are a little small. Food was good, but nothing outstanding.

**KIHEI CHICKEN AND RIBS**   *Local Style/South American*
Kukui Shopping Center (879-2655). SAMPLING: From South America, Evelio and Carmen Mattos have integrated their native recipes with local style meals. Plate lunches run about $4.50 and include potato or macaroni salad, South American fried rice and ribs. Take out available as well.

**McDONALD'S**   *American*
1900 area of South Kihei Rd. at the Kihei Shopping Center. HOURS: Breakfast, lunch, and dinner. Breakfast served only from 6 am-10 am. SAMPLING: There are a few unusual island items added to the menu. For breakfast, you can have Portuguese sausage with rice, and chase it down with chilled guava juice. COMMENTS: Indoor seating available. Prices slightly higher than mainland.

**NEW YORK DELI**   *American*
2395 S. Kihei Rd., (879-1115). HOURS: 10 am-9 pm. SAMPLING: In addition to luncheon meats there is a wide selection of salads and entrees such as lasagna. Bagels, in every variety, are flown in fresh from New York! Sandwiches run $5-6.

**PARADISE FRUIT** ★   *Healthy American*
1913 South Kihei Rd. next to Kihei Town Center (879-1723). HOURS: Market open 24 hours, food served until 10 pm. SAMPLING: "Healthy oriented"

sandwiches, salads, and smoothies. Veggie sandwich runs $2.95, turkey sandwich $3.25, salads priced $3 - $4. Smoothies and yogurt shakes are $1.50 - $2.00. COMMENTS: This is an open air fruit and vegetable market that also sells some sundry items. Tucked in the back is their walk-up snack bar. A few tables in the back of the restaurant. Their yogurt shakes and smoothies are delicious! Try a pizza to take and bake from their deli case.

## PIZZA FRESH   *Italian*
2395 S. Kihei Rd. in Dolphin Plaza (879-1525). HOURS: Daily 3-9 pm. SAMPLING: White or whole wheat crust pizza that they make and you bake. Available in four sizes, small to x-large. One topping small $7.15, in x-large it's $19.95.

## POLLI'S   *Mexican*
101 S. Kihei Rd., Kealia Village (879-5275). HOURS: Lunch 11:30-2:30 and dinner 5-10 pm, happy hour 2:30-5:30, has burgers, free chips and a pupu menu. SAMPLING: Dinner combinations run $8 - $12. Fresh fish prices are quoted daily. COMMENTS: A deck over the beach offers outdoor dining and cocktails. Major credit cards accepted.

## THE SANDWITCH   *American*
145 North Kihei Rd., by Sugar Beach Condos (879-3262). HOURS: Mon.-Sat. 11-11, Sunday noon to 11. SAMPLING: Sandwiches, burgers, salads.

## SHAKA SANDWICH AND PIZZA   *American/Italian*
Located behind Jack 'n the Box in Paradise Plaza on South Kihei Rd. near Azeka's market. (874-0331 or 879-4284) Not open as we go to press.

## THE SPORTS PAGE GRILL AND BAR   *American*
2411 S. Kihei Rd. (879-0602). HOURS: 11 am-11 pm. SAMPLING: New York Yankees Hot Dogs, Fowl Ball Burger, Edmonton Oilers Tuna Melt, Dick Butkus Corned Beef and Cheese, Flo Jo Roast Beef. Also hot dogs and salads. COMMENTS: You probably already have the idea! This is definitely the spot for the sports afficionado. With confident good humor their menu resembles a newspaper tabloid and reads, "You will be served in 5 minutes...or maybe 10 minutes...or maybe even 15 minutes...relax and enjoy yourself." It may take at least 15 minutes to read over the menu. The front page covers exotic beverages and a hearty selection of imported beers, followed on the inside by dugout dogs, champion burgers, sport fishing sandwiches, bowl games (those are salads), and "game favorite" sandwiches. Not much on the menu over $5.95. A big screen TV with remote monitors and satellite reception should ensure plenty of good conversation for the athletic enthusiast!

## SUBWAY SANDWICHES   *American*
Kukui Shopping Center. HOURS: 9 am-midnight. Sandwiches $2.69 - $7.39, salads include roast beef, seafood, chef or tuna $3.29 - $5.99.

## SUDA SNACK SHOP   *Local Style*
By the gas station along N. Kihei Rd. HOURS: 5 am-1 pm. SAMPLING: Plate lunch $3.50, burgers $1.15, chow fun $1.85. The pizza portion of the restaurant is open 3 pm-9 pm. Both closed Sunday. COMMENTS: We were disappointed

that this little "dive" wasn't one of the island's best kept secrets. The burgers were so-so, the french fries were pricey for the portion and the chow fun wasn't a meal, it was snack size. See you in Wailuku for burgers and chow fun!

## MODERATE

**THE BREAKERS**   *American/Seafood*
760 S. Kihei Rd., (879-5898) HOURS: Breakfast 8 am-10:30 am, lunch 11:30-2 pm and dinner from 5 pm. SAMPLING: Breakfast selections include omelettes, steak and eggs $4.50 - $8.95. Lunch offers sandwiches or soup and sandwich combinations in the $7 range. Teriyaki steak, cajun or sauteed or broiled or stuffed fresh fish, tiger prawns, shrimp stir fry, BBQ pork ribs $12.95 - $19.95. COMMENTS: Attractive restaurant with oceanview dining located in the Menehune Shores condominiums. We tried out the Early Bird Specials ($9.95 - $13.95), well prepared, but the price here seems to be an indication of the portions, smallish! No credit cards.

**BUZZ'S WHARF**   *American/Seafood*
Maalaea Wharf area (244-5426 or 661-0964). HOURS: Everyday 11 am-3 pm for lunch, 5-9 for dinner. SAMPLING: Lunch includes sandwiches, salads and hot entrees $3.95 - $12.95. Dinners include shrimp, trout, oysters, spaghetti, BBQ ribs and scallops $9.95 - $26.95. COMMENTS: Recently remodeled with a much, much larger menu, but still has that lovely Maalaea harbor view. The renovations have created a lounge/bar area too. The service remains mediocre to poor, but the food, especially their specialty Tahitian Shrimp is very good.

**CANTON CHEF**   *Cantonese/Szechuan*
Kamaole Shopping Center (879-1988). HOURS: Lunch 11-2 pm and dinner 5-9:30 pm. SAMPLING: Vegetable, chicken, beef, duck, seafood and pork dishes priced $5.50 - $11. COMMENTS: Owned in conjunction with the Hong Kong Restaurant. Orders to go.

**CHUCK'S** ★   *American*
Kihei Town Center (879-4488 or 879-4489). HOURS: Lunch 11:30-2:30 pm, interim menu served 2:30-5:30, Mon.-Sat.; dinner 5:30-10:00 pm daily. SAMPLING: Lunch items include cold beef, ham, turkey, hot BLT, cheeseburger, or French dip sandwiches. Early bird dinners are 5:30-6:30 pm and for $8.95 feature a choice of mahi or chicken thighs. A good salad bar is available a la carte or included with dinners such as teriyaki steak, prime rib, chicken or fish $10.95 - $14.95. Fresh fish is quoted daily. COMMENTS: No reservations taken. Children's menu $6 - $11. Especially popular for its salad bar.

**ERIK'S SEAFOOD BROILER**   *American/Seafood*
2463 S. Kihei Rd., Kamaole Shopping Center (879-8400). HOURS: Lunch 12-4 pm, dinner 5-10 pm. SAMPLING: In addition to fresh island fish try halibut steak, Louisiana catfish, roast duck a la Grand Marnier, veal oscar, scampi. Dinners include soup or salad, bread, potatoes or rice. COMMENTS: If the menu seems familiar, it may be because this is the fourth and newest restaurant operated by Erik Jakobsen. Early bird dinner specials.

191

### GREEK BISTRO ★　*Greek*
Kai Nani Shopping Center, 2511 South Kihei Rd. (879-9330). HOURS: Breakfast, lunch and dinner from 7 am-9:30 pm. SAMPLING/COMMENTS: This little restaurant is a delightful surprise. Breakfasts include three egg omelettes with three fillings of your choice, Belgian waffles with assorted exotic toppings, Eggs Benedict and fruit plates $3.25 - $8.95. Lunch and dinner offers the same menu. Hot and cold sandwiches $4.95 - $5.95 include bistro burger, sausage, avocado and turkey. Salads run $3.95 - $7.95. We sampled the Greek salad and at $5.95 that had ample portions of feta cheese, chunks of cucumbers, Greek olives, tomatoes and was enough for two to share for a dinner salad. Dinner selections are diverse with Mediterranean prawns $10.95, fresh fish (market price), chicken $8.95, and Greek selections of spanakopita (filo with spinach) mousaka (lamb, eggplant and potato), pastichio (Greek lasagna) gyros (lamb sandwiches) and side orders of potato, pilaf, linguini, and vegetables. We tried the Greek Gods Platter which was a combo of all the above for $10.95 and found it excellent. Table seating is limited to a small outdoor area. Formerly from California, the owners have applied for a liquor license and hope to expand their menu in the future. Local folks have quickly been discovering this gem tucked in the rear of this small shopping center. Check this one out! Table seating at this very casual eatery is limited to a small outdoor area.

### HONG KONG　*Oriental*
61 South Kihei Rd. (879-2883). HOURS: Lunch 12:30-4 and dinner 5-9. SAMPLING: Individual dishes from $5 include Szechuan-style cooking.

### ISANA SHOGUN　*Japanese Teppanyaki*
515 S. Kihei Rd., (874-5034). HOURS: Dinner 5:30-10 Tues.-Sat. SAMPLING: Sushi bar available as well as seven teppanyaki tables which serve entrees ranging from $10.95 for vegetable teppanyaki - $23.95 for steak and lobster. All entrees include shrimp appetizer, salad, soup, teppan vegetables, rice and Japanese tea along with two dipping sauces. COMMENTS: Isana brings the first teppanyaki cookery to Kihei. Isana means "big fish" or "whale" in Japanese and a giant whale sculpture by James Hagedorn graces the central stairway. Upstairs in the Mermaid Lounge is a karaoke bar. The sushi bar here is extremely popular too!

### ISLAND FISH HOUSE ★　*American/Polynesian/Seafood*
1945 South Kihei Rd. (879-7771). HOURS: Dinner from 5:30 pm. SAMPLING: Complete dinners include chowder or salad, au gratin potatoes or island rice, fresh vegetables and homemade bread. Chicken teriyaki $12.95, scallops $16.95, shrimp polynesian $17.95, or their daily fresh fish which is offered cooked six different ways. COMMENTS: The fish is consistently excellent, and we are very partial to the way they prepare their carrots! The service is very good and the wine list offers numerous choices. Reservations are a must here unless you plan on arriving by 6:30. You may want to try their Kahului location called Mickey's. Check on their great early bird dinner specials.

### KIHEI PRIME RIB AND SEAFOOD HOUSE ★　*American*
2511 South Kihei Rd., in the Nani Kai Village (879-1954). HOURS: Dinner from 5-10 pm. SAMPLING: Ribs $16.95, polynesian chicken $16.95, prime rib in varied cuts from $18.95 - $21.95. Salad bar a la carte is $8.95. Early bird dinner

special of chicken, prime rib or fish $9.95 - $11.95. COMMENTS: Dinners include a salad bar, Caesar salad, or red snapper chowder, and is served either with fettucini noodles or rice. Homemade bread also accompanies your meal. The salad bar was very good, and the choices included sweet Kula onions. Our beef was tender, however, our lobster somewhat flavorless. The high-beamed ceilings with the hanging plants compliment the gorgeous wood carvings done by Bruce Turnbull and paintings by a German artist named Sigrid. They offer piano entertainment nightly.

**LA FAMILIA** *Mexican*
2511 South Kihei Rd., at Kai Nani Village (879-8824). HOURS: Cocktails begin at 4 pm. SAMPLING: Fajitas $11.95 - $13.95, Mexican entrees $7 - $10. COMMENTS: A very popular spot for locals and visitors alike, probably encouraged by their very inexpensive margaritas during happy hour.

**LUIGI'S PASTA PIZZERIA** *Italian*
Azeka's Shopping Center on S. Kihei Rd. (879-4446). HOURS: Daily 11:30 am until midnight, bar until 2 am. SAMPLING: Lunch selections include pasta or pizza as well as sandwiches such as submarines, French dip or club. Dinner selections $8 - $17 include pizza, calzone, spaghetti, fettucini, shrimp scampi and steak. COMMENTS: Early bird specials 4-6 pm, karaoke entertainment six nights a week 9 pm-1 am.

**MAUI LU LONGHOUSE** *Polynesian*
Maui Lu Resort, 575 South Kihei Rd. (879-5858). HOURS: Breakfast 7-10 am, closed for lunch, dinner 6-9 pm Sun.-Fri. Saturday luau at 8 pm. SAMPLING: Breakfast buffet. An uninspired dinner selection includes salad bar, veggies and potato or rice, ham steak with fresh pineapple, chicken breast $8.95 - $9.25.

**MAUI OUTRIGGER** *Polynesian/American*
2980 South Kihei Rd. (879-1581). HOURS: Breakfast begins at 6 am, lunch 11:30-2:30 and dinner. SAMPLING: Prime rib, polynesian chicken, ahi teriyaki, fish and chips $10.95 - $17.95. COMMENTS: This restaurant is ON the beach. Early bird specials. We would recommend it for cocktails at sunset.

**OCEAN TERRACE** *Polynesian/American*
2960 South Kihei Rd., Mana Kai Hotel (879-2607). HOURS: Breakfast, lunch and dinner served between 7 am and 10 pm. SAMPLING: Breakfasts include specialty pancakes, i.e. coconut or macadamia. Lunches offer salads, sandwiches and hot entrees $3.95 - $10.95. Dinner selections include teriyaki chicken, veal marsala, prawns Tahitian. COMMENTS: Operated by the owners of Island Fish House, this restaurant has a contemporary look enhanced by its oceanfront location. A wonderful spot to enjoy cocktails and the sunset. Pupu items range from local dishes to Polynesian foods. Children 12 and under half price. Early bird specials.

**RAINBOW LAGOON** *American*
2439 S. Kihei Rd., upstairs at the Rainbow Mall (879-5600). HOURS: Dinner 5-10 pm. SAMPLING: Early bird specials from 5-7 pm. The regular menu includes a seafood assortment $8.95 - $16.95, prime rib cuts $12.95-15.95, and two chicken selections $9.95.

**SILVERSWORD GOLF CLUB RESTAURANT**   *American/French*
Located at the Silversword Golf Course (879-0515). HOURS: 10:30 am-3 pm for lunch daily, 6 pm-9 pm for dinner Tues.-Sat. SAMPLING: Sandwiches, hamburgers, soups, salads and lunch entrees are affordably priced $3.25 - $5.75. The dinner menu is limited but diverse. Chicken Cordon Bleu, Fish of the Day, Catfish in garlic butter, shrimp scampi maison, ground sirloin and steak bordelaise run $8.25 - $13.50. COMMENTS: The prices here are comparable with the early bird specials in lower Kihei! Located on a lofty setting with a pleasant view of Kihei and beyond Kahoolawe and Molokini.

## EXPENSIVE

**FERRARI'S**   *Californian/Mediterranian*
1945 S. Kihei Rd. behind Island Fish House (879-1535) HOURS: Dinner only 5 - 9 pm. SAMPLING: Full dinners around $25. SAMPLING: Fresh seafood and freshly made pastas. COMMENTS: Just opening as we go to press, but operated by The Island Fish people who do a great job with all their restaurants.

**WATERFRONT**   *French*
At the Milowai Condo, Maalaea (244-9028). HOURS: Dinner only 5-10 pm, Cocktails 3-11 pm. SAMPLING: Rack of lamb, chicken cordon bleu, seafood mornay, cioppino, crab legs, lobster $17.95 - $24.95. COMMENTS: Early bird dinner specials available. Advance reservations are requested.

# WAILEA/MAKENA

Many new resorts and restaurants are opening in Wailea duirng 1990 and 1991 and we will keep you updated with restaurant reviews in the *MAUI UPDATE*. Be sure and share with us your comments as well!

## LOUNGES

**MOLOKINI LOUNGE** ★
Maui Prince Hotel, Makena. A wonderful opportunity for a sunset view. Entertainment and complimentary pupus served 4:00-6:30 pm daily. Call to check on entertainment available. Currently the entertainment schedule is as follows. Contemporary Hawaiian sounds of Mele Ohana nightly from 4:30-6:30 pm in the Molokini lounge; Out in the atrium, a classic duo, The Sterling Strings, plays Sat. - Thurs. evenings 6:30-8:30 pm; Popular Hawaiian entertainer (our favorite!) George Paoa is at the keyboard Sunday through Tuesday from 8:30-11 pm and during Sunday brunch; The upbeat group, Hau'ula, plays for dancing Wednesday and Thursday 8:30-11 pm and Friday and Saturday until 1 am.

**SUNSET TERRACE** ★
Stouffer Wailea Beach Resort, on the lobby level. Pleasant Hawaiian music and cocktails can be combined with a magnificent Hawaiian sunset.

# INEXPENSIVE

**ED & DON'S**  *American*
Wailea Shopping Center. SAMPLING: Sandwiches and ice cream.

**KIAWE TERRACE**  *American*
Maui Inter-Continental Wailea (879-1922) HOURS: 11 am-4 pm. SAMPLING: Burgers, salads, basket meals and sandwiches $7.50 - $12. COMMENTS: This casual restaurant is located near the central pool. Their burgers are great. The secret may be that the meat is ground fresh from organically grown Maui beef!

**MAKENA GOLF COURSE RESTAURANT**  *American*
At the Golf Course, just beyond Wailea (879-1154). HOURS: Continental breakfast 8:30-10:30., lunch 11-3 pm. Bar open until 5:45. SAMPLING: Salads and sandwiches $5.25 - $8.50. COMMENTS: Furnished in a tan and green theme, this open-air restaurant features a golf and ocean view.

**MAUI ONION**  *American*
Stouffer Wailea Beach Resort, poolside (879-4900). HOURS: Lunch 11 am-6 pm daily, dinner Wednesday and Friday 5:30-8:30 pm. Bar service 9 am-6 pm. SAMPLING: Lunch features scrumptious Maui onion rings are priced at $4.25, jumbo hot dogs $5.75, sashimi $8.50, cheeseburgers $7.00 or cajun grilled chicken breast sandwich $8.75. Twice a week "Dining under the Stars" is offered and at $19.95 it includes an entree choice of NY steak, chicken or ahi, plus salad bar, baked potato, corn on the cob and fruit salad. Live Hawaiian entertainment.

**PIZZA FRESH**  *Italian*
2395 S. Kihei Rd at the Dolphin Shopping Plaza (879-1525). HOURS: Daily 3 pm-9 pm SAMPLING: Pizza prepared for baking at your home or condo. Choice of 20 different toppings, white or whole wheat crust. They grind their own cheese and make their own dough. Also available are mini-pupu pizzas.

**SET POINT CAFE**  *American*
131 Wailea Ike Place atop the pro shop at the Wailea Tennis Center (879-3244). HOURS: Breakfast 7-10 am except Sat. and Sun. which is 8 am-2 pm, lunch 11-2:30. SAMPLING: Breakfast items run $3.75 - $5.50, lunch selections include pasta dishes, sandwiches, soup or salad $5 - $7.

**WIKI WIKI PIZZA**  *Italian*
2411 S. Kihei Rd. (874-9454). HOURS: 11 am-10 pm. SAMPLING: Limited table seating outside. Pizzas are baked and available to go.

# MODERATE

**CABANA CAFE**  *American*
Four Seasons, Wailea. (874-8000). COMMENTS: Not open as we prepare for press, however, we understand it will be a casual atmosphere serving lunch and an afternoon bar with pupu's out of doors.

**CAFE KIOWAI** *Polynesian/American*
Maui Prince Hotel, Makena (874-1111). HOURS: Breakfast, lunch and dinner are served daily with snacks available between regular meal service 6:30 am-10 pm SAMPLING: Breakfasts include Belgian waffles or French toast $4. The snack menu offers sandwiches, salads, desserts and beverages. The lunch menu features unusual selections including grilled chicken sandwich with pepper jack cheese ($7), or mini-rolls with sliced egg and watercress, turkey and prime rib ($6), hot entrees include Korean BBQ ribs with pineapple pickles ($11) or fresh clams and pasta ($12). Dinners begin with appetizers, soup and salad as well as sandwiches for the light eater. A la carte entrees include dungeness crab clusters with honey mustard and lime sauce ($17.50), seared rare tuna with Maui ginger marmalade and sundried cranberries ($21), picatta of veal ($18) or charbroiled lamb loin steaks with Maui wine onion relish ($15). COMMENTS: Keiki menu available. Kiowai pronounced "Key-oh-wy" means "fresh flowing water."

**LANAI TERRACE ★** *American/Polynesian*
Maui Inter-Continental Wailea (879-1922). HOURS: 6 am-11 pm. SAMPLING: Breakfast fare ranges from a Continental breakfast or pancakes to the New York breakfast with salmon lox, bagels and cream cheese or the Japanese breakfast of miso soup, butter fish Nitsuki, Tsukemono, rice and tea. $6.95 - $12. The daily luncheon buffet includes an array of soups, salads, cold cuts and hot entrees served 11:30-2:30 for $12.95. Lunch selections available from the menu include hamburgers, salads or entrees priced $7 - $11. Dinners for the light appetite include salads, sandwiches, egg or pasta dishes. Entrees include catch of the day, teriyaki sirloin, lamb brochettes, or New York steak. Price ranges from $8.50 - $19.50. COMMENTS: An attractive, bright and cheery atmosphere. Check out their "theme" nights, Wednesday being pasta night with a buffet of pastas in a variety of sauces, a salad bar and garlic bread or choose a Mexican buffet on Sat. night. Both are $14.95 and served 5-9:30 pm.

**SANDCASTLE** *American*
Wailea Shopping Center (879-0606). HOURS: Lunch 11 am-3 pm and dinner daily from 4 pm. SAMPLING: Lunch selections includes soup, salads, and sandwiches $4.95-12.95. Dinner choices of fresh fish, shellfish, meat and fowl $15.95-23.95, and includes soup or salad, rice or potato, vegetable and roll. COMMENTS: Currently offering early bird specials 4 pm-6 pm.

**WAILEA STEAK HOUSE**
100 Wailea Ike Drive, Wailea (879-2875). Easy to find sign near the Maui Inter-Continental Wailea indicates turn-off. COMMENTS: Located on the 15th fairway of Wailea's Blue Golf Course this restaurant suffered a major fire in 1989. Reopening date not yet announced.

# EXPENSIVE

**FAIRWAY ★** *American/Continental*
100 Kaukahi St., at Wailea Golf Course Clubhouse (879-4060 or 879-3861). HOURS: Breakfast from 7:30, lunch 11-3:30 pm, and dinner 5:30-9:30 pm. SAMPLING: Breakfast selections include pancakes $4 and eggs benedict $6.95. Lunch offers burgers $5.75 or turkey clubhouse $6.25. Dinners are priced $14.95 -

$25.75 and include a salad bar. Salad bar only $11.95. Entree selections include beef, seafood, veal, duck and chicken. COMMENTS: This restaurant is open-air with outdoor seating available. It offers a beautiful ocean view. Cocktails are available from the adjoining bar, the Waterhole. Their ice cream drinks are richly refreshing any time of day. Here is a sampling to tempt your palate: Fairway Grasshopper - creme de menthe, creme de cocoa, and ice cream all blended together, and topped with chocolate mint liqueur and chocolate sprinkles. Wailea Almond Joy - Amaretto, Kahlua, ice cream, blended and topped with whipped cream and almond slices. Brandy Alexander - Brandy, ice cream, and creme de cocoa blended and sprinkled with nutmeg. These wonderful concoctions run $5-ish. Check for early bird specials. Dinner reservations are suggested.

### HAKONE ★ *Japanese*
Maui Prince Resort, Makena (874-1111). HOURS: Dinner only, 6-10 pm. SAMPLING: Soups, salads and appetizers $2 - $7. A la carte dinners include nikujaga (boiled thinly sliced pork, potatoes and onion) ($10), breaded pork tonkatsu ($15.50) or assorted tempura ($16.50). COMMENTS: Authenticity is the key to this wonderful Japanese restaurant, from its construction (the wood, furnishings and even small nails were imported from Japan) to its food (the rice is flown in as well). A sushi bar is also available. Reservations recommended.

### KIAWE BROILER ★ *American/Seafood*
Maui Inter-Continental Wailea (879-1922). HOURS: Dinner 6-10 pm. SAMPLING: Dinners are served with salad bar, and a choice of baked potato, rice pilaf, or steak fries. Among the options are steamed snapper with chowder, sauteed scampi with tomato, fresh island fish or for the meat lover a choice of chicken breasts with fresh pineapple, filet mignon, lamb or veal chops. $17.50 - $27. Salad bar only $8.50. COMMENTS: A pleasant, quiet restaurant with a very good salad bar.

### LA PEROUSE ★ *International/seafood*
Maui Inter-Continental Wailea (879-1922). HOURS: 6:30-10 pm. SAMPLING: Pacific lobster tail filled with ambrosia mousse wrapped in crispy pastry and surrounded with Maui blush wine butter sauce, succulent guinea hen roasted with garden herbs and Molokai baby vegetables served in its own juices, or island

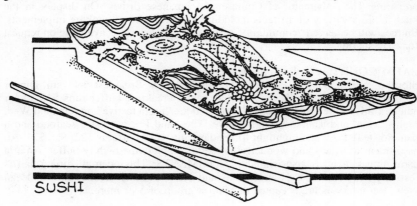

SUSHI

snapper filled with shrimps and wild Haiku mushrooms accompanied by a cream puree of red and green peppers. $23 - $32. Soups and salads $5 - $10. Hot and cold appetizers $7.50 - $18.50. COMMENTS: The decor here has an Oriental theme and is rich with koa wood. Pianist Frank Withalm entertains nightly. The service is excellent. The callaloo soup of crabmeat with Maui taro leaves was richly wonderful. The house green salad was more an edible picture, being artistically arranged to include sprouts, Maui onions, cherry tomato, red cabbage, lettuce, cucumber, and mushrooms. Reservations are recommended and guests are asked to dress in the spirit of elegant evening dining. This Travel/Holiday award-winning fine-dining restaurant is a good choice for that special evening out.

### MAKANI ROOM ★ *American*
Maui Inter-Continental Wailea (879-1922). HOURS: Sunday brunch only, 9 am-1:00 pm. COMMENTS: The buffet is $26 adults, $13 children. Seating is on the large lanai with a scenic view of the neighboring islands or at indoor tables. The feeling at this brunch is enjoyably casual, however, there is nothing casual about their lavishly laden tables. Omelets are cooked to order, as well as other choices which vary but may include fresh seafood, beef or lamb entrees as well as eggs benedict or crepes. Don't forget to save room for at least two trips to the dessert table. The pastry chef here has a wonderful way with the more unusual fruits - we tried a gooseberry pie. Reservations are a really good idea. Major credit cards.

### NICHOLAS, NICKOLAS
Between the Grand Hyatt and The Four Seasons Resort in Wailea. This restaurant is scheduled to open in June 1991 and, following in the footsteps of the O'ahu restaurant of the same name, should offer excellent continental cuisine.

### PACIFIC GRILL *East/West*
Four Seasons Resort in Wailea. (874-8000) HOURS: 6 am - 11 pm for breakfast, lunch, dinner and Sunday brunch. SAMPLING: Breakfast is a la carte or from a buffet. Lunch features an informal setting with sandwiches, hamburgers, hot entrees such as fresh fish or pasta and a salad buffet. Dinner features Pacific Rim Cuisine, Asian-Oriental, complimented by continental fare such as lamb and steak. COMMENTS: In addition to Western style selections, guests can choose Oriental Pacific Rim cuisine. Three Chinese chefs are in view at the restaurant's entrance preparing Thai, Vietnamese, Chinese and Japanese dishes. On display in the kitchen are a variety of artifacts including antique Oriental cooking implements. Entrees will be served "community" style and a lazy susan is available on request for easy sharing.

### PALM COURT ★ *International/buffet*
Stouffer Wailea Beach Resort (879-4900). HOURS: Breakfast 6-11 am, Dinner 6-10 pm. SAMPLING: The daily champagne breakfast buffet runs $13.50 for adults. Daily dinner buffets currently offered are "pasta festiva" on Sun. and Wed. nights, Mon. and Thurs. is English fare, Tues. and Fri. is classic seafood and a paniolo steak fry is the highlight for Friday. Sun.- Fri. buffet is $22, Sat. $24. An assortment of salads and wonderful desserts complete the nightly buffet. An a la carte dinner menu is also available. COMMENTS: This open-air dining hall is festively decorated in reds and greens and offers evening breezes and an ocean view. Reservations are accepted only for a group of 5 or more.

**PRINCE COURT** ★ *Gourmet American*
Maui Prince Hotel, Makena (874-1111). HOURS: Dinner 6-10 pm, Sunday brunch
10 am-2 pm. SAMPLING: "Creative Beginnings" include kiawe broiled duck
sausage with black bean cake and tomato salsa ($9) or sauteed baby Kona abalone
dusted in macadamia nut flour with ocean greens salad $18. "Lighter fare"
selections are designed with a reduction of fat and salt and available in appetizer,
soup, salad and entrees. Entrees such as Salmon Grilled in Ti leaves with roasted
banana and ginger vinaigrette ($26). "Main fare" entrees run $25 - $36 and the
selections are mouth watering. Black Angus Filet Mignon smothered with wild
mushrooms in cabernet sauce or on the "wild side" choose the mixed grill offering
a chef's selection of available game such as elk, venison, rabbit, quail and
antelope. Sunday brunch $25 adults/$15 children. COMMENTS: What do you
say? American cuisine? You may not remember mom's Sunday dinners to be
anything like this! The culinary cuisine of the Prince Court is an incredible blend
of flavors which highlight the best and freshest Hawaiian produce, meats and fish.
Beautifully situated to offer diners a splendid view of the ocean and hotel
grounds. The Sunday champagne brunch is tops on our list for a tasteful extravaganza. Over 160 items are served and might include croissants stuffed with
Portuguese sausages or ham, smoked oysters, octopus or island fish, avocados
stuffed with crab, or an unusual and delicious midora bisque soup. Brunch entrees
during our visit ranged from roast lamb or prime rib to veal piccata. The dessert
table is beyond belief! Reservations recommended.

**RAFFLES'** ★ *Continental/Seafood*
Stouffer Wailea Beach Resort (879-4900). HOURS: Sunday brunch 9-2 pm,
dinner nightly 6:30-10:30 pm. SAMPLING: Brunch includes champagne, and
entrees of chicken, veal, beef and fish which vary weekly. A salad bar, variety of
egg dishes and omelets made to your request. Pastries and desserts to delight any
sweet tooth. The buffet runs $23.50. Raffles' dinner menu also has seasonal
changes, but generally includes rack of lamb or scampi at $28.50 and fresh fish,
such as mahi, ono, or opakapaka (as available) around $25.. The wine list includes
Italian, French, California, Washington State, Australian, and German selections.
COMMENTS: Raffles' bears an Oriental theme, in keeping with its Singapore
origin. Sir Thomas Stamford Raffles (1781 - 1826) was the British founder of the
city where the Raffles' Hotel has become a legend. The buffet is a terrific splurge.
The dessert bar was constantly being replenished and each new offering looked
better than the last. The chocolate mousse merited a second serving and the
chocolate rum cake was rich and moist. Aloha wear is acceptable - an extra large
muu muu might not be a bad idea!

**SAKURA JAPANESE RESTAURANT** *Japanese*
100 Wailea Ike Drive, adjacent to the Wailea Steak House. HOURS: Dinner from
5 pm. SAMPLING: Dinners run $18 - $25.

**SEASONS** *Continental/Californian*
Four Season's Resort in Wailea. (874-8000) HOURS: Dinner only 6 - 10 pm.
SAMPLING: Pan fried opakapaka with steamed fennel and apple raisin curry
sauce, blue Hawaiian prawns and green lip mussels with olive oil and shaved
asiago. COMMENTS: Not yet open as we go to press, but look for terrace seating
with an ocean view combined with background music performed by a jazz trio.

Lobby bar outside of Seasons will be open 11 am - 11 pm. The Four Season's Sunset Lounge will offer nightly entertainment and dancing 6 pm - 1 am. A private dining room for 38 people, with its own entry and fireplace for those intimate gatherings.

# KAHULUI/WAILUKU

Along Lower Main Street in Wailuku are a number of local restaurants which are not often frequented by tourists and may well be one of the island's best kept secrets! Don't expect to find polished silver or extravagant decor, but do expect to find reasonable prices for large portions of food in a comfortable atmosphere.

Dairy Queen, Pizza Hut, McDonald's, Burger King, and Jack in the Box are a few of the restaurants in and around Kahului and the Maui Mall. These don't require elaboration. Orange Julius at Kaahumanu Mall offers top your own waffles and seafood and turkey pita sandwiches.

## INEXPENSIVE

While not restaurants, two of our favorite haunts bear mention here. The Home Made Bakery on Lower Main St is an island institution. More than just donuts, you'll find unusual specialties such as empanadas, manju, and bread pudding. You can also pick up their items in selected grocery stores. Phone 244-7015. Nearby is the Four Sisters Bakery on Vineyard and Hinano in Wailuku. It is run by Melen, Mila, Beth and Bobbie who arrived from the Philippines after helping their father run a Spanish Bakery in Manila for fifteen years. Not a large selection, but delicious and different items. One is a sweet bread filled with a cinnamon pudding, a sponge cake "sandwich," as well as cinnamon rolls and butter rolls. The only place you can purchase these delicacies is at their bakery or at the Saturday Swap meet in Kahului.

### AKI'S HAWAIIAN FOOD AND BAR    *Hawaiian*
309 N. Market, Wailuku (244-8122). HOURS: Mon.-Sat. 11 am-10 pm, Sunday 5 pm-10 pm. SAMPLING: Chicken hekka, Hawaiian favorites such as kalua pig with cabbage and a wonderful octopus soup. The prices have remained about the same as well, $4 - $6 for lunch or dinner. COMMENTS: Formerly the location of the popular Yori's restaurant, this site is now operated by two sisters, Aki Okamuro and Yoshie Hirozawa. The new owners have kept virtually the same menu. The photographs Yori has taken over the years of customers that covered the walls from floor to ceiling are gone, but we understand Yori still stops by to shoot a few for his personal collection.

### ALOHA RESTAURANT    *Hawaiian*
Kahului. COMMENTS: A long time favorite for local residents and visitors alike seeking out true Hawaiian fare has been this family owned business. After 22 years in their current location (and ten years before that at the Kahului Mall) they are currently again looking for a new facility. Our quarterly newsletter, THE MAUI UPDATE, will keep you posted!

**ARCHIE'S** *Japanese*
1440 Lower Main St., Wailuku (244-9401). HOURS: Mon. through Sat. 10:30-2 pm, and 5-8 pm. Closed Sundays. SAMPLING: Their specialty is (Hama'ko) Teishoku $8.50. Don't know what that is? You'll have to stop in and find out. Sandwiches priced $1.30 - $3, saimin $1.90, plate lunches run $3 - $8. Dinners come with soup, rice and tea $4.80 - $8. Sashimi available too! COMMENTS: The food is good and the prices are reasonable.

**BACK STREET CAFE** *American*
335 Hoohana Blvd., #7A, Kahului (877-4088). HOURS: Mon.-Fri. 10:30-2:30. SAMPLING: Daily homemade specials, salads and sandwiches. Take-out lunches available as are catering services. Very popular with the local residents.

**C.D. RUSH'S** *Italian*
Kaahumanu Shopping Center (877-6001) HOURS: Mon.- Sat. 9 am-10 pm, Sun. 10 am-3 pm. SAMPLING: Quiche $3.75, deli sandwiches from $5.25, Salad bar single serving $4.50 or all you can eat for $6.95. Menu features ravioli, lasagne from $8.95. COMMENTS: Big screen TV for sporting events in the lounge.

**CAMI'S** *Korean*
3500 Hoonhan St., Kahului (871-2445). HOURS: 10:30 am - 2:30 pm Mon. - Fri. SAMPLING: Their specialty is jhun, a Korean teriyaki, spicier than the Japanese version. The meat is dipped in flour and egg and then fried on the grill. It is served with rice, macaroni salad, kim chee and mandoo. (Mandoo is a Korean Won Ton which is deep fried but softer.) For those who like a little spice in their lives try the Ko Choo Jang, a chili-sesame sauce.

**CANTO'S**
2902 Vineyard St. (242-9758). HOURS: Vary seasonally - call for exact times SAMPLING: Gourmet dishes to go or for dining in at their limited number of tables. French dip, chicken breast, chef salad $4.75 - $6.75, daily specials might include poached mahi with dill cream, vegetable and wild rice $8.75. COMMENTS: Dinner by reservation ONLY.

**CHUM'S**
1900 Main Street, Wailuku (244-1000). HOURS: 6:30 am-2 pm daily, dinner Tues.-Sun. 5-9 pm SAMPLING: Homemade soups, stew, Hawaiian plate lunches, and chili, priced $2.50 - $5.95.

**THE CLASS ACT ★**
Maui Community College Campus (242-1210). HOURS: Lunch only Wednesday and Friday 11 am - 12:30 pm. SAMPLING: Here are a few examples 1990 luncheon menus! Regular menu: Grilled scallops with tarragon sauce, romaine lettuce, roast bell pepper and anchovy salad, duck breast with wild rice and mango chili salsa, pecan puff pastry with chocolate sauce and sabayon. Healthy heart entree: Oriental buckwheat noodles. COMMENTS: This is one of Maui's best kept secrets. Insiders know they are in for a treat when they stop by for a five-course gourmet lunch for $6.50. The Food Service students of the Maui Community College prepare and wait on the tables as well, with a varied selection of entrees weekly. Two selections are prepared, one is a healthy heart selection

which is low in sodium and fat. The program is only offered during the school year, so be sure and call to check on availability and schedule a reservation.

### FUJIYA'S ★ *Japanese*
133 Market Cafe, Wailuku (244-0206). HOURS: Lunch 11-2 Mon.-Fri., Dinner 5-9 Mon.-Sat. SAMPLING: Tempura $6.50, teriyaki $4.50, chicken $3.90. Five dinner choices include a combination such as Tempura with Yakitori, Fried Ahi, Tsukemono, miso soup and rice for $8. Beer & sake available. COMMENTS: One of our best bets for Japanese food. Sushi lovers will appreciate their new sushi bar where a large variety of selections are available at half the usual resort area price.

### HAZELS ★ *American/Local style*
2080 Vineyard St., (244-7278). HOURS: 6 am-9 pm for breakfast, lunch and dinner. SAMPLING: Beef teriyaki, chicken, pork chops, fresh fish $4 - $6, also saimin, and sandwiches. COMMENTS: Hazel's is the place that the locals go for a good homecooked meal. In addition to the regular dinner menu, there are a half dozen evening specials and a fresh fish entree. Even at these low prices the dinners are generous portions, well prepared and are accompanied by soup or salad, rice or mashed potatoes, dinner roll and coffee, tea or fruit punch. A great place for the family to dine out, fill up and still leave with change in your pocket. 1989 recipient of Best Bet for local dining presented by Paradise Publications!

### ICHIBAN THE RESTAURANT *Japanese*
Kahului Shopping Center (871-6977) HOURS: Breakfast and lunch 7 am-2 pm, dinner 5-9 pm. SAMPLING: Tempura Don, Katsu chicken, teri steak, sukiyaki $6.95. COMMENTS: Located in the older Kahului Center, this restaurant doesn't stand out as memorable for its food or ambiance.

### IMPERIAL TEPPANYAKI *Japanese*
Maui Palms Hotel, Kahului (877-0071). HOURS: Nightly from 5:30 pm-8:30 pm. SAMPLING: A buffet with different items prepared by teppanyaki chefs at the buffet. Entrees might include fried fish, ika tempura, chicken yakitori. From the salad bar sample miso soup, tofu with ginger sauce, sashimi and other local favorites.

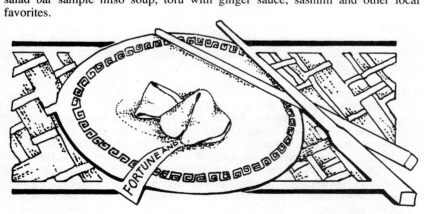

**INTERNATIONAL HOUSE OF PANCAKES**   *American*
Maui Mall (871-4000). HOURS: Sun.-Thurs 6 am-midnight, Fri. and Sat. until 2 am. COMMENTS: A very large facility with a menu that is popular with all family members. Something for everyone and at reasonable prices. Basic dinners begin in the $5 range.

**KALEO'S KAU KAU**   *Local Style/Filipino/Hawaiian*
Lower Main Street (244-2040). HOURS: Lunch 11-2, Dinner from 5:30. SAMPLING: Plate lunches include tripe stew or chop steak $4.50, and a number of Hawaiian dishes include luau stew, squid with coconut or opihi (only place we've every seen this on the menu) and Filipino dishes such as pork sari sari.

**KEN SAN**   *Japanese*
2051 Main Street (242-2971). HOURS: Mon.-Sat. 10:30 am-2 pm. SAMPLING: A self serve counter with a variety of hot and cold selections, all priced under a $1.40. Choose chicken katsu, mahi, noodles or fried shrimp. Scoop out a portion of apple or macaroni salad, fried or white rice too. A sushi tray for $2.40 includes two pieces of shrimp sushi and four pieces of California roll. COMMENTS: Formerly a head chef at a Lahaina restaurant this owner/chef runs a new and highly popular spot for the Wailuku work force. A few seats along the counter up front and four tables in back, or fill up your carton to go. Sat. is the best time to visit when it is not only quieter, but the selection is better. The sushi especially gets pretty well picked over if you don't get there REAL early during the week. But it is no wonder - where else can you get four California rolls for a $1!

**KOHO GRILL AND BAR ★**   *American*
Kaahumanu Shopping Center (877-5588). HOURS: Daily for breakfast 8-11 am, lunch and dinner until 9:30 pm on Mon.-Thurs., 11 pm on Fri. and Sat., and 9 pm on Sun. Happy Hour daily 3-6. SAMPLING: Quiche $4.65, fajitas $6.95 blackened chicken breast sandwich $4.75, soups, salads, burgers and sandwiches $1.95 - $4.95. Dinners are served from 5 pm and include pasta, stir fry and cajun as well as trout almondine, T-bone steak, shrimp scampi $6.45 - $11.95. COMMENTS: A great family dining spot. Dinners are affordable and the menu is broad enough to offer something for everyone with sandwiches available for dinner as well as lunch. One of only a few restaurants in the area open on Sundays! Convenient for a bite enroute to the airport home, with fast and friendly service.

**LOPAKA'S BAR AND GRILL**   *American*
161 Alambra, Kahului Industrial Area (871-1135). HOURS: 11 am-9 pm. SAMPLING: Same menu for lunch and dinner. Burgers and sandwiches $3.95 - $7, plate lunches such as Korean ribs and BBQ beef $6.95 - $7.95. COMMENTS: More a bar atmosphere than restaurant.

**LUIGI'S**   *Italian*
Maui Mall, Kahului (877-3761). HOURS: 11:30 am-4 pm for lunch, 4 pm-10 pm for dinner. SAMPLING: Lunch offers potato skins, saimin, calamari strips, pasta, pizza and burgers $3.99 - $8.99. Dinner offers veal marsala, steamed mussels, more pasta and pizza $10.99 - $17.99. COMMENTS: Daily specials. Now offers Karaoke entertainment with tunes from the 50's to 80's and sometimes disco, call for days and times.

### MA CHAN'S OKAZUYA  *Oriental*
Kaahumanu Mall, Kahului. HOURS: Mon., Tues., Wed., Sat. 6 am-8 pm, Thurs. and Fri. until 9 pm, Sun. until 3 pm. SAMPLING: Okasuya style restaurant with plate lunches ($3.45 - $3.75). COMMENTS: Another small okasuya lunch spot is in the "buffet" corner at the Shirokiya Department store. You'll find bento lunches, sushi $3.25 - $4.25, selections of beef broccoli $1.50, and rice 35 cents.

### MAMA DING'S PASTELES RESTAURANT ★  *Puerto Rican/Local*
255 E Alamaha St., Kahului (877-5796). HOURS: Breakfast 7:30-10:30 am, lunch 10:30 am-3 pm, Mon.-Sat. SAMPLING: Eggs Bermuda (2 eggs whipped with cream cheese and onion) served with potatoes or rice and toast $3.25, fresh fruit pancakes $3.25. Lunch specials include Steak Montego, a steak sauteed with onions, mushrooms in a brandy brown sauce and served with vegetables, rice and salad $6.25. Bacalao salad, a tossed green creation with codfish bits, green banana and dressing for $3.75. For the less adventurous try a grilled chicken breast sandwich! COMMENTS: Ready for a different breakfast? Skip IHOP and try this cozy restaurant tucked away in the Kahului Industrial Area. Or try a pastele which has an exterior of grated green banana and a filling of pork, vegetables and spices that is then steamed. Delicious! We've tried several breakfasts, all were good! Looking forward to going back for lunch! Stop by after the Saturday Swap Meet or order a picnic to go to Hana or Haleakala. No credit cards.

### MAUI BOY  *Hawaiian/Local Style*
2102 Vineyard St., Wailuku (244-7243). HOURS: Tues., Wed., Thurs., Fri. 10 am-2 pm for lunch 5 pm-9 pm for dinner. Sat. and Sun. breakfast 7 am-11 am. Lunch 11 am-2 pm, dinner 5-9 pm. Closed Mon. SAMPLING: Hawaiian favorites such as Kalua pig and poi or lau lau. Local style plate lunches available for lunch and dinner include chicken or pork teriyaki, roast beef or pork priced $5 - $7.

### MAUI GRILL AND CANTINA  *Mexican*
Corner of Main and Market St. Wailuku (244-7776) HOURS: Mon. - Fri. 6 am - 3 pm. Saturdays 10 am - 3 pm. SAMPLING: Homestyle Mexican food such as tamales, enchiladas and daily specials $3.50 - $5. Freshly made rolls for breakfast along with Mexican favorites such as huevos rancheros and skillet egg and potato dishes. COMMENTS: Just opening as we go to press, Tom Kane and Curtis Connors are transplanted Arizonians who are familiar with Mexican cookery. They hope to open in the future for dinner and are also looking into some "theme" dinners to coincide with plays at the nearby Iao Theatre.

### MING YUEN  *Chinese*
162 Alamaha, Kahului (871-7787). HOURS: Lunch daily 11:30 am-5 pm, dinner 5-9 pm. SAMPLING: Cantonese and Szechuan style foods, dishes $5 and up. COMMENTS: A little off the beaten track, you'll find it tucked behind Safeway off Kamehameha Avenue in the industrial area.

### MOON HOE CHINESE SEAFOOD RESTAURANT  *Oriental*
752 Lower Main St., Wailuku (242-7778). HOURS: 11 am-2 pm for lunch Mon.-Sat., daily dinner 5-9:30 pm. SAMPLING: 150 menu items including a variety of preparations for duck, pork, beef, chicken, vegetables and fish. Prices generally run $3.95 - $6.95.

**NAGASAKO FISH MARKET**   *Seafood/Local*
1276 Lower Main St., Wailuku (242-4073). HOURS: Store hours Mon.-Sat. 6 am-8 pm, Sun. 7 am-4 pm. Lunches served Mon.-Fri. 9 am-4 pm. COMMENTS: Operated by Jordan Nagasako (of the Lahaina Nagasako grocery family) this newcomer has lots of potential. As a fisherman, he can buy and sell fish wholesale, thus offering an amazing variety of fish and shellfish for you to cook in your Maui home. Exotic reef fish include opelu, akule, weke or oio. Live clams and crabs too!! Looking for something more unusual? Then try some of their poke (marinated raw fish). A very interesting assortment that might include conch-shell poke with flying-fish eggs. Plate lunches run $3.99 - $5 and include stews, chicken or fresh fish and are served with rice and macaroni salad. As of now, this is a "to go" eatery, but they hope to add some tables in the future. They also plan on extending their hours into the evening in the near future.

**NAOKEES**   *Hawaiian/Local*
1792 Main St. (244-9444) HOURS: Mon.-Sat. 11 am-2 pm. Dinner daily 5:30-9:30. SAMPLING: Complete dinners from $7.95 - $13 include a variety of steaks, Korean ribs, broiled fish, prawns, teri chicken. All are served with soup and salad, rice, beverage and vegetable. Lunches run $4.50 - $6. COMMENTS: What it lacks in ambience, it makes up for in pleasant and prompt service. Their special New York steak was tender and juicy, and as well prepared as many of Maui's finest restaurants. The vegetables were from a can, and the salad was standard fare. A slightly more expensive rib steak required a dose of A-1 to bring it to life.

**NAZO'S**   *Hawaiian/Local Style*
1063 Lower Main St., Wailuku, at Puuone Plaza (244-0529). HOURS: Mon.-Sat. 6 am-9 pm. SAMPLING: Sandwiches include egg salad at $1.40, or grilled ham and cheese $2.25. Entrees include soup or salad, rice or mashed potatoes, coffee, tea or fruit punch. Selections include liver with bacon, shrimp tempura for $4 - $5. A tossed salad adds 80 cents. Luau stew is featured on Wed., pig feet on Thurs. and Sat. COMMENTS: A small, family-owned restaurant which is very affordable and prides themselves on their home-style cooking. No credit cards.

**PIZZA CIRCUS ★**   *Italian*
333 Dairy Rd., Kahului (871-1133). HOURS: Daily for lunch or dinner. COMMENTS: Our best bet for pizza on Maui. What makes this pizza great is their unusual crust which is topped with sesame seeds.

**RED DRAGON CHINESE RESTAURANT**   *Chinese*
Maui Beach Hotel (877-0051). HOURS: Nightly 5:30-8:30 pm, closed Mon. SAMPLING: Cantonese buffet dinner with over fifteen selections which change nightly. Entrees may include haposai, clams in hot sauce, sweet and sour pork or roast chicken. Reservations required.

**SAENG'S THAI CUISINE ★**   *Thai*
2119 Vineyard St., Wailuku, (244-1567). HOURS: Lunch Mon.-Fri. 11 am-2:30 pm, dinner daily 5 pm-9:30 pm. SAMPLING: Basil chicken, shrimps asparagus, sate or tofu with peanut sauce $4.50 - $8. COMMENTS: This wins hands down for the most attractive local restaurant in Wailuku. Owners Toh, Tom and Zach Douangphoumy have created a little Eden with lots of plants providing privacy

between tables. They also know how to cook Thai. Traveling in India, Laos, Vietnam and Thailand in their youth they had an opportunity to sample a diversity of foods. 1989 seemed to be the year of the Thai on Maui and we tried them all. This newcomer is far and away the best of the bunch. Not only was the service attentive, but the portions generous and every dish better than the last. We were especially partial to the peanut sauce. In fact it was so good that enroute to the airport and bound for the mainland we picked up some peanut sauce "to go!" Don't miss this one! They currently operate another restaurant on O'ahu.

**SAM SATO'S**   *Local Style/Hawaiian*
318 North Market St., Wailuku (244-7124). HOURS: Breakfast and lunch 8 am-2 pm, pastries served until 4 pm. Closed Thurs. & Sun. SAMPLING: Noodles are their specialty. Breakfast includes eggs or pancakes. Lunch options include combination plates $3.50 - $4 such as teriyaki beef, stew, chop steak or spare ribs. Sandwiches and burgers. Saimin and chow fun are served in small portions for $1.70/$1.85, or large portions for $2.10 - $3.70. COMMENTS: The homemade pastries are wonderful. The peach, apple and coconut turn-overs were fragrant and fresh. In addition to noodles they specialize in manju, a Japanese tea cake. It may come as a big surprise when you discover that these tasty morsels are actually filled with a mashed version of lima beans!

**SIAM THAI**   *Thai*
125 N. Market St., Wailuku (244-3817). HOURS: Lunch Mon.-Fri. 11-2:30, dinner daily 5-9:30. SAMPLING: Exotic green papaya salad $4.25, eggplant tofu $4.95, Thai ginger beef $5.95. COMMENTS: The white table cloths give this restaurant an elegant air. According to our waiter we dined at the same table Robert Redford used when he visited. Very good Thai food although the competition is now stiff with the new Saeng Thai just around the corner.

**SIR WILFRED'S ESPRESSO CAFE**   *American*
Maui Mall, Kahului (877-3711). HOURS: Mon.-Thurs. 9-6 pm, Fri. 9-9 pm, Sat. 9-5:30 pm, Sun. 10-3pm. SAMPLING: Continental breakfasts with fresh pastries. Bagels or deli-style sandwiches are available for lunch as well as a variety of salads or quiche. Also a selection of gourmet teas, coffees or beer and wine.

SAM SATO'S

**SIZZLER**   *American*
355 E. Kamehameha Hwy, Kahului (871-1120) HOURS: 6 am-10pm, until midnight on Fri. and Sat. SAMPLING: Sirloin steak, sizzler steak, chicken dishes all in the $7 - $10 range. All you can eat salad bar includes soup, salad and tostada bar for only $6.99. A keiki (children's) menu offers a choice of steak, shrimp, burger, fish platter, or salad bar, all for $6 or less. COMMENTS: As Sizzlers go, this is one of the better ones. A good selection of entrees at family prices and a very good salad bar as well!

**TASTY CRUST** ★   *Local Style*
1770 Mill St., Wailuku (244-0845). HOURS: Daily 5:30 am-1:30 pm for breakfast and lunch daily, 5 pm-10 pm for dinner except Mon. SAMPLING: Unusual and delicious crusty hotcakes are their specialty, two are a meal for $1.80, or French toast $1.70. Add an egg for 50 cents. Lunches and dinners a la carte. Spare ribs $4.25, fried shrimp $3.95, roast beef $4.75 and served with rice and a salad. Sandwiches and hamburgers $1.35 and up. COMMENTS: Local atmosphere and no frills, just good food at great prices.

**THE VIETNAM RESTAURANT** ★   *Vietnamese/Chinese*
1063 Lower Main St. in the lower level of the Puuone Tower Bldg. (244-3414) HOURS: Mon.-Thurs. 11 am-9 pm, Fri. and Sat. 11 am-10 pm and closed Sun. SAMPLING: Ga Xoa Thap Caam (a chicken and vegetable dish), Beef Ball Soup, Vietnamese Crepe and numerous Chinese dishes $4.50 - $5.95. COMMENTS: Nyguen Walker and Vuong Chan Thanh moved across the street from their original location. The Vietnamese dishes are a wonderful and exotic blend of flavors and foods. If you haven't tried Vietnamese food, by all means stop here! If you don't care for Vietnamese, then there is a large Chinese menu selection as well! All entrees are prepared without MSG. Try their iced coffee too! We seldom come to Wailuku without stopping in for a meal!

**THE VIETNAM RESTAURANT** ★
Same owner/operators with this smaller location at their original location on 1246 Lower Main Street.

**TIN YING**   *Chinese*
1088 Lower Main St., Wailuku (242-4371). HOURS: Daily 10 am-9 pm. SAMPLING: Selections include Hong Kong or Szechuan style with prices ranging from $5.25 - $7.75. Eat in or take out. COMMENTS: Okasuya style lunches at $2.75 include an entree and fried rice or noodles and are a real good value.

**TIPANAN**   *Filipino*
1276 Lower Main St., Wailuku (244-9466). HOURS: 10:30 am-9 pm Mon.-Sat. SAMPLING: Entrees include bistek tagalog, daing na bangus, paksiw na pata. (These translate to filet of beef tenderloin with lemon juice and soy, milkfish fried to a savory taste, and tender pork stewed in sweet-sour sauce.) Priced $3.00 - $6.50 they are served with rice. COMMENTS: Interesting fare, although we have decided that Filipino food isn't one of our favorites.

**TOKYO TEI ★**   *Japanese*
1063 E. Lower Main St., Wailuku (242-9630). HOURS: Lunch 11-1:30 Mon.-Sat., dinner daily 5-8:30. SAMPLING: Lunch specials run $4 - $5. Teishoku trays $7 - $8.50 include shrimp tempura, sashimi, fried fish, teriyaki pork or steak. Dinner selections such as hakata chicken, seafood platter, and a number of others that are difficult to pronounce, run $4.25 - $8 and include rice, miso soup, namasu and koko. COMMENTS: Small and cozy atmosphere. Great food! Take out meals also available. Cocktails. Very popular with local residents and tourists and deservedly so. Winner of one of our top three awards for best local restaurants.

**WING SING**   *Chinese*
1424 L. Main Street, Wailuku (244-3813). HOURS: Tues.-Sun. 11:30-2 for lunch, 4-8 pm for dinner. Closed Mon. SAMPLING: Chop suey, char sui, shrimp canton. Prices run $2.75 - $5.50. COMMENTS: No checks or credit cards.

## MODERATE

**AURELIO'S**   *Continental*
55 Kaahumanu Ave. across from the Maui Mall (871-7656) HOURS: Lunch 11 am-5 pm, dinner from 5 pm Mon.-Sat. SAMPLING: Lunchtime selections $5.50 - $6.95. Dinner entrees are served with salad and bread, rice or scalloped potatoes and fresh vegetables priced $8.95 - $17.95 for rack of lamb, vegetable pasta, crab claws, beef short ribs.

**CHART HOUSE ★**   *American/Seafood*
500 N. Puunene Ave., Kahului (877-2476). HOURS: Dinner 5:30-10 pm. SAMPLING: Prime rib is their specialty, also fresh seafood and chicken priced $12 - $20. COMMENTS: Large portions, excellent children's menu. You can be sure it will be less crowded on this side of the island. No reservations taken, but a nice ocean view is a consolation. Served with a table service salad bar.

**MICKEYS ★**   *American/Seafood*
Kahului Bldg., 333 Lono Ave. (877-7225). HOURS: Lunch and dinner. SAMPLING: Lunches run $6.50 - $12.95. Dinners range from $12.95 for chicken to $18.95 for seafoods. Their "king" platter for two is $64.95. Fresh fish priced daily, children's portions 1/2 price. COMMENTS: Excellent fresh fish! They have a sister facility in Kihei (Island Fish House) that is very popular as well.

# UPCOUNTRY

## INEXPENSIVE

**BULLOCK'S OF HAWAII**   *Local Style/American*
Just past Pukalani Shopping Center on the right side going up the mountain. HOURS: Breakfast 7:30 am-11:30 am, lunch and dinner 11:30 am-9:30 pm. SAMPLING: Usual breakfast items, omelettes from $2.65, French toast $1.95. A moonburger runs $3.50 (it was lunchtime here when the first astronaut walked on the moon), a guava shake at $2.75, sandwiches from $1.95, and plate lunches $3.50 - $6.25. The lunch and dinner menus are the same. COMMENTS: A landmark in Upcountry, celebrating their 23rd year in business October 1990!

**DAIRY QUEEN - MCDONALD'S** - Pukalani

**FU WAH**   *Chinese/Szechuan*
Pukalani Shopping Center. (572-1341) HOURS: 10:30 am - 3 pm Tues. - Sun.,
5 pm - 9 pm Sat. and Sun. SAMPLING: Dinners entrees from $5.25 or choose
family dinner combinations. Sizzling platters and braised pot courses from $6.

**GRANDMA'S COFFEE HOUSE ★**   *Local Style*
Keoke. HOURS: Daily except Monda 7 am-5 pm. COMMENTS: This is a family
operation run by Alfred Franco, his wife and their two young children (son Derek
is known as "the boss" and a younger sister Alyson). Alfred is encouraging the
return of the coffee industry in Upcountry Maui. Born and raised in Upcountry,
his grandmother taught him how to roast the coffee beans to perfection. He does
this several times a day in his 104 year old coffee roasting machine that was
brought from Philadelphia by his great grandmother. It is then sold by the pound
in a blend known as "Maui coffees." Some of the beans used are grown on
Molokai which is part of Maui County. Prices are high, and there is no decaf
available. (Grandmother never taught Alfred how to do that.) A few tables are an
invitation to visitors to sit down, enjoy a cup of coffee, espresso, cappuccino, or
fresh fruit juice along with cinnamon rolls, muffins, and more all fresh from the
oven. A bit more of an appetite might require one of their fresh avocado sandwiches, a bowl of chili and rice or homemade Portuguese bean soup. With the
popularity of this place among locals and visitors alike, it is tough for them to
keep up with the demand for these goodies. So if you're hungry in Upcountry, be
sure to stop in at Grandma's, just five miles before the Tedeschi Winery. Alfred's
goal is to put Keoke on the maps and minds of everyone. And he just may do it!

**KITADAS RESTAURANT**   *Local style*
3617 Baldwin Avenue, Makawao. PHONE: 572-7241 HOURS: 6 am-1:30 pm
daily except Sun. COMMENTS: Popular local eatery.

**YOUR JUST DESSERTS**   *French/American*
Makawao Ave., Makawao (572-1101) HOURS: 10:30 am - 8 pm weekdays, until
9 pm Sat. and Sun. SAMPLING: Low and non fat yogurt, shakes, and espresso.
Quiche, stuffed potatoes, French onion soup, turkey or reuben sandwiches, garden
or fruit salad. $3 - $4 range. COMMENTS: A cozy little corner with good food
and a comfortable atmosphere, this eatery adjoins a gift shop.

KONA COFFEE                                    JBAYOT

# MODERATE

## CASANOVA ITALIAN RESTAURANT AND DELI    *Italian*

1188 Makawao Ave., Makawao (572-0220). HOURS: Deli from 8:30 am-7:30 pm. Dinner at the restaurant from 5 pm-1 am. SAMPLING: The deli offers hot sandwiches, pastries, cappuccino and fresh juices. Pizza too! Restaurant dining offers pastas $6.95 - $10.95 such as Linguine alle vongole veraci (Clams in wine, olive oil, garlic and shell fish broth) along with catch of the day, ossobuco (braised veal shank on a pasta bed), petto di pollo or filetto ai ferri $14.95 - $17.50. Pizza made in Maui's only wood-burning oven. COMMENTS: We'd read some fairly good reviews and were disappointed by our evening repast. We tried three types of pasta and found them lacking. Perhaps because we aren't Italians and don't know an authentic dish, but the sauces seemed rather thin and runny. The best of the pastas was the spaghetti bolognase. Al dente may be in vogue, but they were so lightly cooked that they were chewy. We were also a little disappointed you can't order pineapple on your pizza, maybe that isn't Italian? The pizza crust was thin and crisp, a little burned around the edges (but forgiveable considering their woodburning oven - the only one of its kind on Maui!) Another question we had were several reports that people enjoying a final cup of espresso at 9:30 following their dinner were asked to ante up for the evening entertainment. The menu does say that the entertainment fee may be charged, but for diners it just doesn't seem right. As for entertainment they have a large dance floor and bring in local and mainland talent. We didn't try any of the fish, chicken, lamb or veal dishes, so we're keeping an open mind and will report back! Let us know your experiences here and anywhere else on Maui that you dine!

## HALIIMAILE GENERAL STORE  ★  *Continental*

Haliimaile (572-2666) HOURS: Lunch 11-3, dinner 6-10 Sunday brunch 10-3. SAMPLING: Lunches include a creative selection of sandwiches and salads $5 - $8 served with unusual salad accompaniments. Dinners run $15 - $25. COMMENTS: This restaurant has put Haliimaile on the map since it opened several years ago. The original structure dates back to the 1920's when it served as the General Store and hub of this community. The 5,000 square foot restaurant has two main dining rooms. Jeremy Realton assisted with the interior design (among his many accomplishments is the set for the Pee-Wee Herman show.) The high ceiling is the original and the floors are refurbished hardwood. The tables are set with cloths in green and peach and there is a huge beautifully designed bar in the front dining room. The menu rotates each Thursday stressing quality in their food preparation from the sourdough bread (which they have shipped in from the mainland for a final baking in their own ovens), to the fresh herbs they cultivate in their own garden. The dinner menu features an interesting selection of dishes with unusual and creative preparations. While the items change, standard items include Paniola ribs, vegetarian lasagna, roasted half duck and half smoked chicken. Dinners run in the $15 - $25 range. Our only criticism is that they could really use a children's menu. Enroute to Pukalani on Hwy. 37, look for the sign that points to Haliimaile. Weekend entertainment is also in the works. It's well worth the stop. Bon appetite!

## KULA LODGE   *American*
Five miles past Pukalani on Haleakala Hwy. (878-1535). HOURS: Breakfast 7-11:30 am, lunch 11:30-5:30 pm; dinner 5:30-9 pm. Dinner is served only on Sat. and Sun. reservations are recommended. SAMPLING: Lunches include sandwiches, salads, and burgers from $5, also hot entrees. Dinners include rice or potato, vegetable, French bread and are priced $9.95 - $15.95 for rack of lamb, sesame chicken, curry bombay, or mahi mahi amandine. COMMENTS: An added benefit here is the fireplace, a warming delight after a cold trip to the mountain top, a panoramic view, and cocktails. Children's portions available.

## MAKAWAO STEAK HOUSE ★   *American*
3612 Baldwin, Makawao (572-8711). HOURS: Dinner nightly 5-9:30, Sunday brunch 9:30-2 pm.   SAMPLING: Chicken Zoie $12.95 (breast stuffed with creamed spinach), Scampi $17.95, N.Y. steak $14.95, fresh fish varies daily. Dinners include salad bar and freshly baked bread. COMMENTS: Cocktails & wine list. We enjoyed fresh Muu, a mild white fish, and have never found or heard of it since!

## POLLI'S   *Mexican*
1202 Makawao Ave., Makawao (572-7808). HOURS: 11:30-2:30, Sunday brunch 10:30-2:30, Happy hour 2:30-5 pm, dinner 5-10 pm. SAMPLING: Dinner combinations run $8 - $12. Vegetarian dishes available.

## PUKALANI TERRACE COUNTRY CLUB   *American/Hawaiian*
360 Pukalani Rd. (572-1325). Turn right just before the shopping center at Pukalani and continue until the road ends. HOURS: Open 10-2 for lunch, 5-9 for dinner.   SAMPLING: Lunch offers a Kalua pig Hawaiian plate for $7.50 or tripe stew for $6.65. Dinner menu offers similar selections at slightly higher prices. Salad bar $6.85, salad bar with dinner $2.85. COMMENTS: Check for early bird dinners and nightly dinner specials. A great view of Maui from here. You might consider a stop on the way down from Upcountry for a drink and tropical sunset.

## SILVERSWORD INN   -   Closed again....

# PAIA

A growing number of restaurants are springing up in Paia and making the reasons to stop here for a meal even better.

## INEXPENSIVE to MODERATE

### CHARLEY'S   *American/Italian*
Hana Hwy in Paia. (579-9453) HOURS: Breakfast and lunch 7 am-2 pm, dinner 4-11 pm. SAMPLING: Cheese omelette, steak and eggs, huevos rancheros $4.25 - $7.95, sandwiches $4.25 - $5.95, dinners include two homemade garlic rolls and a dinner salad. Selections include eggplant parmesan, calzone, fettucini, scampi $6.25 - $10.95. COMMENTS: Renovations will double the size of their sports bar which has a satellite dish for their six televisions, enlarge the eating area and upstairs provide a concert or large dining area.

### DILLON'S ★    *American*
Downtown Paia, 89 Hana Hwy. (579-9113). HOURS: Breakfast, lunch, and dinner 7 am-2 am. SAMPLING: Complete dinners such as pepper steak, pasta or fish run $12.95 - $17.95. COMMENTS: We enjoyed a hearty breakfast with generous portions and some unusual items, such as Kahlua French toast $4.95, Hamburgers $5.95 - $7.25. Opens early for diners desiring breakfast enroute to Hana or Haleakala. Dancing each evening 9 pm.

### KIHATA RESTAURANT    *Japanese*
115 Hana Hwy., Paia (579-9035). HOURS: 10 am-2 pm for lunch Tues.-Sat., dinner 5-9 pm daily except Sunday. SAMPLING: Lunches include chicken katsu, noodle dishes, beef teriyaki $4 - $7. Dinner Teishoku meals include miso soup, rice, shrimp tempura, chicken teriyaki and fish for $11 - $12. Bento lunches and sushi also available. COMMENTS: Good food, but prices are slightly higher than similar restaurants in Wailuku.

### LA VIE EN ROSE ★    *French*
62 Baldwin Ave., Paia. (879-9820). HOURS: Daily for breakfast lunch and dinner 8 am-9 pm. COMMENTS: "Funky" is the word that comes to mind as a description for this local haunt. Entering through an open doorway with hanging beads (Is this 1970?) you are greeted with an odd assortment of tables, chairs and even a sofa for seating. The menu is on the blackboard behind the counter and ordering is done there. For a number of years this has been an especially popular spot with locals or windsurfers tired and hungry from a long day on the beach or in the water. Don't let the surroundings dissuade you from trying their wonderful fresh pasta or baked goods. Omelettes run $3.95 - $4.50 and a freshly made croissant and espresso make an excellent breakfast fare. For lunch they offer sandwiches such as veggie cheese, ham with boiled egg or tuna. For lunch or dinner a huge portion of fettucini or spaghetti runs $5. Each evening they offer a chicken and fresh fish meal as well. The fish dinner is the most expensive, at $9. We tried the Ulua which was a small serving covered with a bernaise sauce and a tasty brown rice. The chicken was a fairly good portion, perfectly seasoned and served with pasta or rice. However, this spot is at its best when it comes to the pasta. Be sure you are REALLY hungry, or have a buddy to share with, and try their spaghetti bolognaise, linguini basquaise or veggie tofu lasagna, all incredibly priced at $4.95. Still hungry? Check out their homemade desserts! Salads run $4.25 and include Greek, Caesar or "dinner." Don't miss stopping by for a hearty meal after a day Upcountry or after windsurfing at Hookipa and have a pasta-rific meal.

### PAIA FISH MARKET RESTAURANT    *American/Seafood*
On the corner of Baldwin Ave. and Hana Highway. (572-8030) HOURS: Lunch noon to 5 pm and dinner 5 pm-9:30 pm SAMPLING: Lunch and dinner selections similar with lunch plates around $7.95, two or three dollars less expensive than the dinner counterpart. A blackboard slate recounts the selections such as fish tacos, fish chowder, shrimp fajitas, ahi burgers. Fresh fish is selected from the case and runs $10.95 - $14.95 for a dinner portion served with rice or home fries and coleslaw. Beer, wine and champagne. COMMENTS: This is the sixth restaurant for owner Warren Roberts from Malibu California. Our trial here was disappointing. Order at the counter and then pick up your meal and seat yourself at a half dozen over-sized picnic tables. A dark interior at night with interesting

and humorous artifacts lining the walls. The fish in the display looks fresh, but filets of such fish as mahi or onaga are dried out when charbroiled. Our fish was accompanied by a pile of fried potatoes that filled up the plate. No good selections for children.

### PEACHES AND CRUMBLE    *American/Healthy*
On Baldwin Ave. just off Hana Hwy. (579-8612). HOURS: 6:30 am-6 pm Mon.-Fri., Sat. and Sun. 9-6. SAMPLING: Unusual baked items such as carrot cake with guava filling, lilikoi cheesecake, "jungle bars" (dried organic bananas, coconut, macademia nuts and passion fruit). Sandwiches such as salmon with cream cheese or Mexican avocado are made on their own freshly baked bread. Specials include pizza. A few counter seats.

### PICNIC'S    *American/Healthy*
30 Baldwin Ave, a few blocks off the Hana Hwy. (579-8021). HOURS: 7:30 am - 3:30 pm, 7 days a week. Closed Thanksgiving and New Years. SAMPLING: The Plantation breakfast includes eggs scrambled with cheddar, coffee and toast $2.95. Their lunch specialty is the spinach nutburger $3.85, Mahi mahi supreme $4.95, turkey $3.85, avacaod and swiss $3.75. Countryside Box lunch at $7.50 includes a choice of sandwich, drink, cookie, chips and fruit. COMMENTS: A very popular place to pick up some lunch goodies for the road to Hana or Haleakala. Anything on their menu is available to go and everything is ready from 7:30 am. No need to call ahead, just stop by enroute.

### THE VEGAN RESTAURANT    *American/Vegetarian*
115 Baldwin Ave., a few blocks off the Hana Hwy. (579-9144). HOURS: 11:30 am-8:30pm. SAMPLING: Serves only vegetarian foods. Tofu omelette $2.50, lasagna tofucci $3.95, sweet and sour tofu $4.50, carrot salad $2.95, sandwich platters $4, complete dinners vary daily $6.95. COMMENTS: It is hard to believe with the wealth of restaurants that this is currently Maui's only vegetarian. It opened in late 1989 with an aim to create foods with tastes and textures to resemble meat products, but without the use of animal products or cholesterol in any of their food. Vegan is a non-profit organization that has been doing vegetari-an nutrition seminars for more than six years before opening this restaurant. Seating for a dozen people. Catering also available.

## EXPENSIVE

### MAMA'S FISH HOUSE ★    *Continental/Seafood*
On Hwy. 36 just 1 1/2 miles past Paia, look for the ship's flagpole and the angel fish sign (579-9672). HOURS: Lunch served 11-2, happy hour and pupus 2:30-4:30, dinner from 5:30 pm. COMMENTS: Mama opened her home, located on a peaceful beachfront in 1973, as Hawaii's first fresh fish house. Her mission was "to serve creative seafood dishes with that elusive taste of Maui island cooking." Mama engages her own fishermen to catch their daily fare. Samples include Ono sauteed in mango butter and Baked Opakapaka Hana. Non-seafood entrees such as Kalbi ribs or chicken in papaya are available. It is a little out of the way, but you can be assured of a quality seafood meal. One of the most expensive seafood restaurants, but worth it. Reservations suggested.

# HANA

## INEXPENSIVE

**HANA RANCH STORE** - Open daily, ready-made sandwiches and hot dogs.

**HASEGAWA GENERAL STORE** - Open daily, a little bit of everything!!

**TUTU'S** - Hana Bay, 8:30-4 pm. Sandwiches, plate lunches.

## MODERATE

### HANA RANCH RESTAURANT   *American*
Downtown Hana (248-8255). HOURS: 6:30-10 am daily for breakfast (take out only), 11-3:30 pm for lunch, dinner served Fri. & Sat. 6-9 pm. Varied. SAMPLING: Entrees include fresh fish, smoked baby back ribs, grilled tiger shrimp, and T-bone steak. The lunch buffet is $8.95 which includes hot foods as well as an assortment of salads. The take-out counter serves hot and cold sandwiches as well as plate lunches. Friday and Saturday dinners are a la carte $18-$25 COMMENTS: Attire is casual. Choose indoor dining or a tree shaded picnic area outside. Full bar service. Dinner reservations recommended.

## EXPENSIVE

### HOTEL HANA MAUI DINING ROOM ★   *Continental*
(248-8211). HOURS: Breakfast 7:30-10 am, a la carte or buffet lunch 11:30-2 pm, dinner 6:30-8:30. SAMPLING: Dinners are a la carte. Wednesday and Sunday evenings is a buffet of seafood, beef, lamb, chicken, fish, or cold seafood. Lunch entrees include grilled ahi, NY steak, or salads. Dinner entrees (vary with specials offered) might include grilled ahi, bamboo steamed mahi mahi, chicken in mango thyme butter, appetizers and dessert. COMMENTS: Wonderful food served with unusually, light, delicious sauces. Non-hotel guests need to call in advance to see if there is space available. Current prices for non-guests are $16 for breakfast, $19 for lunch, and $50 for dinner.

TUTU'S

# SUNSETS AND NIGHTLIFE

Here are a few suggestions as to what to do when and after the sun goes down on Maui. These locations usually offer entertainment, however, call to see what they are offering and which night, as it varies. Also check the Bulletin Board publication, This Week or the Holiday section of the Friday edition of the Maui News, which lists current late night happenings.

## SUNSET WATCHING SUGGESTIONS

On the front lawn of the Pineapple Hill Restaurant near Kapalua
From the lobby bar of the Kapalua Bay Resort with their wonderful pupus
At the Kapalua Grill and Bar
On the promontory at the Bay Club, Kapalua
Atop Black Rock in the Sheraton's Discovery Room
At the Hyatt Regency's Lahaina Provision Company
Enroute down from Haleakala, the Pukalani Country Club
The Maui Outrigger provides your sunset view from surfside.
The Fairway at the Wailea Golf Course (try an ice cream drink)
Enjoy the lobby bar at Stouffer Wailea Beach Resort

## NIGHT SPOTS & ENTERTAINMENT

Consult the *Holiday* section of the *Maui News*, Friday edition, to see who is playing when and where. The following spots generally offer entertainment:

### LAHAINA/KAANAPALI/KAPALUA AREA
*Lahaina:* Moose McGillycuddy's is always hopping for the young crowd. You'll find Jazz at Blackie's Bar four nights a week. Lahaina Broiler offers entertainment several nights a week, dance floor. Pioneer Inn. Whale's Tail has piano entertainment 6 - 10 pm, no dancing and happy hour 3 - 6 pm. Longhi's has live entertainment and a dance floor. Pioneer Inn features happy hour entertainment beginning about 3:30 for several hours and then resuming later in the evening. The 150' Stardancer floating disco livens up the water from 10 pm.

*Kaanapali:* Two of the most popular spots currently are the Banana Moon at the Marriott Hotel, which parties until 2 am with a dress code Friday nights only, no tank tops, shorts or sandals. Spats at the Hyatt Regency is open nightly until 2 am, 4 am on Saturday with a dress code for men of closed toe shoes, and collared shirts and slacks and for women closed toe shoes. Cover charge for non hotel-guests. Also at the Marriott, the Makai Bar features entertainment and a very small dance floor and karoake entertainment in their lobby bar. El Crab Catcher has live Hawaiian style musical entertainment and Nanatomi features Karaoke.

### KIHEI/WAILEA/MAKENA AREA NIGHTSPOTS
The Inu Inu Lounge at the Maui Inter-Continental Wailea, the Lost Horizon at Stouffer Wailea Beach Resort and the Sunset Lounge at The Four Seasons Resort are the prime spots for late night entertainment in South Maui. In Makena, The Maui Prince has something special planned each evening, for more information see the Makena area listing under Molokini Lounge.

# Maui Beaches

# Beaches

## *INTRODUCTION*

If you are looking for a variety of beautiful, uncrowded tropical beaches, nearly perfect weather year round and sparkling clear waters at enjoyable temperatures, Maui will not disappoint you.

With beaches that range from small to long, white sand to black sand or rock, and from well developed to (at least for a little longer) remote and unspoiled, there is something for everyone. The lay-on-the-beach-under-a-palm-tree type, or the explorer-adventurer will not want for the appropriate beach.

Maui's beaches are publicly owned and most have right-of-way access, however, the access is sometimes tricky to find and parking may be a problem! Parking areas are provided at most developed beaches, but are generally limited to 30 cars or less, making an early arrival at the more popular beaches a good idea. In the undeveloped areas you will have to wedge along the roadside. It is vital that you leave nothing of importance in your car as theft, especially at some of the remoter locations, is high.

At the larger developed beaches, a variety of facilities are provided. Many have convenient rinse-off showers, drinking water, restrooms, and picnic areas. A few have children's play or swim areas. The beaches near the major resorts often have rental equipment available for snorkeling, sailing, boogie boarding, and even underwater cameras. These beaches are generally clean and well maintained. Above Kapalua and below Wailea, where the beaches are undeveloped, expect to find no signs to mark the location, no facilities, and sometimes less cleanliness.

Since virtually all of Maui's good beaches are located on the leeward side of East and West Maui, you can expect sunny weather most of the time. This is because the mountains trap the moisture in the almost constant trade winds. Truly cloudy or bad weather in these areas is rare but when the weather is poor in one area, a short drive may put you back into the sun again.

Swells from all directions reach Maui's shores. The three basic swell sources are the east and north-east trade winds, the North Pacific lows, and the South Pacific lows. The trades cause easterly swells of relatively low heights of 2 - 6 feet throughout most of the year. A stormy, persistent trade wind episode may cause swells of 8-12 feet and occasionally 10-15 feet on exposed eastern shores. Since the main resort areas are on leeward West and East Maui, they are protected.

North Maui and Hana are exposed to these conditions, along with strong ocean currents, therefore very few beaches in these areas are considered safe for casual swimming.

Kona winds generated by southern hemisphere storms cause southerly swells that affect leeward Maui. This usually happens in the summer and will last for several days. Surf heights over eight feet are not common, but many of the resort areas have beaches with fairly steep drop offs causing rather sharp shore breaks. Although it may appear fun to play in these waves, many minor to moderate injuries are recorded at these times. Resorts will post red warning flags along the beach during times of unsafe surf conditions. Most beaches are affected during this time causing water turbidity and poor snorkeling conditions. At a few places, such as Lahaina, Olowalu and Maalaea, these conditions create good surfing.

Northerly swells caused by winter storms northeast of the island are not common, but can cause large surf, particularly on the northern beaches, such as Baldwin, Kanaha and Hookipa Beach Parks.

Winter North Pacific storms generate high surf along the northwestern and northern shores of Maui. This is the source of the winter surf in Mokuleia Bay (Slaughterhouse), renowned for body surfing, and in Honolua Bay which is internationally known for surfing.

Land and sea breezes are local winds blowing from opposite directions at different times depending on the temperature difference between land and sea. The interaction of daytime sea breezes and trade winds, in the Wailea-Makena area particularly, produce almost daily light cloudiness in the afternoon and may bring showers. This is also somewhat true of the Honokowai to Kapalua region.

Oceanic tidal and trade wind currents are not a problem for the swimmer or snorkeler in the main resort areas from Makena to Kapalua except under unusual conditions such as Kona storms. Beaches outside of the resort areas should be treated with due caution since there are very few considered safe for casual swimming and snorkeling except by knowledgeable, experienced persons.

Maui's ocean playgrounds are probably the most benign in the world. There is no fire coral, jelly fish are rare, and sharks are well fed by the abundant marine life and rarely come into shore. However, you should always exercise good judgement and reasonable caution when at the beach.

1. "Never turn your back to the sea" is an old Hawaiian saying. Don't be caught off guard, waves come in sets with spells of calm in between.
2. Use the buddy system, never swim or snorkel alone.
3. If you are unsure of your abilities, use flotation devices attached to your body, such as a life vest or inflatable vest. Never rely on an air mattress or similar device from which you may become separated.
4. Study the ocean before you enter; look for rocks, breakers or currents.
5. Duck or dive beneath breaking waves before they reach you.
6. Never swim against a strong current, swim across it.
7. Know your limits.

8. Small children should be allowed to play near or in the surf ONLY with close supervision and should wear flotation devices.
9. When exploring tidal pools or reefs, always wear protective footwear and keep an eye on the ocean. Also, protect your hands.
10. When swimming around coral, be careful where you put your hands and feet. Urchin stings can be painful and coral cuts can be dangerous.
11. Respect the yellow and red flag warnings when placed on the developed beaches. They are there to advise you of unsafe conditions.

> *Paradise Publications has provided current and accurate information on Maui's beautiful beaches, however remember, nature is unpredictable and weather, beach and current conditions can change. Enjoy your day at the beach, but utilize good judgement. Paradise Publications cannot be held responsible for accidents or injuries incurred.*

Surface water temperature varies little with a mean temperature of 73.0 in January and 80.2 in August. Minimum and maximum range from 68 to 84 degrees. This is an almost ideal temperature (refreshing, but not cold) for swimming and you will find most resort pools cooler than the ocean.

## BEST BETS

On South Maui our favorite beaches are Makena for its unspoiled beauty, Maluaka for its deep fine sand and beautiful coral, Wailea and Ulua/Mokapu for their great beaches, good snorkeling and beautiful resorts, and Keawakapu and Kamaole II which offer gentler offshore slopes where swimming is excellent. A good place for small children is the park at the end of Hauoli Street in Maalaea, just past the Makani A Kai condos. There are two small, sandy-bottomed pools protected by reefs on either side of the small rock jetty.

On West Maui, Kapalua offers a well protected bay with very good swimming and snorkeling. Hanakaoo Beach has a gentle offshore slope and the park has lots of parking, good facilities, numerous activities, and is next to the Hyatt. Olowalu has easy access and excellent snorkeling. An excellent place for small children to play in the sand and water is at Pu'unoa Beach, which is well protected by a large offshore reef.

# BEACH INDEX

# MAALAEA TO LAHAINA

The beaches are described in order from Maalaea to Lahaina and are easy to spot from Honoapiilani Highway. They are all narrow and usually lined by Kiawe trees, however, they have gentle slopes to deeper water and the ocean is generally calmer and warmer than in other areas. The offshore coral reefs offer excellent snorkeling in calm weather, which is most of the time. These beaches are popular because of their convenient access and facilities as well as good swimming and snorkeling conditions.

## PAPALAUA STATE WAYSIDE PARK

As you descend from the sea cliffs on your way from Maalaea you will see an undeveloped tropical shoreline stretch before you. At the foot of the cliffs at mile marker 11, Papalaua Park is marked by an easily seen sign. There are picnic tables, BBQ grills, and portable restrooms. The beach is long, (about 1/2 mile) and narrow and lined with Kiawe trees that almost reach the water's edge in places. The trees provide plenty of shady areas for this beautiful beach. Good swimming and fair snorkeling, popular picnicking area.

## UKUMEHAME BEACH PARK

The entrance to the park is near mile marker 12, but there is no identifying sign. There is off-street paved parking for about 12 cars. Five concrete picnic tables. This is also a narrow 1/2 mile long sand beach with lots of Kiawe trees providing shade. Good swimming, fair snorkeling.

## OLOWALU BEACH ★

About 2/10 mile before and after mile marker 14 you will see a large, but narrow stand of Kiawe trees between the road and the beach, followed by a few palm trees, then a few more scattered Kiawe trees. Parking is alongside the road. No facilities. This narrow sand beach slopes gently out to water four or five feet deep making it good for swimming and beach playing. There are extensive coral formations starting right offshore and continuing out a quarter mile or more, and a fair amount of fish expecting handouts. The ocean is generally warmer and calmer than elsewhere, making it a popular snorkeling spot.

## AWALUA BEACH

The beach at mile marker 16 may be cobble stone or sand depending on the time of year and the prevailing conditions. No facilities. At times when Kona storms create a good southern swell, this becomes a very popular surfing spot for a few days until the swells subside.

## LAUNIUPOKO STATE WAYSIDE PARK

This well-marked beach park near mile marker 18 offers a large paved parking area, restrooms, many picnic tables, BBQ grills, rinse-off showers, drinking water, pay phone, and a large grassy area with trees, all of which makes for a good picnic spot. There is a large man-made wading pool constructed of large boulders centered in the park. (Sand has accumulated to the extent that even at high tide there is no water in the pool). To the right is a rocky beach and to the left is a 200-yard dark sand beach with fairly gentle slope. It looks nice, but signs posted warn "Sharks have been seen in the shallow water off this beach. Entry into the water is discouraged." This area is rumored to be a shark breeding ground with shark fishing done here in the past. There is also a no alcohol sign posted. For some reason the beach does not seem to be used for much besides picnicking! However, a couple hundred yards offshore is good snorkeling and you may see snorkel excursions visit this shoreline when the weather prohibits a trip to Lana'i.

### PUAMANA BEACH PARK
Well marked beach park near mile marker 19, just south of the Puamana Resort complex. Parking for 20 cars in paved parking area, with additional parking along the highway. Nice grassy park with seven picnic tables and plenty of shade trees. At the park itself there is no sandy beach, only a large pebble beach. The only beach is a narrow 200 yard long white sand beach just north of the park and fronting Puamana Resort. Fairly gentle slope to shallow water.

## LAHAINA AND KAANAPALI

### LAHAINA BEACH
There is a large public parking lot across from the 505 Front Street shopping center with easy access to the beach through the mall. There is also on-street parking near the Lahaina Shores with public right of way to the beach at the south end of the complex. Restrooms and showers are only available at the resort. The Lahaina Sailing Center is located on the beach. This narrow sand beach fronts the Lahaina Shores and 505 Front St. and is protected by a reef 30 - 50 yards out. The beach is generally sandy offshore with a gentle slope. The water stays fairly shallow out to the reef and contains some interesting coral formations. The area offers fair snorkeling in clear water on calm days. A good place for beginning snorkelers and children, but not good for swimming due to shallow water and abundant coral.

### PUUNOA BEACH ★
The beach is at the north end of Lahaina between Kai Pali Rd. and the old Mala Wharf and can be seen as you leave Lahaina on Front St. Southern access: Take Kai Pali Rd. off Front St. Parking for about 20 cars along the road which is the entrance for the Puunoa Beach Estates. Public Beach access sign with concrete sidewalk to the beach. Mid beach access: Take Puunoa Place off Front St. at the Public Beach access sign. Parking for about four cars at the end of the road which ends at the beach. A rinse off hose here is the only facility for the beach. North Access: Take Mala Wharf off Front St. Parking for approximately 20 cars along the road just before the entrance to the Mala boat launching parking area.

This narrow, dense, darker sand beach is about 300 yards long and well protected by a reef approximately 100 - 150 yards offshore. The beach slopes gently to water only 3 - 4 feet deep. Unfortunately, rock and coral near the surface make swimming unadvised. There are areas of the beach clear of coral 10 - 15 feet out where children can play safely in the calm, shallow water. At high tide there are more fish to see while snorkeling. This continues to be a favorite with our children because of the calm, warm water.

## WAHIKULI STATE WAYSIDE PARK

There are three paved off-street parking areas between Lahaina and Kaanapali. Many covered picnic tables, restrooms, showers, and BBQ grills are provided. The first and third parking areas are marked but have no beach. The second unmarked area has an excellent, darker sand beach with a gentle slope to deeper water. There is some shelf rock in places but it's rounded and smooth and not a problem. With the handy facilities, trees for shade, and the nice beach, this is a good, and popular, spot for sunning, swimming, and picnicking.

## HANAKAOO BEACH PARK ★

Off Honoapiilani Highway, immediately south of the Hyatt Regency, there is a large, well-marked, off-street parking area. The park has rinse-off showers, restroom, and picnic tables. Wide, darker sand beach with gentle slope to deeper water. This is a popular area because of the easy parking, facilities, good beach, shallow water and good swimming, and you are right next to the Hyatt.

## HANAKAOO BEACH ★ (Kaanapali Beach)

The beach fronts the Hyatt Regency, Marriott, Kaanapali Alii, Westin Maui, Whaler's Shopping Center and condos, Kaanapali Beach Hotel, and the Sheraton, and is known as Kaanapali Beach. Access is through the Kaanapali Resort area. Turn off Honoapiilani Highway at either of the first two entrances. This area was not designed with non-guest use in mind, and parking is definitely a problem.

A) The Hyatt end of the beach is only a short walk from the large parking area of Hanakaoo Beach Park.
B) Public right-of-way with parking for 10 cars at the left of the Hyatt's lower parking lot.
C) Public right-of-way between the Hyatt and Marriott, no parking.
D) Public right-of-way between Marriott and Kaanapali Alii with parking for 11 cars only.
E) Public right-of-way between Kaanapali Alii and Westin Maui, no parking.
F) Public right-of-way between Kaanapali Beach Resort and the Sheraton with parking for 11 cars only.
G) The Whalers Shopping Center has a three-story pay parking lot, but with beach access only through the complex.
H) There is no on-street parking anywhere in the Kaanapali Resort complex.


The Hyatt, Marriott, Westin Maui, and Sheraton all have restrooms, showers, bars, and rental equipment. There is a beautiful, long, wide, white sand beach with an abrupt drop-off to deep water. There are small areas of offshore coral from the Hyatt to the Westin Maui at times, but no true offshore reef. Great swimming and good wave playing with the exception of two or three points along the beach where the waves consistently break fairly hard. In the winter, snorkeling can be fair off the Westin Maui when the coral is exposed underwater. The best snorkeling is at Black Rock, fronting the Sheraton Hotel. The water is almost always clear and fairly calm, with many types of nearly tame fish due to the popularity of hand feeding by snorkelers. (Bread, frozen peas and packaged dry noodles seem popular). Not much colorful coral. The best entrance to the water is from the beach alongside Black Rock.

## KAANAPALI BEACH (South End)

This beach begins at the north side of Black Rock and runs for over a mile to the north fronting the Royal Lahaina Resort and the Maui Kaanapali Villas. Turn off Honoapiilani Road at the last Kaanapali exit at the stop light by the Maui Kaanapali Villas. There are a few places to park on the side of the road near the Public Access beach sign. With the airport now closed there is more parking available. This area is being prepared for future hotel and condo developement and should eventually have much improved public access, facilities and parking. The only facilities now are those of the nearby hotels. This wide, (usually) white sand beach has a steep drop-off to deep water and is usually calm - a good place to swim. Snorkeling around Black Rock is almost always good.

# *KAHANA, NAPILI, KAPALUA AND BEYOND*

## KAANAPALI BEACH (North End)

This section of beach fronts the Mahana Resort, Maui Kai, Embassy Suites, Kaanapali Shores, Papakea, Maui Sands and Paki Maui from south to north, and ends at the Honokowai Beach Park. Access is generally only through the resorts. Most of the resorts have rinse-off showers convenient to the beach, however, no other facilities are available. This is a long, narrow, white sand beach which is fronted by a close-in reef. All the resorts except the Kaanapali Shores and Embassy Suites have retaining walls along the beach. The Kaanapali Shores has, over the last couple of years, suffered considerable erosion of its once wide beach and has recently completed an expensive new under-the-sand retaining wall in an effort to stabilize and restore it. There is also a cleared area through the coral in front of the resort. This is the only good swimming area on the north section of the beach and is the only good access through the reef for snorkeling.

The reef comes into shore at the south end of Papakea and again at the Honokowai Beach Park. At low tide the reef fronting Papakea can be walked on like a wide sidewalk. (See GENERAL INFORMATION - Children, for night walking on the reef) The reef is generally only 10 - 20 yards offshore and the area between is very shallow with much coral and rock making it undesirable for swimming and snorkeling. The middle section of beach, fronting the old Kaanapali Airport, is slated for future development.

## HONOKOWAI BEACH PARK

Turn off the Honoapiilani Hwy. on the first side street past the airport (at the Honokowai sign) and get onto Lower Honoapiilani Hwy., which parallels the ocean. The park is across the street from the Honokowai Grocery Store. There is paved off-street parking for 30 cars. There are 11 picnic tables, 5 BBQ pits, restrooms, showers, and a grassy park with shade trees. Grocery store across the street with pay phone outside. The white sand beach is lined by a wide shelf of beach rock. Between the shelf rock and reef there is a narrow, shallow pool with a sandy bottom which is a good swimming area for small children. There is a break in the reef at the north end of the beach where you can get snorkeling access to the outside reef. Water sport equipment for rent at the Honokowai Store.

## KAHANA BEACH
In front of the Kahana Beach Condominiums, Sands of Kahana, Royal Kahana, Valley Isle Resort and Hololani from south to north on Lower Honoapiilani Hwy. There is limited off-road parking at the south end of the beach. Other access would be through the condos. The only facilities available are at the condos, usually rinse-off showers. There are several grocery stores, one at the Valley Isle Resort, the other at the Hololani condos. This white sand beach varies from narrow to wide and its offshore area is shallow with rock and sand, semi-protected by reef. Good swimming, fair snorkeling. Beach may be cool and windy in afternoons.

During the past year (1989), this area has been particularly plagued by the unexplained green algae bloom which tends to concentrate here due to the wind, current and shoreline continues. The beach is fequently unappealing for swimming and beach use due to the amount of slimy green algae on the beach and in the water. One possible cause of this unsightly mess may be the nitrates and other chemicals which are used for agriculture and golf course maintainence, flowing into the ocean. The county is continuing to investigate and may find it necessary to institute some controls.

## KEONENUI BEACH ★
The beach is in front of and surrounded by the Kahana Sunset with no convenient public access. A lovely wide crescent of white sand with a fairly gentle slope to water's edge, then fairly steep slope to deeper water. The beach is set in a small shallow cove, about 150 yards wide, which affords some protection. At times, especially in winter, rough seas come into the beach. When calm (most of the time), this is an excellent swimming and play area with fair snorkeling.

## ALAELOA BEACH ("The Cove")
This miniature, jewel-like cove is surrounded by low sea cliffs. The small, approximately 25 - 30 yard long, white sand beach has a gentle slope with scattered rocks leading into sparkling clear waters. Pavilion and lounge chair area for use by Alaeloa guests. Good swimming and snorkeling with very clear and calm waters except when storm-generated waves come in. Fortunately, or unfortunately, depending on your point of view, this small cove is surrounded by the Alaeloa residential area which has no on or off-street public parking, therefore, no public access to this beach.

## NAPILI BAY ★
There are two public accesses to this beautiful beach. There is a small, easily missed, public right-of-way and Napili Beach sign just past the Napili Shores at Napili Place Street. On-street parking at sign for Napili Surf Beach Resort. The public beach right-of-way sign shows the entrance to the beach. Public telephone in parking lot of Napili Surf. The second entrance is at the public beach right-of-way and Napili Sunset, Hale Napili, and Napili Bay signs on Hui Street. On-street parking and pay phone at entrance to beach walk.

This is a long, wide crescent of white sand between two rocky points. The offshore slope is moderately steep. Usually very safe for swimming and snorkeling except during winter storms when large waves occasionally come into the bay.

At the south end of the beach are a series of shallow, sandy tide pools which make an excellent place for children, but only under close supervision. Coral formations 30 - 40 yards offshore can provide fair snorkeling on calm days especially at the northern end of the beach and decent boogie boarding with mild swells. No facilities other than at the hotels along the beach. There is a grocery store just past the second entrance at the Napili Village Hotel. Look for the Napili Grocery Store sign.

## KAPALUA BEACH ★ (Fleming Beach)
Just past the Napili Kai Beach Club you will see a public beach right-of-way sign. Off-street parking area for about 30 - 40 cars. Showers and restrooms. A beautiful crescent of white sand between two rocky points. The beach has a gentle slope to deeper water, maximum about 15 feet. From the left point, a reef arcs toward the long right point creating a very sheltered bay, probably the nicest and safest swimming beach on Maui. Shade is provided by numerous palm trees lining the back shore area. Above the beach are the lovely grounds of the Kapalua Bay Resort. Swimming is almost always excellent with plenty of play area for children. Snorkeling is usually good with many different kinds of fish and interesting coral. *REMEMBER*, this popular beach has limited parking.

## NAMALU BAY
Park at Kapalua Beach and take the concrete path along the beach, up through the hotel's grounds, and out to the point of land separating Kapalua Bay from Namalu Bay. This small bay has a shoreline of large lava boulders, no beaches. On calm days snorkeling is very good and entry and exit over the rocks is easy. This little known spot is definitely worth the short walk down the trail.

## ONELOA BEACH
Enter at the public right-of-way sign just past the Kapalua Bay Resort. Paved off-street parking for 12 - 15 cars only, no other facilities. Long, straight white sand beach with a shallow sand bar that extends to the surfline. The beach is posted with a warning sign "No swimming at time of high surf due to dangerous currents." This area tends to get windy and cloudy in the afternoons, especially in the winter months. We have usually found this beach deserted.

## D. T. FLEMING BEACH PARK
Off-street parking on both sides of the road. Restrooms, showers, picnic tables, and BBQ's on the grassy dunes above the beach. The long white sand beach is steep with an offshore sand bar which may cause dangerous water conditions when swells hit the beach. The beach is posted "Dangerous Swimming."

## MOKULEIA BEACH ★ (Slaughterhouse)
On Highway 30, past D. T. Fleming Beach Park, look for cars parked along the roadside and the Mokuleia-Honolua Marine Reserve sign. Park your car and hike down one of the steep dirt and rock trails - they're not difficult. There are no facilities. The wide, white sand beach has a gentle slope to deep water and is bordered by two rocky points and is situated at the foot of steep cliffs. The left middle part of the beach is usually clear of coral and rocks even in winter when the beach is subject to erosion.

During the winter this is *THE* bodysurfing spot, especially when the surf is heavy, however, dangerous water conditions also exist. This area is only for the strong, experienced swimmer. The summer is generally much better for swimming and snorkeling. In the past couple of years, this has become a very popular beach. Snorkeling is fair to good, especially around the left rocky point where there is a reef. Okay in winter when the ocean is calm and visibility good. NOTE: The beach is known as Slaughterhouse because of the once existing slaughterhouse on the cliffs above the beach, not because of what the ocean can do to body surfers in the winter when the big ones are coming in! Remember this is part of the Honolua-Mokuleia Bay Marine Life Conservation District - look but don't disturb or take.

## HONOLUA BAY ★

The next bay past Slaughterhouse is Honolua Bay. Watch for a dirt side road on the left. Park here and walk in along the road. There is no beach, just cobblestone with irregular patches of sand and an old concrete boat ramp in the middle. Excellent snorkeling in summer, spring, and fall especially in the morning, but in winter only on the calmest days. In summer on calm days the bay resembles a large glassy pond and in our opinion, this is the best snorkeling on Maui. Note: After a heavy rainfall, the water may be turbid for several days before it returns to its sparkling clear condition again. You can enter at the boat ramp or over the rocks and follow the reefs either left or right. Remember this is a Marine Life Conservation area, so look but don't disturb.

There is an interesting phenomenon affecting the bay. As fresh water runoff percolates into the bay, a shimmering boundary layer (usually about three feet below the surface) is created between the fresh and salt waters. Depending on the amount of runoff it may be very apparent or disappear entirely. It is less prevalent on the right side of the bay. Honolua Bay is also an internationally known winter surfing spot. Storm generated waves come thundering in around the right point creating perfect waves and tubes. A good vantage point to watch the action is the cliffs at the right point of the bay, accessible by car on a short dirt road off the main highway.

KIHIKIHI

# KIHEI BEACHES

The Kihei beaches aren't quite as beautiful as Wailea's. They don't have the nicely landscaped parking areas, or the large, beautiful resort complexes (this is condo country). They do offer increased facilities such as BBQs, picnic tables, drinking water, and grassy play areas. The Kamaole I, II and III beaches even have lifeguards. The beaches are listed in order from Maalaea Bay to Wailea.

## MAALAEA BAY BEACH

This gently curving white sand beach stretches three miles from the Maalaea boat harbor to Kihei. For the most part, the beach is backed by low sand dunes and large generally wet, sand flats. Public access is from many areas along South Kihei Road. There are no facilities. Casual beach activities are best early in the morning before the strong, mid-morning, prevailing winds begin to sweep across the isthmus. Due to the length of the beach and the hard-packed sand near the water, this has become a popular place to jog. Windsurfing is popular in the afternoons.

The beach begins in front of the last three condominiums in Maalaea, the Kana'I A Nalu, Hono Kai and the Makani A Kai. Just past the Makani A Kai on Hauoli Street is a public park and beach access. There is a good section of beach here with a fairly gentle drop off. Also there are several large, sandy-bottomed pools, protected by the reef on either side of the small man-made rock jetty. These are good play areas for kids. The waves remain fairly calm, except at high tide or high surf conditions. The best snorkeling is out from the beach here, but the conditions are extremely variable, from fairly clear to fairly murky, depending on the time of year and prevailing conditions. Snorkeling is usually better in the winter months. The beach from here to North Kihei is generally fronted by shelf rock or reef and is not good for swimming, but excellent for a lengthy beach walk! The beach becomes excellent for swimming and other beach activities in front of the North Kihei condos. Snorkeling is fair. A beach activity center is located on the beach at the Kealia Beach Center.

## MAI POINA OE IAU BEACH PARK

On South Kihei Road, fronting Maui Lu Resort. Paved parking for 8 cars at the Pavilion (numerous other areas to park are along the road). 5 picnic tables, restrooms, showers. This is actually part of the previous beach. In-shore bottom generally sandy with patches of rock, fronted by shallow reef. Swimming and snorkeling are best in the morning before the early afternoon winds come up. Popular windsurfing area in the afternoon.

## KAONOULULU BEACH PARK

Located across the street from the Kihei Bay Surf. Off-road parking for 20 cars, restrooms, drinking water, rinse-off showers, picnic tables, and four BBQ grills. Very small beach, well protected by close-in reef.

## KAWILIKI POU PARK

Located at the end of Waipulani Street. Paved off-street parking for 30 cars, restrooms, large grassy area, and public tennis courts. Fronts Laule'a, Luana Kai and the Maui Sunset Hotel. Tall graceful palms line the shoreline. Narrow sandy beach generally strewn with seaweed and coral rubble. (See GENERAL INFORMATION - Children, for frog hunting information)

## KAWILILIPOA AND WAIMAHAIKAI AREAS
Any of the cross streets off South Kihei Road will take you down toward the beach where public right-of-ways are marked. Limited parking, usually on street. No facilities. The whole shoreline from Kalama Park to Waipulani Street (3 - 4 miles) is an area of interrupted beaches lined by residential housing and small condo complexes. Narrow sandy beaches with lots of coral rubble from the fronting reefs.

## KALAMA BEACH PARK
Well-marked, 36-acre park with 12 pavilions, 3 restrooms, showers, picnic tables, BBQ grills, playground apparatus, soccer field, baseball field, tennis courts, volleyball and basketball courts. Lots of grassy area. There is no beach (in winter), only a large boulder breakwater. Good view of the cinder cone in Makena, Molokini, Kahoolawe, Lanai, and West Maui.

## KAMAOLE I
Well-marked beach across from the Kamaole Beach Club. Off-street parking for 30 cars. Facilities include picnic tables, restrooms, rinse-off showers, rental equipment, children's swimming area, and lifeguard. Long, white, sandy beach offering good swimming, poor to fair snorkeling. NOTE: The small pocket of sand between rock outcroppings at the right end of the beach is known as Young's Beach. It is also accessible from Kaiau Street with parking for about 20 cars. Public right-of-way sign at end of Kaiau Street.

## KAMAOLE II
Located across from the Kai Nani shopping and restaurant complex. On-street parking, restrooms, rinse-off showers, rental equipment, and lifeguard. White sand beach between two rocky points with sharp drop-off to overhead depths. Good swimming, poor to fair snorkeling.

## KAMAOLE III ★
Well-marked beach across from the Kamaole Sands Condominiums. Off-street parking, picnic tables, BBQ's, restrooms, rinse-off showers, drinking water, playground equipment, a grassy play area, and a lifeguard. 200-yard long, narrow (in winter) white sand beach with some rocky areas along beach, and a few submerged rocks. Good swimming, fair snorkeling around rocks at south end of the beach. Kamaole II and III are very popular beaches with locals and tourists because of the nice beaches and easy access.

# WAILEA BEACHES

This area generally has small, lovely, white sand beaches which have marked public access. Parking off street in well maintained parking areas, and restrooms as well as rinse-off showers are provided.

### KEAWAKAPU BEACH ★

There are two convenient public accesses to this very nice but generally underused beach. There is paved parking for 50 cars across the street from the beach, about 2/10 mile south of Mana Kai Resort. Look for the beach access sign on the left as you travel south. There are two small crescent shaped, white sand beaches separated by a small rocky point. Good swimming, off-shore sandy bottom, fair snorkeling around rocks at far north end. There are rinse-off showers and a restaurant at the Mana Kai which is right on the beach. Access to southern end of beach - go straight at left turn-off to Wailea, road says "Dead End." Parking for about 30 cars. Rinse-off showers. Beautiful, very gently sloping white sand beach with good swimming. Snorkeling off rocks on left. Popular scuba diving spot. Four hundred yards off shore in 80 - 85 feet of water there is supposed to be an artificial reef of 150 car bodies.

### MOKAPU BEACH ★

A public access sign (Ulua/Mokapu Beaches) is near the Stouffer Wailea Beach Resort. Small parking area, restrooms and showers. Rental equipment at nearby Wailea Resort Activities Center at Stouffer's. Beautiful white sand beach. Excellent swimming. Good snorkeling in mornings around the rocks which divide the two beaches. The best snorkeling is on the Ulua beach side.

### ULUA BEACH ★

A public access sign (Ulua/Mokapu Beaches) is located near the Stouffer Wailea Beach Resort. Small paved parking area with a short walk to beach. Showers and rest rooms. Rental equipment is only a short walk away at the Wailea Ocean Activities Center. Beautiful white sand beach fronting the Elua Resort complex. Ulua and Mokapu Beaches are separated by a narrow point of rocks. The area around the beaches is beautifully landscaped because of the resorts. The beach is semi-protected and has a sandy offshore bottom. Good swimming, usually very good snorkeling in the mornings around the lava flow between the beaches. Come early to get a parking space!

### WAILEA BEACH ★

One half mile south of the Inter-Continental Resort there is a public beach access sign and a paved road down to a landscaped parking area for about 40 cars. Restrooms and rinse-off showers. Rental sailboats and windsurfing boards are available. Beautiful wide crescent of gently sloping white sand. Gentle offshore slope. Good swimming. Snorkeling is only fair to the left (south) around the rocks (moderate currents and not much coral or many fish).

You won't recognize this area from a few years ago. The Grand Hyatt, Wailea Suites and Four Seasons Resort (currently in various stages of construction) are transforming this undeveloped area into a Kaanapali of the South coast.

## POLO BEACH

Turn right at the Wailea Golf Club-Fairway Restaurant sign and head down to the Polo Beach Resort condominiums. The public access sign is easy to spot. Parking for 40 cars in paved parking area. Showers and restrooms. The beaches are a short walk on a paved sidewalk and down a short flight of stairs. There are actually two beaches, 400 foot long north beach and 200 foot long south beach, separated by 150 feet of large rocks. The beaches slope begins gently, then continues more steeply off-shore and is not well protected. This combination can cause swift beach backwash which is particularly concentrated at two or three points and also a rough shore break, especially in the afternoons. The beach is dotted with large rocks. Fair swimming, generally poor snorkeling. Construction is underway adjacent to the Polo Beach condominiums, where the Wailea Suites resort is expected to be completed by 1991.

# MAKENA BEACHES

This area includes the beaches south of Polo Beach, out to La Perouse Bay (past this point, you either hike or need to have a four-wheel drive). The Makena beaches are relatively undeveloped and relatively unspoiled, for a while longer at least, and not always easy to find. There are no signs, confusing roads, and some beaches are not visible from the road. Plenty of people do find these pristine beaches, however. Generally, no facilities and parking where you can find it. The nearest grocery is at the Wailea Shopping Center. This area has been changing in the last few years. The new road is now completed to the Makena Golf Club and 1 1/2 miles past it to near the end of Makena Beach. Although development is progressing, much more is desired by various commercial interests. Local opposition has managed, so far, to keep this in check. We hope our directions will help you find these sometimes hard-to-find, but very lovely beaches.

## PALAUEA BEACH

As you leave Wailea, there is a four-corner intersection with a sign on the left for the Wailea Golf Club, and on the right for the Polo Beach Condos. 8/10 mile past here turn onto the second right turnoff at the small "Paipu Beach" sign. At roads end (about 1/10 mile), park under the trees. Poolenalena Beach lies in front of you. Walk several hundred feet back towards Polo Beach over a small hill (Haloa Point) and you will see Palauea Beach stretching out before you. A beautiful beach, largely unkown to tourists. The area above the beach at the south end has been developed with pricey residential homes.

If you drive down to the Polo Beach Condos instead, you can continue on Old Makena Road which will loop back to Makena Alanui Road after about a mile. Palauea Beach lays along this road, but is not visible through the trees. There is a break in the fence .35 miles from Polo Beach with a well worn path to the beach. Although this is all private and posted land, the path and the number of cars parked alongside the road seems to indicate that this beautiful white sand beach is getting much more public use than in the past. Good swimming. No facilities. Both Palauea and Poolenalena beaches have the same conditions as Polo Beach with shallow offshore slope then a steep dropoff which causes fairly strong backwash in places and tends to cause a strong shore break in the afternoon.

## POOLENALENA BEACH
See directions for Palauea Beach. This is a lovely wide, white sand beach with gentle slope offering good swimming. This used to be a popular local camping spot, however no camping signs are now posted.

## UPCOUNTRY ROAD
1.4 miles from Polo Beach. Currently closed in dispute over maintenance.

## PAIPU BEACH (Chang's Beach)
Continue another 2/10 to 3/10 miles on Makena Alanui Rd., past Poolenalena and you will come to the Makena Surf Town Houses (about 1.2 miles from the Wailea Golf Club sign). This development surrounds Chang's Beach, however, there is a public beach access sign and paved parking for about 20 cars. It's a short walk down a concrete path to the beach. A rinse-off shower is provided. This small but sandy beach is used mostly by guests of the Makena Surf.

## ULUPIKUNUI BEACH
Turn right just past the Makena Surf and immediately park off the road. Walk down to the beach at the left end of the complex. The beach is 75 to 100 feet of rock strewn sand and is not too attractive, but is well protected.

## FIVE GRAVES
From the Makena Surf, continue down Old Makena Road another 2/10 mile to the entrance of Five Graves. There is ample parking. The 19th century graves are visible from Makena Rd. just a couple hundred feet past the entrance. There is no beach, but this is a good scuba and snorkeling site. Follow the trail down to the shore where you'll see a good entrance to the water.

## MAKENA LANDING - PAPIPI BEACH
Continue another 2/10 mile on Old Makena Rd. to Makena Landing on the right. There is off-street parking for 22 cars. The beach is located at the entrance and is about 75 - 100 feet with gentle slope, sometimes rock strewn. Not very attractive and is used mostly for fishing, but snorkeling can be good if you enter at the beach and follow the shore to the right. Restrooms and showers available.

Instead of turning right onto Old Makena Road at the Makena Surf, continue straight and follow the signs to the Makena Golf Course. About 9/10 mile past the Makena Surf there is another turnoff onto Old Makena Rd. At the stop sign at the bottom of the hill, you can turn right and end up back at Makena Landing or turn left and head for Maluaka Beach. 2/10 mile past the stop sign you will see the old Keawalai Church U.C.C. and cemetery. Sunday services continue to be held here. Along the road is a pay phone.

## MALUAKA BEACH ★ (Naupaka)
3/10 mile past the stop sign there is a turnaround and public entrance to this beach on the right. There are a few parking places near the entrance along the road, however, the main parking lot with restrooms and showers is located a short walk back up the road. The resort above the beach is the Maui Prince. This gorgeous 200-yard beach is set between a couple of rock promontories. The very fine white sand beach is wide with a gentle slope to deeper water. Snorkeling can be good

in the morning until about noon when the wind picks up. There are interesting coral formations at the south end with unusual abstract shapes, and large coral heads of different sizes. Coral in shades of pink, blue, green, purple and lavender can be spotted. There are enough fish to make it interesting, but not an abundance. In the afternoon when the wind comes up, so do the swells, providing good boogieboarding and wave playing.

The road past this point is marked private and proceed at your own risk. It runs through the resort area behind the beach and connects again with the paved road on the other side of the Maui Prince near the entrance to Oneuli Beach. Please drive slowly and cautiously through the resort area.

## ONEULI BEACH (Black Sand Beach)
On Old Makena Hwy. just past the intersection of the old road is a dirt road turnoff. A 4-wheel drive or a high ground clearance vehicle is good idea for the very rutted 3/10 mile to the beach. The beach is coarse black sand and the entire length of the beach is lined by an exposed reef. No facilities.

## ONELOA BEACH ★ (Makena Beach)
The entrance for the north end of the beach is at the second dirt road to the right off Old Makena Hwy. after the intersection of the old road. It is 3/10 mile from the turnoff to the beach and parking area with room for quite a few cars. The old, very rutted dirt road has been replaced by a graded and somewhat graveled road.

This very lovely white sand beach is long (3/4 mile) and wide and is the last major undeveloped beach on the leeward side of the island. Community effort is continuing in their attempt to prevent further development of this beach. The 360-foot cinder cone (Pu'u Olai) at the north end of the beach separates Oneloa from Puuolai Beach. The beach has a quick, sharp drop off and rough shore break particularly in the afternoon. Body surfing is sometimes good. Snorkeling around the rocky point at the cinder cone is only poor to fair with not much to see, and not for beginners due to the usually strong north to south current.

### PUUOLAI BEACH (Little Makena)

Take the first Oneloa Beach entrance, and park at Oneloa Beach. From there, you hike over the cinder cone. There is a flat, white sand beach, with a shallow sandy bottom which is semi-protected by a shallow cove. The shore break is usually gentle and swimming is good. Bodysurfing sometimes. Snorkeling is only poor to fair around the point on left. Watch for strong currents. Although definitely illegal, beach activities here tend to be au naturel.

### AHIHI-KINAU ★ (NATURAL RESERVE AREA)

About 3/4 mile past Makena Beach, a sign indicates the reserve. There is a small, 6-foot wide, sandy beach alongside the remnants of an old concrete boat ramp. Although it's located in a small cove and is well protected, the beach and cove are very shallow with many urchins. There is also very limited parking here. Up around the curve in the road is a large parking area. It's a short walk to the shore on a crushed lava rock trail. Another couple hundred feet to a very small (3 foot) and partially hidden sand and pebble beach that makes a better entrance to the water than over the rocks. There is excellent snorkeling directly off shore to the right and left. Remember, this is a marine reserve - look, but don't disturb. No facilities.

### LA PEROUSE BAY

2 miles past Ahihi-Kinau over a fairly rough road carved through Maui's most recent lava flow is the end of the road unless you have a 4-wheel drive. The "road" is extremely rough and we would recommend a hike rather than a ride. It's about 3 or 4 miles from road's end at La Perouse Bay to the Kanaio beaches. If you hike, wear good hiking shoes as you'll be walking over stretches of sharp lava rock. There are a series of small beaches, actually only pockets of sand of various compositions, with fairly deep offshore waters and strong currents.

# WAILUKU - KAHULUI BEACHES

Beaches along this whole side of the island are usually poor for swimming and snorkeling. The weather is generally windy or cloudy in winter and very hot in summer. Due to the weather, type of beaches, and distance from the major tourist areas on the other side of the island, these beaches don't attract many tourists (except Hookipa, which is internationally known for wind surfing).

### WAIHEE BEACH PARK

From Wailuku take Kahekili Highway about three miles to Waihee and turn right onto Halewaiu Road, then proceed about one-half mile to the Waihee Municipal Golf Course. From there, a park access road takes you into the park. Paved off-street parking, restrooms, showers, and picnic tables. This is a long, narrow, brown sand beach strewn with coral rubble from Waihee Reef. This is one of the longest and widest reefs on Maui and is about one thousand feet wide. The area between the beach and reef is moderately shallow with good areas for swimming and snorkeling when the ocean is calm. Winter surf or storm conditions can produce strong alongshore currents. Do not swim or snorkel at the left end of the beach as there is a large channel through the reef which usually produces a very strong rip current. This area is generally windy.

### KANAHA BEACH PARK

Just before reaching the Kahului Airport, turn left, then right on reaching Ahahao Street. The far south area of the park has been landscaped and includes BBQs, picnic tables, restrooms, and showers. Paved off-street parking is provided. The beach is long (about one mile) and wide with a shallow offshore bottom composed of sand and rock. Plenty of thorny Kiawe trees in the area make footwear essential. The main attraction of the park is its peaceful setting and view, so picnicking and sunbathing are the primary activities. Swimming would appeal mainly to children. Surfing can be good here.

### H. A. BALDWIN PARK

The park is located about 1.5 miles past Spreckelville on the Hana Highway. There is a large off-street parking area, a large pavilion with kitchen facilities, picnic tables, BBQs, and a tent camping area. There are also restrooms, showers, a baseball and a soccer field. The beach is long and wide with a steep slope to overhead depths. This is a very popular park because of the facilities. The very consistent, although usually small, shore break is good for bodysurfing. Swimming is poor. There are two areas where exposed beach rock provides a relatively calm place for children to play.

### HOOKIPA BEACH PARK

Located about two miles past Lower Paia on the Hana Highway. Restrooms, showers, four pavilions with BBQ's and picnic tables, paved off-street parking, and a tent camping area is provided. Small, white sand beach fronted by a wide shelf of beach rock. The offshore bottom is a mixture of reef and patches of sand. Swimming is not advised. (The area is popular for the generally good and, at times (during winter), very good surfing). Hookipa is internationally known for its excellent wind surfing conditions. This is also a good place to come and watch both of these water sports.

## HANA

### WAIANAPANAPA STATE PARK

About four miles before you reach Hana on the Hana Highway is Waianapanapa State Park. There is a trail from the parking lot down to the ocean. The beach is not of sand, but of millions of small, smooth, black volcanic stones. Ocean activities are generally unsafe. There is a lava tunnel at the end of the beach that runs about 50 feet and opens into the ocean. Other well marked paths in the park lead to more caves and fresh water pools. An abundance of mosquitos breed in the grotto area and bug repellent is strongly advised.

### HANA BEACH PARK

If you make it to Hana, you will have no difficulty finding this beach on the shoreline of Hana Bay. Facilities include a pavilion with picnic tables, restrooms and showers, and also Tutu's snack bar. About a 200-yard beach lies between old concrete pilings on the left and the wharf on the right. Gentle offshore slope and gentle shore break even during heavy outer surf. This is the safest swimming beach on this end of the island. Snorkeling is fair to good on calm days between the pier and the lighthouse. Staying inshore is a must, as beyond the lighthouse the currents are very strong and flow seaward.

### KAIHALULU BEACH (Red Sand Beach)

This reddish sand beach is located in a small cove on the other side of Kauiki Hill from Hana Bay and is accessible by trail. At the Hana Bay intersection follow the road up to the school. A dirt path leads past the school and disappears into the jungle. The trail almost vanishes as it goes through an old cemetery, then continues out onto a scenic promontory. The ground here is covered with marble-sized pine cones which make for slippery footing. As the trail leads to the left and over the edge of the cliff, it changes to a very crumbly rock/dirt mixture that is unstable at best.

You may wonder why you're doing this as the trail becomes two feet wide and slopes to the edge of a 60 foot cliff in one place. The trail down to the beach can be quite hazardous. Visitors and Hana residents alike have been injured seriously. It is definitely not for the squeamish, those with less than good agility or young-sters. And when carrying beach paraphernalia, extra caution is needed. The effort is rewarded as you descend into a lovely cove bordered by high cliffs and almost enclosed by a natural lava barrier seaward. The beach is formed primarily from red volcanic cinder, hence its name. Good swimming, but stay away from the opening at the left end because of rip currents. Although definitely illegal, beach activities here may be au naturel at times. The Hotel Hana Maui has plans to improve the access to this beach sometime in the future.

### KOKI BEACH PARK

This beach is reached by traveling 1.5 miles past the Hasegawa Store toward Ohe'o Gulch. Look for Haneoo Road where the sign will read "Koki Park - Hamoa Beach - Hamoa Village." This beach is unsafe for swimming and the signs posted warn "Dangerous Current."

### HAMOA BEACH ★

This gorgeous beach has been very attractively landscaped and developed by the Hotel Hana Maui in a way that adds to the surrounding lushness. The long white sand beach is in a very tropical setting and surrounded by a low sea cliff. To reach it, travel toward Ohe'o Gulch after passing through Hana. Look for the sign 1.5 miles past Hasegawa store that says "Koki Park - Hamoa Beach - Hamoa Village." There are two entrances down steps from the road. Parking is limited to along the roadside. The left side of the beach is calmer, and offers the best snorkeling. Because it is unprotected from the open ocean, there is good surfing and bodysurfing, but also strong alongshore and rip currents are created at times of heavy seas. The Hana Hotel maintains the grounds and offers restrooms, changing area, and beach paraphernalia for the guests. There is an outdoor rinse-off shower for non-hotel guests. Hay wagons bring the guests to the beach for the hotel's weekly luau.

# Recreation and Tours

## INTRODUCTION

Maui's ideal climate, diverse land environments, and benign leeward ocean has led to an astounding range of land, sea and air activities. With such a variety of things to do during your limited vacation time, we suggest browsing through this chapter and choosing those activities that sound most enjoyable. The following suggestions should get you started.

### BEST BETS:

To see and experience the real Maui, take a hike with guide Ken Schmitt.

For spectacular scenery and lots of fresh air, try the 38-mile coast down the world's largest dormant volcano on a bicycle.

For great snorkeling try Honolua Bay, Namalu, Ahihi Kinau, or Olowalu.

Take a helicopter tour and get a super spectacular view of Maui.

Golf at one of Maui's four resort courses.

Sail to Lana'i and snorkel Hulopoe Beach with the Trilogy Cruise.

If the whales are in residence, take a whale-watching cruise with the Pacific Whale Foundation.

For an underwater thrill consider an introductory scuba adventure, no experience necessary.

For a wet and wild tour of Lana'i or Molokini, with snorkeling, try a Zodiac raft trip.

If you're really adventurous consider parasailing (during the summer when the whales have gone back north!), or sea kayaking.

For great scenery at a great price, drive yourself to Hana and the O'heo Pools or to Upcountry and Haleakala.

# OCEAN ACTIVITIES

## *SNORKELING*

Maui offers exceptionally clear waters, warm ocean temperatures and abundant sea life with safe areas (no sharks or adverse water conditions) for snorkeling. If you are a complete novice, most of the resorts and excursion boats offer snorkeling lessons. Older folks can enjoy this sport that needs little experience and there is no need to dive to see all the splendors of the sea. If you are unsure of your abilities, the use of a floatation device may be of assistance. Be forewarned that the combination of tropical sun and the refreshing coolness of the ocean can deceive those paddling blissfully on the surface, and result in a badly burned backside. Water resistant sunscreens are available locally and are recommended.

Equipment is readily available at resorts and dive shops, and as you can see, much less expensive at the dive shops (even better are the weekly rates). For a listing of dive shops see Scuba Diving.

**TYPICAL RENTAL PRICES -**
**MASK/FINS/SNORKEL FOR 24 HOURS:**
Maui Dive Shop in Kihei - $7.50
Fun Rentals in Lahaina - $5.00
Frog Man - $7.50
Hyatt Regency Resort at Kaanapali - $20.00

*Snorkel Bobs* in Kihei and Napili has about the best weekly rates for mask, fin and snorkels at $15 per week. Silicon mask sets at $27 per week and prescription masks at $39.

Good snorkeling spots, if not right in front of your hotel or condo, are only a few minutes' drive away. The following are our favorites, each for a special reason.

## WEST MAUI

*Black Rock* - At the Sheraton in the Kaanapali Resort. Park at the Whaler's Shopping Center and walk down the beach. Clear water and a variety of tame fish - these fish expect handouts!
*Kapalua Bay* - Public park with off street parking, restrooms and showers. A well protected bay and beautiful beach amid the grounds of the Kapalua Resort. Limited coral and some large coral heads, fair for fish watching.
*Namalu Bay* - At the Kapalua Resort, walk over from Kapalua Bay. Very good on calm days.
*Honolua Bay* - No facilities, parking alongside the road and walk a 1/4 mile to the bay, but the best snorkeling on Maui, anytime but winter.
*Olowalu* - At mile marker 14, about 5 miles south of Lahaina. Generally calm and warmer waters with ample parking along the roadside. Very good snorkeling. If you find a pearl earring, let us know, we have the match!

## EAST MAUI

***Ulua/Mokapu Beach*** - Well-marked public beach park in Wailea with restrooms and showers. Ocean Activities Center amid grounds of Stouffer Wailea Beach Resort offers rental equipment. Good snorkeling on the Ulua side of the rocky point separating these two picturesque and beautiful beaches.

***Maluaka Beach*** - Located in Makena, no facilities and along the road parking. Good coral formations and a fair amount of fish at the left end of the beach.

***Ahihi Kinau Natural Reserve*** - Approximately five miles past Wailea. No facilities. This is not a very crowded spot and you may feel a little alone here, but the snorkeling is great with lots of coral and a good variety of fish.

Generally at all locations the best snorkeling is in the morning until about 1 p.m., when the wind picks up. For more information on each area and other locations, refer to the BEACHES chapter.

A good way to become acquainted with Maui's sea life is a guided snorkeling adventure with Ann Fielding, marine biologist and author of *Hawaiian Reefs and Tidepools*. She takes small groups (minimum 2, maximum 4) to Honolua Bay in summer and Ahihi Kinau in winter. These morning (8:30 am - 12:30 pm) excursions begin with a practical seminar on reef systems and marine life followed by snorkeling. Floatation devices, snorkel gear and refreshments are provided for the $35 fee. Phone 572-8437 for information.

You may feel the urge to rent an underwater camera to photograph some of the unusual and beautiful fish you've seen, and by all means try it, but remember, underwater fish photography is a real art. There are two new video tapes of Maui's marine life available at the island bookstores if you want a permanent record of the fish you've seen.

There are two other great places to snorkel, however, you need a boat to reach them. Fortunately, a large variety of charter services will be happy to assist.

***Molokini Crater*** - This small semicircular island is the remnant of a volcano. Located about 8 miles off Maalaea Harbor, it affords good snorkeling in the crater area. These waters are a marine reserve and the island is a bird sanctuary. Molokini is usually a 1/2 day excursion with a continental breakfast and lunch provided. Costs are $40 - $50 for adults, $25 - $35 for children under 12. (You may find rates even lower with current price wars among the heavy competition.)

***Hulopoe Beach, Lanai*** - This is our favorite. Located on the island of Lanai, it's worth the trip for the beautiful beach and the abundance of coral and fish. We saw a school of fish here that was so large that from the shore it appeared to be a huge moving reef. After swimming through the school and returning to shore we were informed that large predatory fish like to hang out around these schools! Lanai is usually a full-day excursion with continental breakfast, BBQ lunch and a possible tour of the island. Costs run $90 - $110 for adults. Some half-day trips are available.

A variety of snorkel/sail/tour options are available for snorkeling along East and West Maui's coastline, Molokini, Molokai or Lana'i. Your first decision is choosing between a large or small group tour. Large groups go out in substantial monohull or catamaran motor yachts of 60 - 90 feet in length. They get you there comfortably and fast, but without the intimate sailing experience of a smaller, less crowded boat. There are also many sleek sailboats (monohull, catamaran or trimaran) that you can share with 4 to 8 people or privately charter. A fairly new addition to the Maui sea excursions are the Zodiac type rafts that use 20 - 23 foot inflatable rafts powered by two large outboards. These rides can be rough, wet and wild. All tours provide snorkel gear with floatation devices, if needed, and instruction. Food and refreshments are provided to varying degrees. For a list of outings, see the section on Sea Excursions.

## MOLOKINI... again

Molokini is a 10,000 year old dormant volcano with only one crescent shaped portion of the crater rim now providing a sanctuary for marine and bird life. The crater on the inside of the island offers a water depth of 10 - 50 feet, a 76 degree temperature and visibility sometimes as much as 150 feet on the outer perimeter creating a fish bowl effect.

While our last report was not encouraging, we are happy to report that the waters of Molokini are making a comeback. For several years after the detenation of some submerged bombs by the Navy, the aquatic life was sparce. While it hasn't been restored to the way it was during our first excursions a number of years ago, it is now returning to a much improved condition. Fortunately, concerned boat operators pushed for and were finally granted semi-permanent concrete mooring anchors, thereby preventing further reef damage. There is even talk that the number of boats may be somehow limited.

We recommend *Blue Water Rafting* ★ for those die-hard snorkelers who would enjoy their early bird arrival to the crater and the opportunity to explore three different dive locations at Molokini. They arrive at the crater first and snorkel the best spot before the big boats come in. Then it's a stop at the far crater wall for a second snorkeling opportunity. A third stop takes you to a spot over the underwater crater rim where the water on the crater side is shallow, but drops off out of sight dramatically on the other side.

REEF DWELLERS                                                                J. BAYOT

However, not everyone is comfortable traveling in a 6-passenger inflatable raft. We recently went aboard the *Excalibur* and found it a comfortable and fun trip. Large enough (50') to have an indoor cabin with seating/standing room for a number of people, it isn't too large to get lost in the crowd. The crew was very personable and the trip did include two dive spots. The first stop was in the Molokini Crater where the fish were most enthusiastic upon our arrival. (Some unusual fall Kona weather had kept the boats out of the area for four days and the fish had certainly missed the handouts to which they had become accustomed). After the first dive there was time for some fresh fruit and buffalo wing appetizers while we motored to the second dive spot at La Perouse Bay. Here the fish, not tamed and accustomed to feeding, can be viewed in a completely different setting. And a delight for the "land lubber" is their warm, fresh water rinse-off shower. No kidding! A hose with a spray nozzle gushes out a warm refreshing rinse. A deli style lunch with make-your-own sandwiches was followed by some champagne refreshment for the ride home.

# SCUBA DIVING

Maui, with nearby Molokini, Kahoolawe, Lanai and Molokai, offers many excellent diving locations. A large variety of dive operations offer scuba excursions, instruction, certification and rental equipment. If you are a novice, a great way to get hooked is an introductory dive. No experience is necessary. Instruction, equipment and dive, all for $34.95 - $95 with the average about $60. Dives are available from boats, or less expensive from the beach. A beginning beach dive may be advisable for the less confident aquatic explorer. For those who are certified but rusty, refresher dives are available.

The mainstay of Maui diving is the two-tank dive, two dive sites with one tank each. Prices depend on location: Maui coast $40 - $55, Lanai or Molokini $70 - $80, which includes all equipment. If the bug bites and you wish to get certified, the typical course is five days, eight hours each day, at an average cost of $300 plus books. One dive shop suggested that visitors who do dive "PADI" dive preparation on the mainland can be certified on Maui in just two days. Classes are generally no more than 6 persons, or if you prefer private lessons, they run slightly more. Advanced open water courses are available in deep diving, search and recovery, underwater navigation and night diving (at a few shops). If you wish to rent equipment only, a complete scuba package including wet suit runs about $25 per day.

The larger resorts also offer instruction and some offer certification courses and arrange for excursions. Many of the dive operators utilize boats specifically designed for diving. Information, equipment, instruction and excursions can be obtained at the following dive shops and charter operators. As you can see by the number of listings, diving is very popular around Maui.

**SNUBA:** The newest in underwater recreation combines snorkeling with scuba diving. No heavy tanks since the air supply is provided by a hose which is connected to the surface. If you're ready to try something new check this out at Captain Nemo's, located at 150 Dickenson St. in Lahaina. (661-5555).

## WEST MAUI

Aloha Voyages
667-6284

Aquatic Charters &
Underwater Video
879-0976

American Institute
of Diving
278 Wili Ko Pl. #18
Lahaina, 96761
667-5129

Beach Activities
Kaanapali, 661-5500
Kapalua, 669-4664

Captain Nemo's
700 Front St.
Lahaina, 661-5555
1-800-367-8088

Central Pacific
Divers ★
Lahaina, 661-8718
1-800-551-6767

Dive Maui, Inc.
Lahaina, 667-2080

Extended Horizons
P.O. Box 10785
Lahaina, 667-0611

Dive Lanai
Frogman
(44' catamaran)
Kaanapali 667-7622

Hawaiian Reef Divers
129 Lahainaluna
Lahaina, 667-7647

Lahaina Divers
710 Front St.
Lahaina, 667-7496
1-800-657-7885

Maui Dive Shops
Azekas in Kihei
Lahaina Cannery
879-3388

Maui Marriott
Ocean Activities
667-1200

Sundance
Lahaina, 661-4126

Underwater
Adventures
Lahaina, 661-8957

## EAST MAUI

Ed Robinson's
Diving Adventures
Kihei, 879-3584
1-800-635-1273

Maui Dive Shops
Kihei Town Center
Kihei, 879-1919

Makena Coast Charters
Kahului, 874-1243

Maui Dive Shops
Azeka's Shopping
Center, 879-3388

Maui School of Diving
Kahului, 879-7681
or 877-9145

Maui Sun Divers
Kihei, 879-3631

Mike Severns
Kihei, 879-6596

Ocean Activities
Center
Wailea Shopping
Center, 879-4485

Ocean Safaris
101 N. Kihei Rd.
Kihei, 879-7242

Skin Dive Maui
2411 South Kihei Rd.
Kihei, 879-1502

Molokini Divers
1993 S. Kihei Rd
879-0055

The Dive Shop of
Kihei, 879-5172
1-800-367-8047
ext. 368

Underwater Habitat
Groups of 6
Lanai and Molokini
244-9739

Scuba Shack
Valley Isle Divers
Kihei, 879-3483

Books of interest available at dive shops or area bookstores:
*Comprehensive Guide to Scuba Diving in Hawaii* by Phil Hoffman. 111 pgs.
*Diver's Guide to Hawaii* by Chuck Thorne and Lou Zitnik. 248 pages, $9.95.
*Skin Diver's Guide to Hawaii* by Gordon Feund. 72 pages, $2.50.
*Hawaii Diver's Manual*, 162 pages, $3.00.

# SEA EXCURSIONS

Maui offers a bountiful choice for those desiring to spend some time in and on the ocean. Boats available for sea expeditions range from a three-masted schooner, to spacious trimarans and large motor yachts, to the zodiac type rafts for the more adventurous. Your choice is a large group trip or a more pampered small group excursion with a maximum of six people. Two of the most popular snorkeling excursions are to Molokini and Lanai. Most sailboats motor to these islands and, depending on wind conditions, sail at least part of the return trip. All provide snorkel equipment. Food and beverage service varies and is reflected in the price. Many sailboats are available for hourly, full day or longer private charters.

Excursion boats seem to have a way of sailing off into the sunset. The number of new ones is as startling as the number of operations that have disappeared since our last edition. As mentioned previously, competition to Molokini has become fierce. Twenty to thirty boats a day now arrive to snorkel in this area. Many more boats now take trips to Lanai as well. Currently there is only one company offering Molokai excursions.

In the following list, phone numbers of the excursion companies are included in case personal booking is desired, however, most activity desks can also book your reservation. The best deal with an activity operator is ***Tom's Cashback Tours*** who can book most boats and offers a 10% refund. They are located in Lahaina and can be reached at 661-8889.

PRICES PER PERSON WILL RUN YOU ACCORDINGLY:
 1/2 day trip to Lanai $40 - $60
 Full day trip to Lanai $90 - $125
 1/2 day trip to Molokini (3 - 6 hours) $35 - $65
 1/2 day Maui coastline (3-4hrs.) $40 - $65
 Full day Maui coastline $80 - $90
 Sunset sails (1 1/2 - 2 hrs.) $30 - $35
 Whale watching (3 hrs./seasonal) $30
 Private charters $75 per hour and up, $400 per day and up

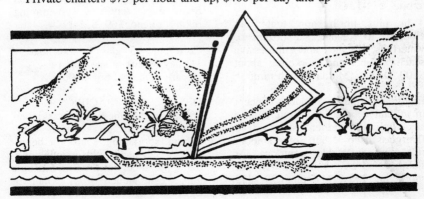

## EXCURSION - CHARTER LISTING

**ADVENTURE ONE**
*Maika'i makani II* 50' catamaran
BBQ sunset dinner sail, Molokini
snorkel, whale watch. Departs
Maalaea, 667-7683, 1-800-356-8989

**ALIHILANI YACHT CHARTERS**
*Makani Wiki Too, Makani Wiki,* 36'
MacGregor trimarans
1/2 and full day trips to Lanai
Maui coastline sail/snorkel Sunset
sail, private charters. Departs
offshore near Lahaina or Lahaina
Harbor, 661-3047 or 1-800-544-2520

**ALOHA VOYAGES**
*Machias* 80' full rigged schooner
All day (12 hr) snorkel/scuba to
Molokai, $105 adults. Includes tour
of leper colony, continental breakfast,
BBQ lunch. Half day $69. 667-6284

*Between the Sheets*, Morgan 42 MKII
3 & 5 hr. snorkel/sail
Sunset sail, whale watching
Max. 6 persons, 661-4095
Departs from Lahaina Harbor

**BLUE WATER RAFT** ★
3 1/2 hr. Molokini snorkel, (3 different
sites in the crater)
2 hr. Molokini, snorkel 1 site
$39-49. Departs Kihei launch ramp

4 hr. plus, north shore tour from
Hookipa to Keanae, see dozens of
waterfalls, jungle and impressive
seacliffs, lunch and snorkeling, about
$100. Departs Maliki launch ramp

Enjoy your ride on the latest concept
in inflatable boats, the Novurania, a
sleek Italian hybrid featuring a ridged
fiberglass deep V hull for a smooth
ride, and inflatable pontoons for
stability and easy entry and exit for
snorkeling. Max. 6 people. 879-7238

**CAPT. KIRK'S ENTERPRISE**
Zodiac-type boats, snorkel wherever
the conditions are best
Lahaina, 661-5333 or 667-9740

**CAPT. NEMO'S**
*Seasmoke*, 58' catamaran. Morning
snorkel, sunset sail. Introductory or
certified dive to Lanai. Seasonal
whalewatch. Departs Kaanapali
Beach. 661-5555, 1-800-367-8088.
$66 snorkel, $79 intro scuba.

**CAPTAIN ZODIAC RAFT**
23' ocean zodiacs, snorkeling, whale-
watching. Depart Mala Wharf. Maximum 15. 7 hr. Lanai snorkel $95, 1/2
day snorkel $55. 667-5351, 667-5862

**CINDERELLA YACHT CHARTERS**
*Cinderella*, 50' Columbia Sloop
3/4 day sail/snorkel, max. 6 people
Kealia Beach Plaza, 879-0634

**CLASSIC VENTURES**
*Alihilani*, 40' cutter. Snorkel cruise,
whalewatching. Private charters
Departs Kihei Cove, 879-7986

**CLUB LANAI**
*Kaulana*, 70' power catamaran, 149
passengers. Departs Lahaina Harbor
daily for private beach on Lanai.
Snorkeling, kayaking, swimming,
biking, picnic. $69. 871-1144

*Coral Sea*, 65' glass bottom boat
1/2 day Molokini snorkel/cruise
Departs Lahaina, 661-8600

**DESTINATION PACIFIC**
26' diesel. Marine biologist, skipper-
owner Ted Mickowski offers sportfishing, snorkeling, scuba for beginners or advanced, professional photography, underwater video. 6 people
max., can launch from Kihei or
Lahaina. 242-5004 or 244-5611

## E'ALA - NATIVE HAWAIIAN OCEAN ACTIVITIES
(808) 667-4355 or 667-5362
186 Malanai St.
Lahaina, Maui, HI 96761
Traditional Hawaiian double hulled canoe (only commercially certified sailing canoe in Hawaii) 45', 26 passenger. 2 1/2 hour snorkel/sail includes gear and instruction, Hawaiian narration, live Hawaiian music, refreshments $45. Seasonal whale watch, adults $29, children under ten $24. Can arrange for private groups.

## EXCEL CHARTERS ★
50' Delta *Excalibur*. Molokini/La Perouse snorkel w/champagne lunch $58. 879-3333

## FRIENDLY CHARTERS
*Maalaea Kai*, 45' catamaran Molokini snorkel/cruise and 2-4 hr. snorkel/sail, max. 26. Private charters available. Maalaea Harbor. 871-0985

## GENESIS YACHT CHARTERS
*Genesis*, 48' ketch. 20 passenger 1/2 day snorkel/sail to offshore Lanai $56. Sunset dinner cruise, private charters. Departs Lahaina. 667-5667

## IDLEWILD CHARTERS
34' Hawaiian sailing cat. Max. 24 persons, snorkel/sail, whale watching, private charter also. Departs Maalaea. 572-8964

## KAMEHAMEHA SAILS, INC.
*Kamehameha*, 40' catamaran Snorkel/sail, sunset sail, whale watching, private charter, max.15 Departs Lahaina harbor, 661-4522

## KIELE V
Contact Hyatt Regency, Kaanapali (808) 661-1234 ext. 3104
*Kiele V.*, 55' catamaran, 4 hr. snorkel/sail $60. Afternoon sail $29

## KIHEI SEA SPORTS
*Kihei Sea Sport*, 55' motor yacht 1/2 day snorkel Molokini Lg. groups, Maalaea, 879-1919

## LAHAINA BEACH CENTER
Lahaina, 661-5762
Lahaina Expeditions 661-3756. Ferry service to Lanai twice daily.

## LELANI CRUISES
49' Motor vessel. 1/2 day Molokini snorkel $59, seasonal whale watching 661-8397 or 1-800-833-5800 U.S.

## LIN WA
*Lin Wa*, 65' glassbottom boat resembling a Chinese junk. 1 1/4 hour coastline cruise, sunset dinner cruise Lahaina, 661-3392

## MAUI CLASSIC CHARTERS
*Lavengro*, 60' gast rigged Schooner built in 1926 is currently in dry dock. *Four Winds*, 53' double deck glass bottom cat., bbq grills, waterslide, daily learning snorkel, afternoon whale watch. Maalaea, 879-8177

## MAUI-MOLOKAI SEA CRUISES
*Prince Kuhio*, 92' motor yacht Whale watching, private charters 1/2 day Molokini, departs Maalaea 242-8777 or 1-800-468-1287

## MICHELLE MARINE
*Michelle II*, 38' power cat. Snorkel/introductory scuba Lahaina, 667-2085

## MOWEE WINDS CHARTERS
42' trimaran. Sail/snorkel to Molokini/Lanai. Sunset sail Private charter, max. 22 people Departs Kaanapali, 669-6445

## OCEAN ACTIVITIES
*Ka Kanani*, 46' catamaran departs at Maui Prince Hotel for Molokini, 879-7218

## OCEAN ACTIVITIES
*No Ka Oi IV & No Ka Oi III*
37' polycraft, scuba trips, 12 max.
Deep sea fishing, 6 max.
*Wailea Kai*, 65' catamaran
Snorkel, whale watch, dinner cruise
*Maka Kai*, 65' catamaran
Lanai trip, whale watching,
snorkeling, party boat fishing
*Manute'a*, 50' catamaran, Lanai
trip, cocktail sails, whale watching
All boats depart Maalaea except
Manute'a which departs Lahaina
879-4485 or 1-800-367-8047 ext. 448

## OCEAN ENTERPRISES
6 passenger cabin cruisers
Snorkeling, scuba, sportfishing.
879-7067 or 242-5096

## OCEAN RIDERS
"Adventure Rafting" 3 unusual destinations (depends on daily weather
conditions). Reefs of Kahoolawe,
Molokai's cliffs or Lanai. 10 persons
max for all day trip. 661-3586

## PACIFIC WHALE
## FOUNDATION CRUISES ★
*Whale One*, 53' motor vessel
*Whale Two*, 50' sailing ketch
Whale watching, Molokini snorkeling,
Lanai BBQ, sunset cruise. Research
scientists on board. All profits benefit
marine conservation.
Departs Maalaea, 879-8811

## PARDNER SAILING CHARTERS
*Pardner* 46' Cal ketch (6 passenger)
Sunset sail, West Maui snorkel, Lanai
(808) 661-3448

## SAIL HAWAII
*Fiesta* and *Mele Kai* 37' or 40' sailing yachts, max. 6. 5 hr. Molokini
snorkel, sunset sail. Trip destinations
flexible. Kihei Cove Park, 879-2201

## SEA ESCAPE    You-Drive zodiac
boat rentals, 879-3721

## SEALINK OF HAWAII
*Maui Princess* 118' excursion ferry
150 passenger. Service between Maui
and Molokai daily. $21 adults one
way. 1-800-833-5800, (808) 661-8397

## SEABERN YACHTS
42' cutter or 42' sloop, Max. 6 for
snorkeling, whale watching, overnight
inter-island charters. 661-8110

## SCOTCH MIST CHARTERS
*Scotch Mist II*, 50' sloop
*Santa Cruz* 50', max 19
West Maui 1/2 & full-day snorkel/sail
Champagne sunset sail, max. 6, char-
ters. Lahaina, 661-0386

*Spirit of Windjammer*
65', 3-masted schooner
Full-day trip to Lanai, 2-hour Maui
coastline dinner cruise, lg. groups
Lahaina, 667-6834, 1-800-843-8113

*Suntan*, 50' Santa Cruz. Molokini
snorkel, sunset sail. Kealia Beach
Center, max. 25, 874-0332

*Stardancer*, 150' yacht
Dinner cruises and nightclub
See Lahaina restaurants

## TRILOGY EXCURSIONS ★
*Trilogy*, 50' trimaran
*Kailana*, 40' trimaran
*Manele Kai*, 40' catamaran
Full-day snorkel/picnic/sightseeing
to Lanai, a definite best bet
1/2-day sail/snorkel
Departs Lahaina Harbor
Also Molokini trips departing
from Maalaea Harbor
661-4743 or 1-800-874-2666

*White Wings*, 35' trimaran
6-hr. sail/snorkel/fishing
Available for overnight and
inter-island excursions, max. 6
Maalaea Harbor, 572-8457

## BLUE WATER ADVENTURES!

While Molokini continues to be a much touted snorkeling spot, those seeking something a little different should check with ***Blue Water Rafting*** about their East Maui trip. The area past Makena is geologically one of Maui's youngest. The coastline is only accessible by foot, however, a trip on the Novurania (a sleek Italian inflatable boat) will get you up close to see the beautiful and unusual scenic wonders of Mother Nature. Natural lava arches, pinnacles and caves are explored with the picturesque slopes of Haleakala providing a magnificent backdrop. Our chosen day for the expedition proved to be an exhilarating and wet one! The ocean conditions were somewhat rougher than desired, but our hearty group of six agreed to push forward. With spray from the ocean drenching us, one of the more witty members of our group donned his snorkel and mask. It performed admirably at keeping the water out of his eyes! The scenic vistas were fabulous and the boat was able to maneuver through one of the arches and up close to the cliffs which appeared to have been sculpted by a fine artisan. A brief stop for some mid-morning nourishment and a snorkel at La Perouse before returning to the Kihei Boat Ramp. This is a trip that can be best experienced in only in this manner. To the best of our knowledge, only Blue Water Rafting is offering this trip. So for an unusual and exciting Maui adventure check this one out!  Phone 879-7238.

## A TRIP TO LANAI

We hadn't done a trip aboard ***Trilogy*** for a few years and wanted to see how things had changed over the course of time. It appears that the Coon family knows not to mess with a good thing. The morning boat trip over to Lanai still starts earlier than most would like, but once underway with warm (yes, still homemade by the Coons) cinnamon rolls and a mug of hot chocolate or coffee, it seems all worth the effort. Don't forget to bring the camera! Two boats bring about 60 guests to the island each day for snorkeling, sun and fun at Manele Bay. With the additional number of people on the trip, they have now divided the snorkeling and Lanai island tour in two. One boat load walks to the beach and snorkels, tours the island and eats. The beginning snorkelers are carefully instructed in a tide pool before entering the ocean. The other group tours the island, snorkels and then eats at a second "sitting." If you would prefer, you can also skip the tour of Lanai City and snorkel even longer. The chicken is cooked on the grill by the ship's captain

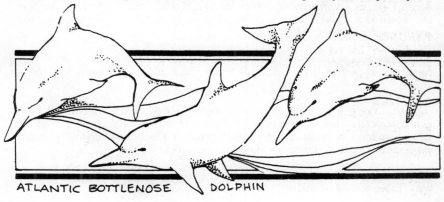

ATLANTIC BOTTLENOSE   DOLPHIN

and served on china plates. Mrs. Coon still isn't giving out the secret ingredients to her salad dressing to anyone. Some work has been done to beautify the eating area with additional plants and an awning. Development of the new resort on Hulopoe Beach is proceeding at a snail's pace (scheduled to open in 1991) and the sand is as white as ever. Just about the only change seemed to be the way they cooked the noodles and they are even better! Unfortunately the Coon brothers don't get out of the office much any more to skipper the boats, but our Captain was energetic and really seemed to enjoy the trip as much as the rest of us, although the rest of the crew weren't as congenial as a previous trip. This is a very special outing and well worth considering as a part of your island holiday. Phone (808) 661-4743 or 1-800-874-2666. $125 for this all day trip (7 am to 4 pm) includes continental breakfast, lunch and snorkel gear.

The Coon family's Trilogy operation has now expanded to a half day Molokini snorkel aboard their new 44 foot trimaran cutter Trilogy IV. We haven't tried it, but if its even half as good as their Lanai venture, it bears looking into. The trip features a continental breakfast, with those same good cinnamon rolls, snorkel lessons and equipment, and an all you can eat hot lunch. Departs Maalaea Harbor 6:30 am from slip 99, behind the Coast Guard Station, returns at noon. Adults $65, Children $32.50. Same phone numbers as above.

# DINNER CRUISES

Dinner cruises are quite popular on Maui with their free flowing Mai Tai's, congenial passengers, tropical nights, and Hawaiian music which entertains while the boat cruises along the coastline. Dinner aboard most is definitely not haute cuisine, but usually quite satisfactory, especially after a few Mai Tai's. The live Hawaiian entertainment varies from amateur to very good. Cruises typically last about two hours and prices run about $50. Samples of dinners listed may vary. Most dinner cruises accommodate 50 - 100 people. New on the waters off Lahaina is the Stardancer. This ship is far and away the best of the bunch.

*Aloha Voyages* - offers dinner cruises for $42.24 adults, $21.12 children on their 80' retired U.S. Coast Guard schooner Machias. Includes BBQ dinner, open bar and live entertainment. (667-6284)

*Frogman* - Enjoy a dinner of filet mignon, fresh island fish, aboard this 44 foot catamaran, departing from Kaanapali Beach. Beer and wine served. $49. (667-7622)

*Genesis* - Enjoy a sunset dinner cruise aboard this 44' ketch, entertained by a guitarist/vocalist. Entrees are either fresh island fish or N.Y. steak served on special lap trays. A maximum of 18 people ensures a more intimate and romantic experience. Departs Lahaina six nights a week. (669-5667)

*Lin Wa* - Sunset Polynesian dinner cruise. Depart Lahaina in this glass bottom boat. Table seating. (661-8397)

***Prince Kuhio*** - Departs from Maalaea Harbor. This 92' cruise boat features a dinner of fresh Mahi, beef ribs and an open bar. Hawaiian entertainment accompanies your cruise. Table seating. (242-4575)

***Seabird Cruises*** - Departs from Lahaina Harbor. This 65' catamaran offers a two-hour sail featuring a prime rib dinner. Table seating. (661-3643)

***Stardancer*** - This 150' vessel accommodates up to 149 guests and offers a complete buffet each evening. Main entrees include prime rib, fresh fish, pasta and teriyaki chicken. The meal is prepared on board, also unique to Maui dinner cruises. Two cruise seatings are offered at 5 pm and 7:30 pm ($50 per person). We recommend the early seating which tends to be less crowded and offers the benefit of a Maui sunset! A ten minute ride on the Kaulana, their shuttle ferry, and we were greeted with a welcome drink and a choice of seating. An open bar, a plentiful buffet, a ship that is sleek and quiet, attentive service, a perfect sunset and the talented performers "Raw Silk" got our evening off to a good start. The boat is dramatically appointed. Only several years old, it had been a dinner cruise ship on the East coast before arriving in the Pacific waters. A complete refurbishing included two painters laboring six months to create the unusual hand painted walls and furniture in an array of purples and fushias. As dinner is winding down, the band moves up one level to the dance floor. A black marble floor gleams invitingly while tunes of the 40's to 80's are rendered by the band. If dancing isn't your style, there are plenty of panoramic ocean views, starlit nights and fresh air on the top deck. First seating diners can receive a pass to come back later and second seating diners can stay on board when the dinner club transforms itself into the High Spirits night club. Starting at 10 p.m. the shuttle will bring guests out for a $5 round trip fare. (10 pm-2 am). (871-1144).

***Wailea Kai*** - Departs Maalaea Harbor. Enjoy the sunset on this 65' catamaran. Dinner is Mahi or chicken. Open bar and live entertainment. Free bus transportation is available. Table seating. (879-4485)

***Windjammer*** - An open bar, live entertainment and a dinner of scampi and steak are served on this 75' three-masted schooner. Table seating. Sails nightly. (667-6834)

LAHAINA HARBOR

# WHALE WATCHING

Every year beginning in November and continuing until April, the humpback whales arrive in the warm waters off the Hawaiian Islands for breeding, and their own sort of vacation! The sighting of a whale can be an awesome and memorable experience with the humpbacks, small as whales go, measuring some 40 - 50 feet and weighing in at 30 tons. The panoramic vistas as you drive over the Pali and down the beachfront road to Lahaina afford some excellent opportunities to catch sight of one of these splendid marine mammals. However, PLEASE pull off the road and enjoy the view. Many accidents are caused by distracted drivers. For an even closer view, there are plenty of boat trips. Although most every boat operator does whale watching tours in season, you may want to check into the one sponsored by Pacific Whale Foundation 879-8811. As they are a research group, they are very well informed and knowledgeable about the whales. You can report your sightings by calling Whale Watch Hotline at 879-6530.

# DEEP SEA FISHING

Deep sea fishing off Maui is among the finest in the world and no licenses are required for either trolling or bottom fishing. All gear is provided. Fish that might be lured to your bait include the Pacific Blue, Black or Striped Marlin (Au) weighing up to 2,000 lbs., Yellow Fin Tuna (Ahi) up to 300 lbs., Jack Crevalle (Ulua) to 100 lbs., Bonita-Skipjack (Aku) to 40 lbs., Dolphin Fish (Mahi) to 90 lbs., Waho (Ono) to 90 lbs., Mackerel (Opelu), Amerjack (Kahala), Grey Snapper (Uku), Red Snapper (Onaga), and Pink Snapper (Opakapaka).

Boats generally offer half or full-day fishing trips on a share or private basis with prices running from $65 - $75 shared or $300 - $350 private for a half day (4 hr.), and $100 -$110 shared or $450 -$550 private for a full day (8 hr). Some are willing to take non-fishing passengers along at half price. Most boats take 4 - 6 on a shared basis, however several, such as Excel and Sport Diver, can handle larger groups. *FINDING A CHARTER:* Your local activity center may be able to direct you to a particular boat that they favor, or you could go down to the docks at the Lahaina or Maalaea Harbor in the afternoon and browse around. There are also a number of activity booths at both harbors that can be consulted. When reserving a spot, be aware that some boats will give full refunds only if 48-hour notice is given for cancellation. If you want to take children fishing, many have restrictions for those under age 12. If you are a serious fisherman, you might consider entering one of the numerous tournaments. Some charters offer tournament packages. Following are a list of just some of the charter fishing boats. While not a scientific study, Finest Kind and Exact have managed to be consistantly mentioned in the local newspaper for regular catches of very large fish.

*A WORD OF ADVICE:* The young man had a grin that reached from ear to ear as he stood on the pier in Maalaea, holding up his small ahi for a snapshot of his big catch. The surprise came when the deck hand returned it to the ice chest and continued on with his work. The family all stood, unsure what to do. It appeared that the fish was to remain on board, while the family had envisioned a nice fresh fish dinner. Finally one family member spoke up and a very unhappy crew member sliced a small filet, tossed it into a sack and handed it to the young man. Unlike sportsfishing charters in some parts of the country, the fish caught on board generally remain the property of the boat. The pay to captain and crew is minimal and it is the selling of the boat's catch that subsidizes their income. Many vacationers booking a fishing excursion are unaware of this fact. There seems to be no written law for how fishing charters in Hawaii handle this, at least everyone we talked to had different answers. Many of the brochures lead one to believe that you keep your fish, they neglect to mention that it is only a filet of fish. Occasionally you may find a head boat which operates under a different sort of guideline. In this situation, you pay for your bait, gear, and boat time and then keep the fish. In any case, be sure you check when you book your trip about just what and how much fish will be yours to keep. If the person at the activity desk assures you that you keep your catch, don't leave it there, also check with the captain when you board. Communication is the key word and have a mahi mahi day!

## LAHAINA HARBOR

AERIAL SPORTSFISHING
CHARTERS
PO Box 12, Lahaina, 667-9089

BALI HA'I CHARTERS
Lahaina, 667-6672, 667-9237

FINEST KIND, INC.
*Exact*, 31' Bertram
*Finest Kind*, 37' Merritt
P.O. Box 10481, Lahaina
Maximum 6 people, 661-0338

HINATEA SPORTFISHING
*Hinatea*, 41' Hatteras
PO Box 1238, Lahaina, 667-7548
Full day only

ISLANDER II SPORTSFISHING
36' Uniflite, 667-6625

LAHAINA CHARTERS
*Broadbill*, 36' Hardcraft
*Escape*, 26' Bertrum
PO Box 12, Lahaina
maximum 6 people, 667-6672

LUCKEY STRIKE CHARTERS
*Luckey Strike*, 45' custom
*Kanoa*, 31' Uniflite, max. 6
Luckey Strike has a large cockpit
with full sunshade that can
accommodate up to 22 people
PO Box 1502, Lahaina, 661-4606

MANO KELA SPORTSFISHING
*Mano Kela*, 48' deluxe sportfisher
6 passenger, includes a/c, microwave,
TV, VHS. Exclusive full day $988,
half day $572. Charters available
(808) 661-3446

## MAALAEA HARBOR

CAROL ANN CHARTERS
33' Bertram, max. 6
877-2181 or 242-4575

RASCAL SPORTFISHING
CHARTERS
*Rascal*, 31', 874-8633

EXCEL FISHING CHARTERS
*Excel*, 48' Delta
*Excite*, 35' Bertram
PO Box 146, Makawao, 96768
877-3333 or 1-800-462-4103

OCEAN ACTIVITIES CENTER
Departs Maalaea, 879-4485

# SMALL BOAT SAILING

Small boat sailing is available at a number of locations with rentals, usually the 14', sometimes 16' and 18', Hobie Cat. Lasers are also available. Typical rental prices are $35 - $40 per hour, and lessons are available. The Maui Inter-Continental, Stouffer Wailea Beach Resort, Hyatt Regency Maui, Marriott and Kaanapali Alii also offer free sailing clinics for their guests.

## FOR MORE INFORMATION ON RENTALS CONTACT:

*Fun in the Sun* - West Maui Sailing School. Also windsurfing and snorkel equipment. (667-5545)
*Sea Sails* - Located at the Sheraton, Royal Lahaina and Kaanapali Beach Hotel, Westin, Sands of Kahana, Kaanapali Shores (661-0927)
*Maui Sailing Center* - Kealia Beach Plaza (879-5935)
*Ocean Activities Center* - Stouffer Wailea Beach Resort (879-9969), Maui Inter-Continental (879-8022)

# WINDSURFING

Windsurfing is a sport that is increasing in popularity astronomically. Hookipa Beach Park on Maui is one of the best windsurfing sites in the world. This is due to the consistently ideal wind and surf conditions, however, this is definitely NOT the spot for beginners.

For the novice, group beginner lessons runr $15 - $25 an hour, which generally involves instruction on a dry land simulator before you get wet with easy to use beginners equipment. Equipment and/or lessons are available from the following:

*Hawaiian Island Windsurfing* - Pro shop, sales, service, rentals and insturction. Three hour group beginner lessons or advanced water start $44. Private instruction $35 per hour. Toll free 1-800-231-6958 or locally 871-4981.

WINDSURFING

**Hawaiian Sailboarding Techniques** - Alan Cadiz offers advanced instruction on Maui's Kanaha Beach (871-5423)
**Hi-Tech Sailboards** - 51 Baldwin Ave. (579-9297) or (877-2111)
**Kaanapali Boating Center** - At Kaanapali Villas (661-5424)

**Kaanapali Windsurfing School** - Hanakaoo Beach by Hyatt Regency. Two hour beginner group lessons $40/ Private beginner $30 per hour. (667-1964)
**Maui Magic Windsurfing School** - Group & private lessons for all levels. Beginning group 2 1/2 hours $55; private 1 1/4 hours $59. Rentals & retail (877-4816)
**Maui Windsurf Co.** - offers group instruction for novice or the experienced and rental equipment. 520 Keolani Place, Kahului (872-0999)

**Maui Sailing Center** - Free windsurfing demonstration is featured 8 am daily at Kealia Beach Plaza as does a two hour class, first on land and then on the water. No reservations necessary, but call ahead to verify time. Kealia Beach Center (879-5935), or Kai Nani Village (879-6440)
**Second Wind** Pro shop rental, used and new sales. Private lessons $45 per hour, $60 for 2 1/2 hour class. (Toll free 1-800-852-SHOP) U.S. (808-877-7467).

**Indian Summer Surf Shop** Rental Equipment.
**Fun Rentals** (661-3053), Lahaina, rents equipment ($40 for 24 hrs.) and can recommend an instructor (661-3794)
**Ocean Activities Center** - At Stouffer Wailea Beach Resort (879-9969), Maui Inter-Continental (879-8022), Central office 1-800-367-8047 ext. 448 or 879-4485

**Sea Sails** - Located at the Sheraton, Royal Lahaina, Westin, Sands of Kahana, Kaanapali Shores and Kaanapali Beach Hotel. (661-0927)
**Windsurfari** - offers packages which include accommodations, rental car, windsurfing equipment nad excursions with Alan Cadiz's of Hawaiian Sailboarding. (1-800-367-8047 ext. 334. or 808-871-7766).

Some resorts offer their guests free clinics. Rental by the hour can get expensive at $15 - $20 per hour and $40 - $65 per four hours. A better rate is $35 for all day. If you are interested in renting equipment for longer periods or desire more advanced equipment try:

**Freedom Maui** - (877-0202) 55 Kaahumanu Avenue, Kahului
**Hawaiian Island Windsurfing** - (871-4981), 460 Dairy Road, Kahului
**Sailboards Maui** - (871-7954) 210 Dairy Road, Kahului
**Pacific Ocean Sports** (871-1374)

Typical costs are $40 for a full day, $200 for seven days. Deposit may be required. Rental includes board, universal mast, boom, sail and soft car rack. There are no reservations and no refunds unless returned within one hour. You also pay for broken, lost or stolen equipment.

Books of interest: *How to Windsurf Hawaiian Style* by Thomas J. Cunningham, 18 pages, softcover $2.50.

# SURFING

Honolua Bay is one of the best surfing spots in Hawaii, and undoubtedly the best on Maui, with waves up to 15 feet on a good winter day and perfect tubes. A spectacular vantage point is on the cliffs above the bay. In the summer this bay is calm and, as it is a Marine Reserve, offers excellent snorkeling.

Also in this area is Punalau Beach (just past Honolua) and Honokeana Bay off Ka'eleki'i Point (just north of the Alaeloa residential area). In the Lahaina area there are breaks north and south of the harbor and periodically good waves at Awalua Beach (mile marker 16). On the north shore of Hookipa Beach Park, Kanaha Beach, and Baldwin all have good surfing at times. In the Hana area there is Hamoa Beach. There are a couple of good spots in Maalaea Bay and at Kalama Beach Park.

Conditions change daily, and even from morning to afternoon around the island. Check with local board rental outlets for current daily conditions.

***Surfing Maui*** - (242-7572) offers group lessons (1 hour for $25 per person) or Private (1 hour for $40). Beginning, intermediate or advanced levels are available and equipment is provided.

***Maui Surfing School*** - (877-8811) Andrea Thomas originated the "Learn to Surf in One Lesson" and is reported to have taught thousands of people between the ages of 3 and 70. She's been teaching on Maui since 1980 and offers private and group lessons specializing in the beginner and coward. Board rentals available. Lahaina Harbor.

***Maui Beach Center*** - (661-4941 or 667-4355) surfing tours are offered by Maui Beach Center. Free hotel pickup, all equipment, instructors and refreshments. Hotel pick up at 9 a.m. $45 per person for board surfing, $35 for body boarding. Discounts for multiple days.

Books of interest available at local bookstores:
*Surfing Hawaii* by Bank Wright, 96 pages softcover $3.95.
*Hawaii Surfing Map*, descriptions of 97 surfing spots.

## BODY SURFING

Mokuleia (Slaughterhouse) Beach has the best body surfing especially in the winter. This is not a place for weak swimmers or the inexperienced when the surf is up. The high surf after a Kona storm brings fair body surfing conditions, better boogieboarding, to some beaches on leeward Maui.

Books of interest available at local bookstores:
*How to Body Surf* by Nelson Dewey, 10 pages softcover $1.00.

# JET SKIIS

**CONTROVERSY!** A ban restricting *thrillcraft* activities on Maui, scheduled to go into effect in 1989 for five months every year, was thrown out of court. The regulations for Oʻahu and Hawaiʻi are expected to be rewritten. Five parasailing companies filed a suit against the State Department of Transportation stating that such a ban would cause them severe economic hardship and discrimination since this ban would not affect fishing or dive boats. This ban was spurred by the reports of a number of incidents, particularly with jet skiiers harassing whales. The proposed ban would have affected all the Hawaiian Islands and was designed to be a protective measure for the humpback whales which winter there each year in the warm Pacific waters. We'll update this issue in our quarterly newsletters.

The areas off Sugar Beach in Kihei and Hanakaoo Beach Park in Kaanapali offer excellent jet skiing conditions, usually calm but with enough waves to make it interesting. Prices are usually $50 for one hour and $35 for a half hour.

*Pacific Jet Ski* - On the beach in front of Whaler's Village. They also have two passenger wave runners. Call for reservations. (667-7851)
*Pacific Jet Ski Rental* - South end of Kaanapali Beach at Hanakaoo Beach Park. (667-2066)
*Maui Sailing Center* - Located in Kihei. Kawasaki 440 and 550's. (879-5935)

# PARASAILING

Some flights start at the offshore floating platform off Lahaina. You are helped into a life jacket and special harness-like seat and instructed in takeoffs and landings. Then you are hooked up to the tow line and in four or five steps you're off and soon at 200 feet enjoying the ride and view. The flight lasts 8 to 10 minutes which seems neither too long nor too short. The landing is really no more traumatic than stepping off a high curb, unless a last second wind gust sends you off the platform's edge and into the drink (the reason you wear swimming suits). Prices range from $35 to $45 for a 8 to 10 minute ride. Some charge for an "observer" (friend) to go along, others allow them free.

UFO Paracruiser and Wailea Para Sail use a new wrinkle. You get started standing on the boat and as your parachute fills, you are simply reeled out 200 - 400 feet. When it comes time to descend, you're simply reeled back in.

*Parasailing Hawaii* - Lahaina, reservations required. Boat trip one hour. $10 charge for observer. (661-5322)
*Wailea Para Sail* - Kihei, $10 observer charge, early morning discount. (879-1999)
*West Maui Para-Sail* - Lahaina Harbor, booth #4. Honeymoon shute for two. Dry landing. Boat departs every 30 minutes with six passengers. Call for reservations. (661-4060)
*UFO Paracruiser* - Observers can go free. Departs in front of the Westin Maui. (661-7836)

# WATERSKIING

***Challenger Waterski*** - Kihei's only licensed waterski company offers free instruction for beginners or advanced levels. Children's skiis also available. Private charter for six passengers also available. (879-3271)

***Lahaina Water Skiing Inc.*** -Hanakaoo Beach Park in front of Hyatt Regency. $35 for 15 minutes. Charters available. Also a new feature are their "Screemin' Weanie Rides"! (661-5988)

***Kaanapali Water Ski*** - Departs Mala Wharf. $25 one person for 15 minutes. Hourly rate for one to three skiers. (661-3324)

# KAYAKS

***Maui Beach Center*** - has two man kayaks $20 per hour, 1 man $10 per hour. Kaanapali 667-4355 or Lahaina at 505 Front Street. (661-4941)

***Maui Kayaks*** - The novice or experienced paddler can join a half or full day guided trip along the Kihei coast. Hourly rentals also available. Contact: Bill Pray, Maui Kayaks, 50 Waiohuli St., Kihei, Maui 96753. (874-3536)

***Sea Sails*** - Locations at the Maui Westin, Sands of Kahana, Kaanapali Shores, Sheraton, Royal Lahaina and Kaanapali Beach Hotel have surf ski rentals. (661-0927)

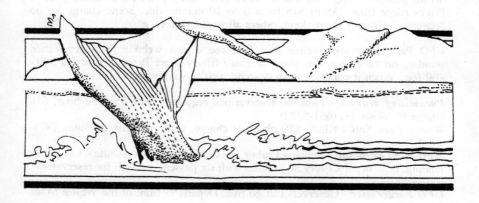

# LAND ACTIVITIES

## *LAND TOURS*

Land excursions on Maui are centered upon two major attractions, Hana and the Seven Pools, and Haleakala Crater. Lesser attractions are trips to the Iao Valley or around West Maui. You can do all of this by car (refer to the WHERE TO STAY - WHAT TO SEE chapter), however, with a tour you can sit back and enjoy the scenery while a professional guide discourses on the history, flora, fauna and geography of the area. The single most important item on any tour is a good guide/driver and, unfortunately, the luck of the draw prevails here.

Another, somewhat expensive option, is a personalized custom tour. A local resident will join you in your car for a tour of whatever or wherever you choose. You can do the driving or sign on your guide with your rental car company to do the driving. This may allow you the opportunity to linger at those places you enjoy the most, without following the pace of a group. Your guide may also be able to take you to locations the tour vans don't include.

Driving to Hana and back requires a full day and can be very grueling, so this is one trip we recommend you consider taking a tour. A Haleakala Crater tour spans 5-6 hours and can be enjoyed at sunrise (3 am departure), mid day or sunset. The West Maui and Iao Valley trips are half-day ventures. Only vans travel the road to Hana, however, large buses as well as vans are available for other trips. (Be aware that some vans are not air conditioned.) Prices are competitive and those listed here are correct at time of publication. Some trips include the cost of meals, others do not.

Also available are one day tours to the outer islands. The day begins with an early morning departure to the Big Island, Oʻahu or Kauaʻi. Some excursions provide a guided ground tour, others offer a rental car to explore the island on your own.

*Akamai* - Trips include Haleakala ($39), Hana ($48). (871-9551, 1-800-922-6485)

*Aloha Nui Loa Tours* - Haleakala ($60), Hana ($65). They also provide a special horseback tour. A drive to Hana is followed by a ride over the ranchlands of the Hotel Hana, then a lunch at a quiet beachfront. (879-7044)

*Arthur's Limo Service* - They provide tours to Haleakala and Hana in their luxury limousine or their new super stretch van. (661-5466)

*Ekahi Tours* - Herbie Watson and his crew provide tours of Hana ($60), the Keanae Peninsula ($45), or Haleakala ($30 -$35). (572-9775)

*Grayline* - Provides a variety of large bus and small van tours to Halekala and Hana. (877-5507)

257

# RECREATION AND TOURS
*Land Tours*

*Historic Lahaina* - This one is a FREE 30 minute tour provided courtesy of the Lahaina area merchants. Pickups are on Papalaua St. behind the Post Office, or at the intersection of Lahainaluna Road and Front Street at the seawall. They provide an open sided tram and a narrated tour of this picturesque and historic sea port.

*Guides of Maui* - Discover Maui in your car with an island resident as your guide. Explore the destinations of your choice. $150 for two people (plus gas). (877-4042)

*Jesse's Maui Adventures* - Pick up from the Kihei and Wailea area. Hana ($52), Haleakala sunset ($42). (879-1329)

*No Ka Oi Scenic Tours* - Takes you to Hana ($40), Iao Valley and Haleakala ($35-$38). (871-9008)

*Polynesian Adventure* - Haleakala ($42/children $30), Hana ($60/children $45), as well as one day trips to the other islands (from $125 each island). (877-4242)

*Roberts-Hawaii Tours* - Offers land tours in their big air conditioned buses or vans. They depart to all scenic areas from Kahului, Kihei, Wailea, and the West Maui Hotels. (871-6226)

*Scenic Air Maui* - Trips from Maui to Moloka'i, Hawai'i, Kaua'i or O'ahu for a one day flightseeing-sightseeing, $99 - $149. (871-2555)

*Sugar Cane Train* - The Sugar Cane Train makes six round trips daily with one way fares for adults $5, two way is $8. Children 2-12 years are $2.50 one way, $4 round trip. Their main depot is located just outside of Lahaina, turn at the Pizza Hut sign. The Kaanapali Station is located across the highway from the resort area. The free Kaanapali trolley picks up at the Whaler's Shopping Center and drops off at this station. The Puukolii boarding platform and parking lot is located on the Kapalua side of Kaanapali.

They also have several package options which include the "Orient Express," a round trip train ride plus scenic voyage on the Lin Wa II at $20 for adults, $10 for children. A "Historic Lahaina" excursion includes a train ride and a self-guided tour of historic landmarks and admission into the Baldwin House and the Carthaginian II for $13 adults and $4 children. The newest adventure is to combine the train ride with the Omni Experience Theater $14 adults $7.75 children. If you plan on a full day round trip excursion, buy your return tickets early as they often sell out quickly. (661-0089)

*Trans Hawaiian* - A day trip to Hana ($45) departs 7 am and returns about 5:30, lunch is not included. Haleakala sunrise or morning tour ($28-$32). Also available are day trips to Hawai'i, Kaua'i, and O'ahu via jet. Guided tours on the outer islands or a car can be included for self-guided exploration. (871-1180)

# ART TOURS

Maui Art Tours offers visitors an opportunity to visit the homes and studios of Maui's finest artists, meet them and watch them work. The tour begins with a lei greeting at your hotel or condominium and an escorted tour to meet the artisans of your choice. A catered lunch may be followed by a stroll through a tropical arboretum where you can enjoy a hot tub and fresh island tea. P.O. Box 1058, Makawao, Maui 96768. (808-572-8374)

# BIKE TOURS

The Hawaiian Islands offer an endless array of spectacular air, sea and land tours, but only on Maui is there an experience quite like the bicycle ride down from the 10,000 foot summit of the world's largest dormant volcano. Bob Kiger, better known as Cruiser Bob, was the originator of the Haleakala downhill. (Cruiser Bob is reported to have made 96 individual bike runs himself to thoroughly test all aspects of the route before the first paying customers attempted the trip).

Each tour company differs slightly in its adaptation of the trip, but the principal is the same, to provide the ultimate in biking experiences. For the very early riser (3 am) you can see the sunrise from the crater before biking down. Later morning expeditions are available as well. Your day will begin with a van pickup at your hotel for a narrated trip to the Haleakala summit along with safety information for the trip down. The temperature at the summit can be as much as 30 degrees cooler than sea level, so appropriate wear would include a sweater or sweatshirt. General requirements are for riders to wear closed-toe rubber soled shoes, sun glasses or prescription lenses (not all helmets have visors), a height requirement of 5 feet is requested by some and no pregnant women are allowed on the trip. Bikers must also sign an acknowledgement of risk and safety consideration form. For the descent, riders are equipped with windbreaker jackets, gloves, helmets and specially designed bicycles with heavy duty brakes.

A leader will escort you down the mountain curves with the van providing a rear escort. Somewhere along the way will be a meal break. Some tours provide picnics, others include a sit-down meal at the lodge in Kula or elsewhere. Actual biking time will run about 3 hours for the 38-mile downhill trip. The additional time, about 5 hours, is spent commuting to the summit, meals, and the trip from the volcano's base back to your hotel. Prices for the various tours are competitive and reservations should be made well in advance.

We biked down with Maui Downhill and opted for the "late" 7 am trip. We found them to be very careful, courteous and professional. Unfortunately they don't have control over the weather and the day we chose was clear on the drive up, fogged in and misty at the summit and a torrential downpour for more than half of the 38 miles down. Due to the weather, we couldn't enjoy much of the scenery going down, but probably wouldn't have had much time to gander as it is important to keep your eyes on the road! The leader set a fairly slow pace, not much of a thrill for the biking speedster, but safe and comfortable for most. At any time we were invited to hop in the van, but ours was a hearty group and after a stop to gear up in rain slickers, we all continued on. Our leader also advised that if the weather posed any kind of risk, he would load us on the van. The weather broke long enough for us to enjoy sandwiches or salads at the Sunrise Market and to bask in the sun's momentary warmth. In radio contact with the group just ahead of us we were advised that the rain promised to await us just a little farther down the volcanic slope. As predicted, the drizzle continued as we biked down through the cowboy town of Makawao. We arrived in Paia only a little wetter for the experience.

*Cruiser Bob's* - (667-7717) (1-800-654-7717) This is the original Haleakala downhill trip. You can choose between an early (3 am) sunrise trek or the regular excursion for $89. Both include continental breakfast and sit down lunch at Kula Lodge.

*Maui Downhill* - (871-2155 or 1-800-535-BIKE in U.S.) Transportation from Lahaina, Kihei, Kahului and Paia. The sunrise trek includes a brunch after the ride. The day trip begins with a continental breakfast at the base yard before departure up the mountain and a picnic lunch at the Sunrise Market (and protea gardens) on the way down. $99 plus tax.

*Maui Mountain Cruisers* - (572-0195 or 1-800-232-MAUI in U.S.) Pickup provided from Kaanapali, Lahaina, Kahului and Kihei. Sunrise cruise includes a champagne brunch and continental breakfast. Midday trip includes continental breakfast and gourmet picnic lunch at Sunrise Market. $89.50 plus tax.

# BIKE RENTALS

Bikes and mopeds are an ambitious and fun way to get around the resort areas, although you can rent a car for less than a moped. Available by the hour, day or week, they can be rented at several convenient locations.

*A & B Rentals* - 3481 Honoapiilani Hwy. at the ABC store. They have mopeds, bicycles, beach equipment, surfing and boggie boards, snorkel gear, fishing poles and underwater cameras. Mopeds run $25 for 24 hours, bikes $10 day or $50 week. (669-0027)

*Fun Rentals* - 193 Lahainaluna Rd., Lahaina. In addition to bikes they have boogie boards, fishing rods, beach chairs and even baby strollers! (661-3053)

*Go-Go Bikes* - Located just past Kaanapali on Lower Honoapiilani Hwy. 30-B Halawai Drive #5. Bikes, scooters and mopeds can be rented. Choose your moped from a selection of one, two or variable 10 speed. Mopeds $5 hour, $25 for 24 hours. Mountain bikes, touring bikes, 6 and 12 speed $25 for 24 hours. Scooters also. Free pickup from Kaanapali area with 3-hr. minimum rental. Hawaii State law is 18 years minimum age, no helmet or special license. (661-3063)

# GOLF

Maui's golf courses have set for themselves a high standard of excellence. Not only do they provide some very challenging play, but they also offer distractingly beautiful scenery. Most of the major resorts offer golf packages. For the avid player, this may be an economical plan.

Sunseeker Golf Schools at Kaanapali offers golf instruction for the beginner or the advanced. Each Wednesday and Saturday at 10 am they conduct a free golf clinic. They offer private lessons ($35 for 45 minutes), playing lessons (one person $100, couple $150) and special one day classes ($175) classes. Phone 667-7111.

## KAANAPALI
The Kaanapali Resort offers two championship courses. Green fees are $90 for 18 holes, cart included from 7 am - 2:30 pm, 2:30 pm - 6 pm $55. Also found at the southern entrance to the Kaanapali resort area is the Royal Kaanapali driving range. (661-3691)

*The North Course* has been attracting celebrities since its inaugural when Bing Crosby played in the opening of the first nine holes. This is now home of the LPGA's Women's Kemper Open. Designed by Robert Trent Jones, this 6,305 yard course places heavy emphasis on putting skills. At par 72, it is rated 70 for men and 71.4 for women.

*The South Course* first opened in 1970 as an executive course and was reopened in 1977 as a regular championship course after revisions by golf architect Arthur Snyder. At 6,205 yards and par 72 it requires accuracy as opposed to distance, with narrower fairways and more small, hilly greens than the North Course. As an added distraction, the Sugar Cane Train passes by along the 4th hole.

## KAPALUA
The Kapalua Resort features the Bay Course and the Village Course. Eighteen holes for hotel or villa guests is $45 plus $15 for the required cart. $75 plus $15/cart for non-guests. Twilight play $50. Guests may reserve tee-off times up to 7 days in advance. Non-guest reservations 2 days in advance. (669-8044)

*The Bay Course* under the design of Arnold Palmer, opened October 13, 1975 and sprawls from sea level to the mountain's edge. This beautiful and scenic par 72, 6,850 yard course has a distinctly Hawaiian flavor and is the home of the internationally televised Kapalua International Championship of Golf. Imagine a 530-yard par four that travels downhill doglegging past the ancient stones of a pre-missionary fishing village to a green positioned on a black lava peninsula surrounded by white sand beaches. Or how about a 158-yard par 3 where the tee shot has to loop over a small bay. No wonder this course demands a controlled and knowledgeable game.

*The Village Course* opened in 1981 and sweeps inland along the pineapple fields and statuesque pine trees. At par 71 and 6,858 yards designer Arnold Palmer and course architect Ed Seay are reported to have given this course a European flavor.

261

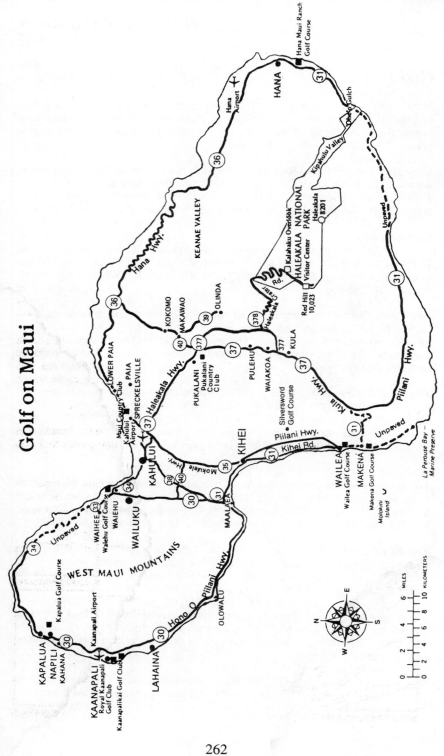

# Golf on Maui

Hana Maui Ranch Golf Course

HANA

31

Hana Airport

36

KEANAE VALLEY

Kipahulu Valley

Oheo Gulch

Unpaved

HALEAKALA NATIONAL PARK

Kalahaku Overlook

Haleakala 8201

Visitor Center

Red Hill 10,023

31

Piilani Hwy.

Hana Hwy.

KOKOMO

MAKAWAO

OLINDA 39

40 377

378

Haleakala Crater

Rd.

Crater Rd.

LOWER PAIA

PAIA

SPRECKELSVILLE

Maui Country Club

Kahului Airport

37

377

PUKALANI

Pukalani Country Club

PULEHU

37

WAIAKOA

KULA

377

Silversword Golf Course

Kula Hwy.

Piilani Hwy.

KIHEI

Piilani Hwy.

35

31

Kihei Rd.

31 Unpaved

La Perouse Bay – Marine Preserve

WAILEA

Wailea Golf Course

MAKENA

Makena Golf Course

Molokini Island

Mokulele Hwy.

KAHULUI

38

40

30

31

MAALAEA

34

Unpaved

WAIHEE 33

Waiehu Golf Course

WAIEHU

WAILUKU

WEST MAUI MOUNTAINS

Kapalua Golf Course

Kaanapali Airport

KAPALUA

NAPILI

KAHANA

30

KAANAPALI

Royal Kaanapali Golf Club

Kaanapalikai Golf Club

LAHAINA

OLOWALU

Piilani Hwy.

Hono

30

N E S W

MILES

KILOMETERS

0 2 4 6

0 2 4 6 8 10

Resembling the mountainous countryside of Scotland, this course is reputed to be the most difficult and demanding in Hawaii and one of the most challenging in the world.

Coore and Crenshaw of Austin Texas have designed the newest third golf course at Kapalua. The 18-hole championship *Plantation Course* will be located on 400 acres north of the Village Course. Kapalua has plans to make this newest premier course the location of the Kapalua International tournament. The course is scheduled to open in January 1991.

### KIHEI
*The Silversword Golf Course* is the newest on Maui. This non-resort course is a 6,800 yard par 71, 18-hole course and is located off Piilani Highway near Lipoa Street. Green fees are currently $50 which includes a shared cart. Twilight play $25. Maui residents receive a discounted rate. (874-0777)

### MAKENA
The Maui Prince Resort features one 18-hole course with plans for another 18 in the near future. Ultimately they plan to have two 36-hole courses. Their greens fees change throughout the year so it is best to call for current rates. (879-3344)

*The Makena Course* is a Trent Jones designed 6,800 yard par 71, 18-hole course which opened in 1981. It winds among the hillsides and down along the coastline, where remnants of early Hawaiian rock boundary walls have been preserved. The large cactus which abound in this area were imported to feed the cattle which were once ranched in this area.

### WAILEA
The Wailea resort offers the challenging Orange and Blue Courses. Greens fees for Wailea guests and Maui residents are $45 plus $15 cart, non-guests $90 plus $15 cart. (879-2966)

*The Orange Course* is a par 72, 6,810 yard championship course designed by Arthur Jack Snyder. It opened in 1978. With more trees, ancient stone walls, narrow fairways and dog legs at half the holes, it is a considerable challenge.

*The Blue Course* is par 72 and 6,700 yards from the championship tees. It opened in 1972. Four lakes and 72 bunkers provide added hazards along with the exceptional scenery. The 16th hole is especially lovely with numerous people stopping to snap a picture from this magnificent vantage point.

A third golf course for Wailea is now in the development stage.

### PAIA
*The Maui Country Club* is a private course which invites visitors to play on Mondays. Call on Sunday after 9 am to schedule Monday tee times.It originally opened in 1925. The front 9 holes have a par 37 as do the back nine. Greens fees are $35 cart included for 9 or 18 holes. (877-0616)

### PUKALANI
*Pukalani Country Club and Golf Course* is nestled along the slopes of Haleakala and affords a tremendous panoramic view of Central Maui and the ocean from every hole. Designed by Bob Baldock, the first 9 holes opened in 1980. Nine additional holes have been added making a par 72, 6,692 yard course. Greens fees are $45 cart included. (572-1314)

### WAILUKU
*The Waiehu Municipal Course*, which is north of Wailuku, opened with nine holes in 1929 and an additional 9 holes were added later. This a par 72 course. Greens fees are $25 weekdays and weekends and holidays. A cart is optional at $12.50, pull carts available for $1. Phone (243-7410)

# TENNIS

Tennis facilities abound on Maui. Many condos and major hotels offer tennis facilities, also, there are quite a few very well kept public courts. They are, of course, most popular during early morning and early evening hours.

## PUBLIC COURTS:

*Hana* - Hana Ball Park, one lighted court.
*Kahului* - Maui Community Center (Kaahumanu and Wakea Avenue) has two unlighted courts. Kahului Community Center (Onehu and Uhu St.) has two lighted courts.
*Kihei* - Kalama Park has two lighted courts. Six unlighted courts in park fronting Maui Sunset condos.
*Lahaina* - Lahaina Civic Center has two lighted courts and there are four lighted courts at Malu-ulu-olele Park.
*Makawao* - Eddie Tam Memorial Center has two lighted courts.
*Pukalani* - Pukalani Community Center has two lighted courts, located across from the Pukalani Shopping Center.
*Wailuku* - Maui Community College has four lighted courts available after school hours. (244-9181)

## PRIVATE COURTS WITH FACILITIES OPEN TO PUBLIC:

*Hyatt Regency*, Kaanapali. Five courts. FEE CHARGED for guests and non-guests. (661-1234 ext. 3174)

*Kapalua Bay Hotel*, Kapalua. Offers the Tennis Garden with 10 courts. Tennis attire required at all times. FEE CHARGED. (669-5677)

*Makena Tennis Club*, 5415 Makena Alanui, Makena.Six courts. FEE CHARGED. (879-8777)

*Maui Marriott Resort*, Kaanapali. Guests and non-guests can play on five unlighted courts. FEE CHARGED. (667-1200)

*Napili Kai Beach Club*, Napili. Has two courts which are not lighted. FEE CHARGED. (669-6271)

*Royal Lahaina Hotel*, Kaanapali. Has the 2nd largest tennis facility on the island with 11 courts, 6 lighted and one is a stadium court. FEE CHARGED. (661-3611)

*Sheraton Maui Hotel*, Kaanapali, . Has three lighted courts. FEE CHARGED. (661-0031)

*Wailea Tennis Center*, Wailea. Has fourteen courts, 3 lighted, 3 grass. FEE CHARGED. (879-1958)

## RESORT COURTS RESTRICTED TO GUESTS:

Hale Kamaole, Hotel Hana Maui, Kaanapali Alii, Kaanapali Plantation, Kaanapali Shores, Kaanapali Royal, Kahana Villa, Kamaole Sands, Kihei Akahi, Kihei Alii Kai, Kihei Bay Surf, Kuleana, Maalaea Surf, Mahana, Mahinahina, Makena Surf, Maui Hill, Maui Islander, Maui Lu Resort, Maui Vista, Papakea, Puamana, Royal Kahana, Sands of Kahana, Shores of Maui, The Whaler.

# HORSEBACK RIDING

Historically, the first six horses arrived on the islands in 1803 from Baja California. These wild mustangs were named "Lio" by the Hawaiians, which means "open eyes wide in terror." They roamed and multiplied along the volcanic slopes of Maui and the Big Island until they numbered 11,000. They adjusted quickly to the rough terrain and had a reputation for terrific stamina. Today these ponies, also known as Kanaka ponies or Mauna Loa ponies, are all but extinct with fewer than a dozen purebreds still in existence.

Lush waterfalls, pineapple fields stretching up to the mountain's peaks, cane fields, kukui nut forests and Haleakala's huge crater are all environments that can be enjoyed on horseback. Trips are for the beginner, intermediate or experienced rider and last as little as an hour or two or up to three days. Most stables have age restrictions.

*Adventures on Horseback* - A 5 hour waterfall ride outside Haiku is $125 per person and includes lunch and gear. Enjoy the cliffs of North Maui, the slopes of Haleakala, the old Hanan Hwy, rainforest streams and secluded waterfalls. breakfast, picnic lunch, and swimming. Maximum 6 riders. (242-7445)

*Aloha Nui Loa Tours* - Drive to Hana for a tour of the town, lunch on a quiet beach, and a ride across the scenic pastureland of the Hotel Hana Ranch. Children over age 7 with experience can ride alone. (879-0000)

*Charley's Trailride and Pack Trips* - Features overnight trips to Haleakala with guide arranging cabin and supplies. Rates for 4 - 6 people are $150 per person, food provided; $125 per person if you bring your own. Rates for 2 - 3 people are $200 each with food provided. $75 non-refundable deposit unless rain cancels trip. Write Charles Aki, c/o Kaupo Store, Hana, Maui, HI 96713. (248-8209)

*Holo Lio Stables* - Located near the Maui Prince Hotel at Makena. Shoreline, sunset or moonlight rides $20 - $35. All day trips $65. (879-1085)

*Hotel Hana Maui* - Offers guided trail rides around the 4,500 acre working cattle ranch on open range, shoreline, rain forest and mountains. $25/hr non-guests, $22 hotel guests. Private rides $50, custom rides available. (248-8284)

*Kau Lio Stables* - Offers three hour excursions with refreshments for $53 per person. Kaanapali Resort area pick up for the half-hour drive up the slopes above Kaanapali to the stables. Pick up times 10:00 am, and 1:00 pm. A four-hour picnic excursion includes the drive, a three-hour ride and a one-hour picnic for $67. You ride up mountain sides, through cane fields, and the kukui nut forests above Kaanapali. (667-7896)

*Makena Stables* - Located on Makena Rd., rides are along the King's Hwy. and La Perouse Bay. (879-0244)

***Pony Express Tours*** - Has trips into Haleakala Crater, weather permitting, Mon. - Fri. The Kapalaoa Cabin Tour is 12 miles and 8 hours starting at the craters rim for $120 plus tax per person and includes lunch. Half day trips $90. One hour rides $25, two hour $45. (667-2202)

***The Rainbow Ranch*** - Follow Honoapiilani Hwy. 11 miles north of Lahaina, entrance near exit for Napili. They tour the mountain area. Trips are available for beginning, intermediate or advanced riders in either English or Western style. 1-hour mountain trip $25, 2-hour $40, 2-hour sunset trip $40, 3-hour picnic ride $50, and 3-hour trip $65. (669-4991)

***Thompson Riding Stables*** - Located on Thompson Rd. in Kula. Trail rides and crater tours. Day and overnight camping. Sunset and picnic rides. (879-1910)

# HIKING

Maui offers an array of excellent hiking opportunities for the experienced hiker or for a family outing. Comprehensive hiking information is available from several excellent references (see ORDERING INFORMATION). The following are descriptions of several hiking experiences which we have enjoyed.

Among the most incredible adventures to be experienced on Maui is one of the fifty hikes with your personal guide ***Ken Schmitt.*** These hikes, for 2 - 6 people only, can encompass waterfalls and pools, ridges with panoramic views, rock formations, spectacular redwood forests (yes, there are!), the incomparable Haleakala Crater or ancient structures found in East Maui.

Arriving on Maui some 10 years ago, Ken has spent much of that time living, exploring and subsisting out-of-doors and experiencing the "Natural Energy" of this island. This soft spoken man offers a wealth of detailed knowledge on the legends, flora, fauna and geography of Maui's many diverse areas. Ken has traversed the island nearly 400 times and established his fifty day hikes after considerable exploration. His favorites are the 8 and 12 mile crater hikes which

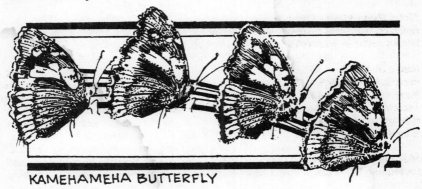

KAMEHAMEHA BUTTERFLY

he says offer a unique, incredible beauty and magic, unlike anywhere else in the world. The early Hawaiians considered Haleakala to be the vortex of one of the strongest natural power points on earth.

The hikes are tailored to the desires and capabilities of the individual or group and run 1/2 or full day (5 - 12 hours). They range from very easy for the inexperienced to fairly rugged. Included in the $45 - $85 fee (children are less) are waterproof day packs, picnic lunch, specially designed Japanese fishing slippers, wild fruit and, of course, the incredible knowledge of Ken. Also available are overnight or longer hikes by special arrangement. Ken can be reached at 879-5270, or by writing **Ken Schmitt**, P.O. Box 330969, Kahului, Maui, HI 96733.

## WEST MAUI

*Escape!* Just a few miles beyond Kapalua is a world far removed from the tourist haunts. With naturalist **Ken Schmitt**, some unique sights are in store for the hiking enthusiast. The hike is a fairly easy one. There are a few places requiring the footing of a mountain goat and a moderate climb up a rather steep slope, but a hike most people could enjoy.

This West Maui hike leads first to an unusual Hawaiian archeological site. While visitors are familiar with Haleakala and the role it played as the islands most sacred spot, the circle of stones in West Maui is unknown to most and is an area veiled with mystery and filled with beauty. The first portion of the hike is less than a mile over private land (which Ken has permission to cross). A dirt path used by cattle twists and turns with low hanging bows of Australian pine (false ironwood) trees. Emerging on a grassy bluff with the crystal blue Pacific below and the warm sunny skies above is a delightful surprise. Except for an overly friendly horse, there is no sign of people for as far as the eye can see. The circle of stones dominates Kanounou Point with a large center stone in the middle. Resembling the Indian medicine wheels of Colorado and Wyoming this area is still revered by the Hawaiians as a source of power. Evidence of this was apparent with the small, beautiful purple crystals that had been left as an offering in the center alter stone. Ken explains that there is a fairly straight ley line through the islands that emits a subtle force field. On each of the islands that cross their path you will find a similar Hawaiian structure. A Great Frigatebird, which land only to nest, soars overhead. It is a place full of beauty and peace.

As with each hike, Ken introduces the group to the tropical flora of the area. At this location we discover the beautiful, low growing, multi-blossomed Lantana, the Beach Naupaka, the half blossom with a legend of lost love.

The second portion of the hike also covers about a mile in distance. While we begin our trek less than a mile from the first site, the environs are dramatically different. Descending a natural path along the side of the cliff we enter into what appears to be an area more lunar than terran. When the lava of West Maui's volcano, Mauna Kahalawai, spewed forth this area was covered by water. The lava erupted from a localized vent during the third series (and probably the last, occuring approximately 5,000 - 10,000 years ago) of West Maui eruptions and poured into the ocean where it was rapidly cooled, creating unusual and intriguing formations, caverns and blow holes. The highlight of this portion of the sojourn was standing next to one of the island's most spectacular blow holes and being sprinkled (or drenched depending on the tide and wave activity) with salty spray. Many a visitor view this powerful natural phenomena from afar, but few experience the excitement so close at hand. This half day hike ends with the return back up the bluff to a picnic lunch of sandwiches and fresh fruit on the summit with a panoramic view of the deep blue Pacific. This hike can be conveniently combined with a snorkeling trip to Honoloa Bay (in summer) where Ken explains the reef and marine life.

## CENTRAL MAUI
A wonderful hike for the family is to the **Twin Falls** area on Maui's windward side. This easy trail passes through cattle pastures and then woodlands to three beautiful and tantalizingly cool fresh water pools formed by cascading falls. It is a popular hike enjoyed by many visitors and island residents over an easy to follow trail described in many hiking books. However, it is Ken that makes the trip special. The average hiker would pass right over the fallen kukui nuts while Ken stops, opens several and passes them around for a taste test, warning that they have a strong laxative quality! A small purple blossom, an herbal blood purifier, is sampled, a juicy guava and liliokoi are tasted and the base of one of the pandanus blossoms with a caramel-like flavor can be chewed on. Passing over a canal, Ken explains the history of the early cane industry when this series of amazing water channels was built one day short of a two year deadline.

## UPCOUNTRY MAUI
Poli Poli is ideally situated on the leeward slopes of Haleakala. Cool crisp mountain air provides a temperate climate for hiking and the trails are suitable for the entire family. Since the weather can be cool, warm attire and rain apparel should be included in your day or night pack, however, the clouds often clear and treat visitors to a sunny and very mild afternoon. To get to the turn off, go just past the Kula Botanical Gardens and turn on Waipoli Rd., or go 3/10 mile past the junction of Hwy 37 and Hwy 277. Follow the paved, steep and windy road approximately 3.9 miles (about 26 minutes), then continue another 5.8 miles or about 20 minutes on gravel road. The last 1/4 mile is downhill and especially rutted and can be very muddy and too slippery for anything but a four wheel drive when wet. *WARNING: Rental car agencies are not responsible for damage done*

COMMON 'AMAKIHI

269

*to cars that travel this road.* We have found it passable in a car with high clearance if the road has been recently graded and is dry. After recent rains, the rutted road becomes treacherous and only a four wheel drive vehicle is advised. Once you reach the park there is a graveled parking area and a grassy camping area. Two BBQ's and the luxury of a flush toilet in a small outhouse. Running drinking water is available. Trail options include a .8 mile trek to the Redwood Forest, a 6 mile Haleakala Trail, 4.8 mile loop trail, 1.0 mile to the cave shelter and 1.5 miles to the Plum trail.

The 4.8 mile loop is a very easy trail and with frequent snack stops, even our three year old was able to make it the entire distance. There is an array of lush foliage and plums may be ripe if you arrive during June and July. The clouds can roll quickly in, causing it to be pleasant and warm one minute and cool the next as well as creating some interesting lighting effects amongst the trees. The cave shelter, is a bit of a disappointment. It is a shallow cavern and reaching it meant a descent down a steep incline of loose gravel that was too difficult for our young ones. The Eucalyptus was especially fragrant as the fallen leaves crunched underneath our tennis shoes. An area along the trail that had been freshly rutted by wild boars demonstrated the incredible power of these animals. On one trip we heard a rustling in the bushes nearby followed by grunting sounds. We have been told that while you definitely want to avoid the wild boars, they are accustomed to being hunted and will also choose to avoid you. Apparently we were down wind of them and since they have poor eyesight we passed by quietly without them noticing us and without seeing them.

## HALEAKALA

A hike, once again with Ken Schmitt, is a thrill for all the senses. Not only does he pack a great lunch and yummy snacks, but he provides beauty for the eyes, fresh air for the nose, peace and serenity for the ears, and an opportunity to touch and get in touch with Maui's natural beauty. We chose a trip to Haleakala to see the awesome crater up close. The trip was an 8-mile hike down Switchback (Halemauu) Trail to the Holua cabin and back. The first mile of the trek was over somewhat rocky, but fairly level, terrain. The next mile seemed like three as we descended seemingly endless hairpin twists down the side of the crater with changing panoramic vistas at each turn. Sometimes fog would eerily sweep in, hovering around and obscuring the view completely, only to soon move away. The vegetation (following a period of heavy rains) was exceptionally lush. All the greenery seemed quite out of place in the usually rather desolate crater.

We continued along the crater floor and past the Holua cabin to reach the Holua lava tube. The small opening was not marked and could be easily overlooked. A small sign advised the use of lights inside the cave. We prepared our flashlights and bundled up for the cooler temperatures to be encountered below. A ladder set by the park service provide access. Once at the base of the ladder, the cavern was large, cool and dark. While there were several directions that lead quickly to dead ends, Ken took us further down the main tube. The cavern was so large that seldom did we have to do more than occasionally duck. Once inside we turned out the lights to enjoy a few moments of the quiet darkness. The tube travels about 100 yards with a gradual ascent to daylight. As daylight peeks down through the dark shaft, it appears the end is in sight. Another turn in the tunnel reveals not the end, but an altar-like flat rock piled with assorted stones. Light cascading through a hole in the ceiling casts an almost supernatural glow to eyes now accustomed to the darkness. The effect is to create a luminescence on the stones making them appear to be statues set in a natural cathedral. Returning to the cabin, we picnicked on the grounds while very friendly nene geese begged for handouts. Ken advised against feeding them as the park rangers would prefer the geese not become dependent on human handouts. After a rest in the warm sun, we retraced our path back up Switchback Trail to the van and continued on to enjoy the summit before heading back to town.

Other more strenuous and lengthy crater ventures include the Sliding Sands Trail and one that traverses down the side of the volcano through the Kaupo Gap.

Another hiking tour company is *Outer Island Adventures*, P.O. Box 996, Makawao, Maui, HI 96768, (808-572-6396), which also offers guided hikes to Haleakala, Iao Valley and other areas.

The public is invited to join the guided hikes of the *Maui Chapter of the Sierra Club*. This is a wonderful and affordable way to enjoy this beautiful island with a knowledgeable group of people. The group has weekend outings twice a month. There is an optional donation. At the Sierra Club contact Mary Evanson (572-9724).

Hiking off established trails without a guide is not advised, however, there are several good resources. *Hawaiian Hiking Trails* by Craig Chisholm, 128 pages, $12.95, is an excellent and accurate book with good maps and color photographs. It includes nearly 50 hiking trails throughout the islands and seven of these are on Maui. Robert Smith's book *Hiking Maui* 160 pages, $7.95, will guide you on 27 fairly accessible trails throughout Maui. The author also has guides to Hawai'i, Kaua'i and O'ahu. Both of these publications are available from Paradise Publications, see ordering information at the back of the book. A good trail companion book might be *Trailside Plants of Hawaii's National Parks* by Charles Lamoureux, 77 pages with great pictures for $4.95. Available at Maui bookstores.

PANDANUS TREE

271

# CAMPING

County and state camping permits are required for camping in many of the parks. *County permits* are available at the Maui War Memorial in Wailuku (243-7230). *State permits* are issued 8 am - noon and 1 - 4:15 pm, Mon. - Fri. at the State Office Building, Dept. of Land and Natural Resources, 54 High St., P.O. Box 1049, Wailuku, Maui 96793 (244-4354). The maximum stay at state parks is five days. *Haleakala National Park permits* are issued at the park headquarters for tent camping on a first-come basis on the day you plan to stay, beginning at 7:30 am -4 pm daily. The three cabins can each accommodate up to 12 people. Water, stove, wood and cooking utensils are provided. Bring your own bedding. The reservations are set up at a monthly lottery. Send requests to Haleakala National Park, PO BOX 369, Makawao, Maui, Hi 96768 (572-9177 or 572-9306).

*Baldwin Beach Park* - This county park is a grassy fenced area near the roadside. It is located near Lower Paia on the Hana Hwy. and has tent camping space, restrooms and outdoor showers.

*Honomanu Bay* - This county park in the Hana area has tent camping facilities.

*Hookipa Beach Park* - This county Park offers tent camping, restrooms and outdoor showers.

*Hosmer Grove* - Haleakala National Park. Tent camping. No permit required. Located at the 7,000 foot elevation on the slope of Haleakala. Three night maximum stay. Cooking area with grill, pit toilets, water, picnic tables.

*Kaumahina State Wayside* - Tent camping. Located in the rain forest area 28 miles from Kahului Airport on the Hana Highway. Restrooms, picnic tables, outdoor BBQ.

*Ohe'o Gulch* - Haleakala National Park, Hana. Tent camping. No permit required. Chemical toilets, picnic tables, BBQ grills, bring your own water. Three-day maximum stay.

*Poli Springs Recreational Area* - Located in Upcountry, this state park has one cabin and offers tent camping. This is a wooded, two-acre area at the 6,200 foot elevation on Haleakala's west slope and requires four wheel drives to reach. Extensive hiking trails offer sweeping views of Maui and the other islands in clear weather. Seasonal bird and pig hunting. Nights are cold, in winter below freezing. No showers. Toliets, picnic tables. The single cabin sleeps 10 and has bunk beds, water, cold shower, kitchenware. Sheets and towels can be picked up along with the key. See additional description in the hiking section which follows.

*Wainapanapa State Park* - Located near Hana. Tent camping, 12 cabins. More information on this location can be found under WHERE TO STAY - HANA. Restrooms, picnic tables, outdoor showers. This is a remote volcanic coastline covering 120 acres. Shore fishing, hiking, marine study, forests, caves, blow holes, black sand beach. The park covers 7.8 acres. BRING MOSQUITO REPELLENT!

Currently there are no rental companies offering camping vehicles. There have been a few companies in recent years, but they have come and gone quickly. Car rental agencies generally prohibit use of cars or vans for camping. Camping equipment is available for rent at Silversword Stoves in Makawao. (572-4569)

Camping tours are offered from *Pacific Quest Outdoor Adventures*. Camping, hiking and activity packages offered on Kaua'i, Moloka'i, The Big Island and Maui. Trips are offered monthly and include inns, meals, local air and ground transportation, natural and cultural history tours. Trips begin at $1,495. P.O. Box 205, Haleiwa, HI 96712. (1-800-367-8047 ext. 523 in U.S. (808) 638-8338)

*Charley's Trail Rides* offers guided overnight horseback trail excursions from Kaupo up the Haleakala slopes to the crater. Parties of 4 - 6 persons $150 each includes meals, cabin or campsite equipment, parties of 2 - 3 are $200 each. Advance notice and deposit required. Write c/o Kaupo Store, Kaupo, HI 96713. (248-8209)

# *PHOTOGRAPHY*

Photo safaris to bamboo jungles, cascading waterfalls, fern laden paths amid redwood forests are among the options offered by *Maui-Anne's Island Photograph Tours* ★. Professional freelance photographer, Marianne Pool, offers excursions along short, but scenic two to three mile trails where you receive photo instruction, tips and techniques along the way. Tours include transportation from Kihei, rain ponchos and hiking footwear, water resistant bags for equipment and deli lunch. Depending on the length of the hike and destination, trips run $75 - $250. Also available are photography workshops of up to 10 days, photo and hiking tour of neighbor islands or choose to have you and your family portrayed at a scenic location for personalized postcards. Write P.O. BOX 2250, Kihei, Maui, HI 96753. (808) 874-3797.

WILIWILI

# HUNTING

Contact **Hunting Adventures of Maui, Inc.**, 645B Kapakalua Rd., Haiku, Maui, HI 96708 (808-572-8214). Year round hunting season. Game includes (Kao) Spanish Goat and (Pua'a) Wild Boar. Hunting on 100,000 acres of privately owned ranches with all equipment provided. Rates $400 first person, $150 each additional person, 3 max. (non-hunters free) includes sunrise-sunset hunt, food, beverages, four wheel drive transportation, clothing, boots, packs, meat storage and packing for home shipment, hotel or airport pick up. Rifle rentals and taxidermy available. Also available are sightseeing safaris and overnight lodgings for small groups.

Hotel Hana Maui offers guided wild boar hunting trips over its 4,500 acre ranchland in the hills above Hana. Pheasant hunting and trap shooting available. Boar $340 plus license, pheasant $240 plus license. c/o Hotel Hana Maui, PO Box 8, Hana, HI 96713. (808-248-7238)

# ARCHERY

Valley Isle Archers holds weekly meetings in Kahului at the National Guard Armory each Wednesday, visitors are welcome. An annual competitive shoot is held each year in June on Kamehameha Day.

# RUNNING

Maui is a scenic delight for runners. **Valley Isle Road Runners** can provide you with up-to-date information on island running events. They can be reached at 242-6042. For a free running map, write "Hawaii Safe Running Council," P.O. Box 23169, Honolulu, HI 96822, and tell them which islands you will be visiting. Stouffer Wailea Beach Resort also provides guests with a running guide to the Kihei/Wailea area. Also available at area bookstores is *HAWAII: A Running Guide* by Noel Murchie and Paul Ryan, softcover. $6.95.

WILD BOAR

# FITNESS CENTERS AND HEALTH RETREATS

If you are interested in keeping in shape while you are on Maui and you have no fitness center at your resort, check with the following facilities which welcome drop in guests.

*Nautilus Fitness Center*, Dickenson Square, Lahaina (667-6100)
*Nautilus Fitness Center*, 1325 Lower Main St., Wailuku (244-3244)
*Valley Isle Fitness Center*, Wailuku Industrial Park, (242-6851)
*Lahaina Health Club*, at Maui Kaanapali Beach Villas, (667-6684)

The Hotel Hana Maui offers a *Wellness Center*. The hotel is offering a seven day Hana Health and Fitness Retreat at the beginning of each month conducted by Director Susan Thomson. The retreat is limited to twelve guests and includes hiking, exercises, and a 1,200 calorie daily menu. Arrangements are made through the hotel number 248-8211 and the charge is $2,00 and up which includes room and meals.

*The Strong, Stretched and Centered Fitness Body/Mind Institute* is offered by Gloria Keeling. The six week health and fitness retreats are designed for personal wellness as well as fitness instruction certification. Gloria explains that this is a program where people learn completely new lifestyles and learn how to teach those lifestyles to others. The $4,000 cost includes meals, land transportation, accommodations in the Kihei area and classes. P.O. Box 758, Paia, HI 96779. (808) 575-2178.

*Pualani Fitness Retreat*, PO Box 1135, Makawao, Maui, HI 96768 (808-572-6773) 1-800-PUALANI. Pualani (which means heavenly flower) is located on the slopes of Haleakala. A maximum of eight guests experience a personalized program of swimming, aerobics, weight lifting, crater hikes and beach runs. Stress management, yoga, nutrition and cooking classes round out the schedule with gourmet vegetarian meals served. Owners Susan and William Linneman spent several years traveling throughout the U.S. and Europe researching this fitness retreat design. Rates are $1,800 per person per week for private, $1,600 per person shared, $1,500 per person for a couple sharing a room. Bed and Breakfast accommodations are also available $55 - $85 per night.

# AIR TOURS

## SMALL PLANE FLIGHTSEEING

Flightseeing trips are available via helicopter or small plane and include Hana and Haleakala as well as island flights to Mauna Loa on the Big Island, Oʻahu, Kahoolawe, Lanaʻi or Molokaʻi. Small plane trips are less expensive, but you won't get as close to the scenery. Prices are from $39 for half-hour flights to $125 for full-day trips combined with ground tours.

**Akamai Tours** - Operates daily from Maui to Oʻahu or Hawaiʻi for a one-day sightseeing trip. On the Big Island a 260 mile "see it all" ground tour includes a visit to a Kona fishing village, Volcanoes National Park and the Parker Ranch. On Oʻahu, visitors will enjoy the sites of Pearl Harbor, downtown Honolulu and Sealife Park. Hotel pickup, approximately 5:30 am, returning 9 pm. Cost $119.71 plus tax. (871-9551) 1-800-922-6485

**Maui Air Sports** -Twin engine planes fly commuter service to Maui/Oʻahu. (877-4253)

**Panorama Air Tour** - Air tour of all 8 islands with ground tour of Hawaiʻi and Kauaʻi (with a river boat ride on Wailua River). Transportation to and from Kahului Airport. Day trips to Oʻahu and the Big Island via twin engine Piper Chieftain. (On Maui 1-800-352-3732, from mainland 1-800-367-2671)

**Paragon Air** - Charter service, five passenger, $325/hour. (878-6412)

**Scenic Air Maui** - Ten-passenger twin engine Beechcrafts provide flightseeing to outer islands of Lanaʻi, Kahoolawe, and the Big Island. (871-2555)

# HELICOPTER TOURS

The price of an hour helicopter excursion may make you think twice. After all it could be a week's worth of groceries at home. We had visited Maui for more than 7 years before we finally decided to see what everyone else was raving about. It proved to be the ultimate island excursion. When choosing a special activity for your Maui holiday, we'd suggest putting a helicopter flight at the top of the list. (When you get home you can eat beans for a week!)

Adjectives cannot describe the thrill of a helicopter flight above majestic Maui. The most popular tour is the Haleakala Crater/Hana trip which contrasts the desolate volcanic crater with the lush vegetation of the Hana area. Maui's innermost secrets unfold as the camera's shutter works frantically to capture the memories (one roll is simply not enough).

Ken Rankin (Sunshine Helicopters) loves his work so much that he even stops by on his days off to fly visitors around Maui and the neighboring islands in a Bell 206B Jet Ranger helicopter. Bearing a quick wit and a resemblance to Kris Kristopherson, pilot Ken takes you right next to waterfalls cascading down to cool mountain pools and provides an opportunity to see first-hand the unique qualities of the world's largest volcano. While scenic landmarks are narrated by the pilot, mellow music provides the perfect backdrop to this outstanding experience. Ken is just one of several excellent and knowledgeable pilots at Sunshine Helicopters (see below).

Currently all the helicopters depart from Kahului. One helipad near Kapalua has been closed. Keeping up with the prices is impossible. Listed are standard fares, and we hope you'll be delighted to learn of some special discount rates when you call for reservations.

*Alexair* - Four passenger. Hana/Haleakala 45 minute special $105. Departs Kahului. (871-0792) (1-800-462-2281)

*Blue Hawaiian* - Four passenger Bell 206 . (871-8844 or 1-800-247-5444) West Maui (30 minutes, $80) Hana/Haleakala (45 minutes $99), Hana/Haleakal Deluxe (60 minute $1356), Complete island $185 (1 hour 45 minutes)

*Cardinal Helicopters* - (661-0092) 70 minute flight includes video for $140.

*Hawaii Helicopters Inc.* - West Maui 30 min. ($90), All of Maui (2 1/2 hours $220) Departs Kahului. (877-3900)

*Kenai Helicopters* - Departs from Kahului. Maui No Ka Oi Deluxe for $195/person, West Maui and North Shore of Molokai ($180), Hana/Haleakala ($140). (871-6463 or 1-800-622-3144)

*Maui Helicopter* - These are the look-alike helicopters to the ones used on the Magnum PI television show. Maui unlimited tour ($200), Hana, Upcountry and Keanae ($105), Hana and Haleakala Crater ($135), or West Maui ($110). (879-1601 or 1-800-367-8003)

*Papillon* - Trips include Haleakala Sunrise Flight with a continental breakfast 1 1/2 hrs. ($245), West Maui ($99), West Maui/Molokai 1 hr., ($185), and Circle Island Special, 1 hr., ($205), All day Hana excursion $395. (669-4884 or 1-800-367-7095)

*Sunshine Helicopters Inc.* ★ - Their fine ground and flight crew will make sure that your flight in one of their four passenger Bell 206 Jet Rangers is one of your most memorable island experiences. The West Maui trip is filled with cascading waterfalls, but given the choice, our favorite is still the Hana/Haleakala flight! They also provide a free video tape of your trip. (871-0722) (1-800-544-2420)

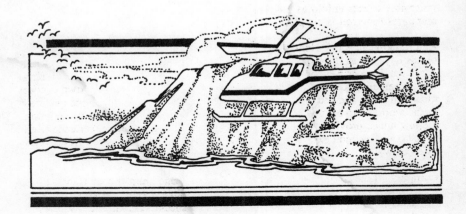

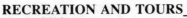

# *HANG GLIDING*

A different sort of air tour! USHGA certified instruction at Waipoli Pasture and Waihee Sand Dune. Beginner lessons $48 for 3-4 hours; beginner certification 3-4 days; 2 day lesson $84; Crater hang gliding available for advanced gliders; equipment rentals also. Check with *Maui Soaring Supplies* (808-878-1271) RR II, Box 780, Kula, HI 96790.

# RECOMMENDED READING

Ashdown, Inez. *Ke Alaloa O Maui. Authentic History and Legends of the Valley Isle*. Hawaii: Kama'aina Historians. 1971.

Ashdown, Inez. *Stories of Old Lahaina*. Honolulu: Hawaiian Service. 1976.

Barrow, Terence. *Incredible Hawaii*. Vermont: Charles Tuttle Co. 1974.

Begley, Bryan. *Taro in Hawaii*. Honolulu: The Oriental Publishing Co. 1979.

Bird, Isabella. *Six Months in the Sandwich Islands*. Tokyo: Tuttle. 1988

Boom, Bob and Christensen, Chris. *Important Hawaiian Place Names*. Hawaii: Bob Boom Books. 1978.

Chisholm, Craig. *Hawaiian Hiking Trails*. Oregon: Fernglen Press. 1989.

Christensen, Jack Shields. *Instant Hawaiian*. Hawaii: The Robert Boom Co. 1971.

Clark, John. *Beaches of Maui County*. Honolulu: University Press of Hawaii. 1980.

Daws, Gavan. *The Illustrated Atlas of Hawaii*. Australia: Island Heritage. 1980.

*Echos of Our Song*. Honolulu: University of Hawaii Press.

Fielding, Ann. *Hawaiian Reefs and Tidepools*. Hawaii: Oriental Pub. Co.

Haraguchi, Paul. *Weather in Hawaiian Waters*. 1983.

*Hawaii Island Paradise*. California: Wide World Publishing. 1987.

Hazama, Dorothy. *The Ancient Hawaiians*. Honolulu: Hogarth Press.

Judd, Gerrit. *Hawaii, an Informal History*. New York: Collier Books. 1961.

Kaye, Glen. *Hawaiian Volcanos*. Nevada: K.C. Publications, 1987.

Kepler, Angela. *Maui's Hana Highway*. Honolulu: Mutual Publishing. 1987

Kepler, Cameron B. and Angela Kay. *Haleakala, A Guide to the Mountain*. Honolulu: Mutual Publishing. 1988.

Kyselka, Will and Lanterman, Ray. *Maui, How it Came to Be*. Honolulu: The University Press of Hawaii. 1980.

*Lahaina Historical Guide*. Tokyo: Maui Historical Society. 1971.

Lahaina Restoration Foundation, *Story of Lahaina*. Lahaina: 1980.

London, Jack. *Stories of Hawaii*. Honolulu: Mutual Publishing. 1965.

Mack, Jim. *Haleakala and The Story Behind the Scenery*. Nevada: K.C. Publications, 1984.

Mrantz, Maxine. *Whaling Days in Old Hawaii*. Honolulu: Aloha Graphics. 1976.

*Na Mele O Hawai'i Nei*. Honolulu: University of Hawaii Press. 1970

Nickerson, Roy. *Lahaina, Royal Capital of Hawaii*. Hawaii: Hawaiian Service. 1980.

*On The Hana Coast*. Hong Kong: Emphasis Int'l Ltd. and Carl Lundquist. 1987.

Pukui, Mary K. et al. *The Pocket Hawaiian Dictionary*. Honolulu: The University of Hawaii Press. 1975.

Randall, John. *Underwater Guide to Hawaiian Reef Fishes*. Hawaii: Treasures of Time. 1981.

Smith, Robert. *Hiking Maui*. California. 1990.

Stevenson, Robert Louis. *Travels in Hawaii*. Honolulu: University of Hawaii Press. 1973.

Tabrah, Ruth. *Maui The Romantic Island*. Nevada: KC Publications. 1985.

Thorne, Chuck. *50 Locations for Scuba & Snorkeling*. 1983.

Titcomb, M. *Native Use of Fish in Hawaii*. Honolulu: University of Hawaii Press. 1952.

Twain, Mark. *Letters from Mark Twain*. Hawaii: University of Hawaii Press. 1966.

Twain, Mark. *Mark Twain in Hawaii*. Colorado: Outdoor Books. 1986.

Wallin, Doug. *Exotic Fishes and Coral of Hawaii and the Pacific*. 1974.

Westervelt, H. *Myths and Legends of Hawaii*. Honolulu: Mutual Publishing. 1987.

Wisniewski, Richard A.. *The Rise and Fall of the Hawaiian Kingdom*. Honolulu: Pacific Basin Enterprises. 1979.

*"One cannot determine in advance to love a particular woman,*

*nor can one so determine to love Hawaii.*

*One sees, and one loves or does not love.*

*With Hawaii it seems always to be love at first sight.*

*Those for whom the islands were made,*

*or who were made for the islands,*

*are swept off their feet in the first moments of meeting,*

*embrace and are embraced."*

Jack London

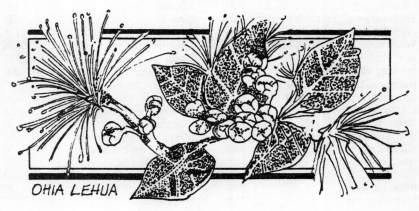

OHIA LEHUA

# INDEX

GINGER &
ANTHURIUMS

# Kauai
## A Paradise Guide

by Don and Bea Donohugh

Second Edition
Expanded & Updated

# Hawaii The Big Island
## A Paradise Guide

By John Penisten

# Oahu
## A Paradise Guide

By Ken Bierly

**UPDATE NEWSLETTERS!** Each Paradise Guide features a companion newsletter. These information filled quarterly publications highlight the most current island events along with late breaking tips on the newest restaurants, island activities or special, not-to-be missed events. Special feature articles are also included. *THE MAUI UPDATE, THE KAUA'I UPDATE, THE O'AHU UPDATE,* and *HAWAI'I: THE BIG ISLAND UPDATE* are available at the single issue price of $1.50 or a yearly subscription (four issues) price of $6 per year each.

# READER RESPONSE

# ORDERING INFORMATION

Dear Reader:

We hope you have had a pleasant visit to Maui. Since our book expresses primarily our own opinions on accommodations, restaurants, and recreation, we would sincerely appreciate hearing of your experiences. Any updates or changes would also be welcomed. Please address all correspondence to the publisher.

*FREE!* To keep you current on the most recent island changes, Paradise Publications has introduced *THE MAUI UPDATE*. A complimentary copy of this quarterly subscription newsletter is available by writing the publisher (Newsletter Dept.) and enclosing a self-addressed, stamped, #10 size envelope.

Traveling to another island? A Paradise Guide is available for each of the major Hawaiian Islands.

*MAUI, A PARADISE GUIDE* by Greg & Christie Stilson. Packed with information on over 150 condos & hotels, 200 restaurants, 50 great beaches, sights to see, travel tips, and much more. "A down-to-earth, nuts-and-bolts companion with answers to most any question." L.A Times. 256 pages, multi-indexed, maps, illustrations, $9.95. Fourth edition.

*KAUA'I, A PARADISE GUIDE* by Don & Bea Donohugh. Island accommodations, restaurants, secluded beaches, recreation and tour options, remote historical sites, an unusual and unique island tour, this guide covers it all. "If you need a 'how to do it' book to guide your next to Kaua'i, here's the one."..."this guide may be the best available for the island. It has that personal touch of authors who have spent many happy hours digging up facts." Hawaii Gateway to the Pacific Magazine. 256 pages, Multi-indexed, maps, illustrations, $9.95. Second Edition.

*O'AHU, A PARADISE GUIDE* by Ken Bierly. Let this exciting book be your personal tour guide to a Hawaii that you'll always remember. Discover why O'ahu is today's best vacation bargain, enabling the visitor to enjoy three wonderfully different vacations all on this one tropical Hawaiian isle. This guide features restaurants, accommodations, beaches, sight-seeing, and recreational and tour opportunities. Discover what awaits you beyond Waikiki! 320 pages, Multi-indexed, maps, illustrations. $11.95.

*HAWAI'I: THE BIG ISLAND, A PARADISE GUIDE* by John Penisten. Outstanding for its completeness, this well-organized guide provides useful information for people of every budget and lifestyle. Each chapter features the author's personal recommendations and "best bets." Comprehensive information on more than 70 island accommodations and 150 restaurants. Sights to see, recreational activities, beaches, and helpful travel tips. 256 pages. Multiple indexes, maps and illustrations. 256 pages. $9.95.

**See ordering information on the following page.**

**OTHER TITLES!** Also available from Paradise Publications are interesting Hawaiian titles which may be difficult to locate in mainland bookstores.

*MAUI, THE ROMANTIC ISLAND*, and *KAUA'I, THE UNCONQUERABLE* by K.C. Publications. These books present full color photographs depicting the most magnificent sights on each island. Brief descriptive text adds perspective. Highly recommended. 9 x 12, 48 pages, $5.95 each, paperback.

*HALEAKALA* and *HAWAII VOLCANOES* by K.C. Publications. Fascinating and informative with vivid photographs depicting these natural wonders. A great gift or memento. 9 x 12, $5.95 each.

*WHALES, DOLPHINS, PORPOISES OF THE PACIFIC*, by K.C. Publications. Enjoy the antics of these beautiful aquatic mammals through full color photographs and descriptive text. 9 x 12, paperback, $5.95

*ON THE HANA COAST*, published by Emphasis International Ltd., and Carl Lindquist. This work captures in rich color photographs and descriptive text the history of a people who arrived in double-hulled canoes to create a new life on the wind-ward side of Maui. 6 x 9, 164 pgs., $12.95, paperback.

*HAWAIIAN HIKING TRAILS*, by Craig Chisholm. This very attractive and accurate guide details 49 of Hawaii's best hiking trails. Includes photography, topographical maps, and detailed directions. 6 x 9, paperback, 152 pgs., $12.95.

*HIKING MAUI*, by Robert Smith. Discover 27 hiking areas all around Maui. 5 x 8, paperback, 160 pages. $8.95. Also by Robert Smith *HIKING HAWAII, THE BIG ISLAND*, $8.95; *HIKING KAUAI*, $8.95.

*COOKING WITH ALOHA,* by Elvira Monroe and Irish Margah. Discover the flavors and smells of the Hawaiian islands in your own kitchen with this beautiful illustrated and easy-to-follow cookbook. Delicacies range from exotic pickled Japanese seaweed or taro cakes to flavorful papaya sherbet or chicken 'ono niu. Drinks, appetizers, main courses and desserts are covered. 9 x 12, paperback, 184 pages, $7.95.

**For a full listing of current titles send request to Paradise Publications. Prices are subject to change without notice.**

**SHIPPING:** In the mainland U.S. add $2 per book title, maximum of $4 to same address. Books shipped out UPS or first class mail. For orders to Canada, Alaska or Hawaii add $3 per book, maximum $6. A gift? Just supply us with the name add address. Orders promptly shipped.

**PARADISE PUBLICATIONS**
8110 S.W. Wareham, Suite 101
Portland, Oregon 97223.
(503) 246-1555

Mauna Kia

500

Aurel Timken
Travel Agent

Read & Falcon